Texts and Readings in Mathematics

Volume 88

The **Texts and Readings in Mathematics** series publishes high-quality textbooks, research-level monographs, lecture notes and contributed volumes. Undergraduate and graduate students of mathematics, research scholars and teachers would find this book series useful. The volumes are carefully written as teaching aids and highlight characteristic features of the theory. Books in this series are co-published with Hindustan Book Agency, New Delhi, India.

Arup Bose

Elements of Measure and Probability

HINDUSTAN BOOK AGENCY

Springer

Arup Bose
Statistics and Mathematics Unit
Indian Statistical Institute
Kolkata, West Bengal, India

ISSN 2366-8717 ISSN 2366-8725 (electronic)
Texts and Readings in Mathematics
ISBN 978-981-95-2757-1 ISBN 978-981-95-2758-8 (eBook)
https://doi.org/10.1007/978-981-95-2758-8

Jointly published with Hindustan Book Agency
The printed edition is not for sale in India. Customers from India please order the print book from: Hindustan Book Agency, P 19 Green Park Extension, New Delhi 110016, India.
ISBN of the Co-Publisher's edition: 978-81-988317-6-7

This Springer imprint is published by the registered company Springer Nature Singapore Pte Ltd.
The registered company address is: 152 Beach Road, #21-01/04 Gateway East, Singapore 189721, Singapore

Ishani and Anupama

bundles of joy

Contents

Abbreviations, notation, and conventions

a.s./a.e. : almost surely/almost everywhere.
cdf: cumulative distribution function.
cf: characteristic function.
mgf: moment generating function.
iff: if and only if.
iid: independent and identically distributed.
pdf: probability density function.
pmf: probability mass function.
rcd: Regular conditional distribution.
ui: uniformly integrable.
CLT: Central limit theorem.
DCT: Dominated convergence theorem.
MCT: Monotone convergence theorem.
SLLN: Strong law of large numbers.

$\mathbb{N}$: set of positive integers $\{1, 2, \cdots\}$, sometimes with 0 included.
$\mathbb{Q}$: set of rational numbers.
$\mathbb{R}$ and $\mathbb{R}^+$: set of all, real and non-negative real numbers.
$\bar{\mathbb{R}}$: $\mathbb{R} \cup \{\infty\} \cup \{-\infty\}$ (extended real numbers).
$\bar{\mathbb{R}}^+$: $\mathbb{R}^+ \cup \{\infty\}$.
ι: $\sqrt{-1}$, the imaginary square root of -1.
$a \wedge b$: minimum of a and b.
$B(x, r)$: ball of radius r around x, $\{y \in \mathbb{R} : |x - y| < r\}$.
$\mathcal{R}e(z)$: real part of the complex number z.

$\#(A)$: the number of elements of A.
$\mathbf{1}_A$: indicator function of the set A.

A^c, A': complement of the set A.
A^o and $\bar{A}$: interior and closure of the set A (in a metric space).
∂A: boundary of the set A (in a metric space).
$A \cup B$ and $A \cap B$: union and intersection of the sets A and B.
$A \Delta B$: $(A \setminus B) \bigcup (B \setminus A).=(A \cap B^c) \cup A^c \cap B)$.
$\limsup A_n$: $\cap_{n=1}^{\infty} \cup_{k=n}^{\infty} A_k$, $\liminf A_n$: $\cup_{n=1}^{\infty} \cap_{k=n}^{\infty} A_k$.

$\mathcal{P}(\Omega)$: the *power set* of Ω, which contains all subsets of Ω.
$\mathcal{M}$: monotone class. $\mathcal{M}(\mathcal{C})$: smallest monotone class containing the class of sets $\mathcal{C}$.
$\mathcal{F}$: field. $\mathcal{F}(\mathcal{C})$: smallest field that contains the class of sets $\mathcal{C}$.
$\mathcal{I}_{\mathcal{F}}$: class of all sets that are increasing limits of sets from $\mathcal{F}$.
$\mathcal{A}$: σ-field. $\sigma(\mathcal{C})$: smallest σ-field containing the class of sets $\mathcal{C}$.
$\mathcal{B}(M)$, Borel σ-field: smallest σ-field containing all open sets of M.
$\sigma(\mathcal{X})$: smallest σ-field that makes all variables in the class $\mathcal{X}$ measurable.
$\mathcal{G} \vee \mathcal{H}$: smallest σ-field containing $\mathcal{G}$ and $\mathcal{H}$.
$(\Omega, \mathcal{F}, \mu)$: measure space where $\mathcal{F}$ is a field.
$(\Omega, \mathcal{A}, \mu)$: measure space where $\mathcal{A}$ is a σ-field.
$\mathcal{L}$: Lebesgue σ-field (completion of Borel σ-field wrt Lebesgue measure).
$\mathcal{N}_\mu$: null sets of a measure μ.
$\mathcal{A}_\mu$: completion of the σ-field $\mathcal{A}$ with respect to the measure μ.

$||f||_p$: $(\int |f|^p d\mu)^{1/p}$. $L^p(\mu)$: $\{f : ||f||_p < \infty\}$, $0 < p < \infty$.
λ and λ_d: Lebesgue measures on $\mathbb{R}$ and $\mathbb{R}^d$.
$\nu \ll \mu$: ν is absolutely continuous with respect to μ.
$\mu_1 \perp \mu_2$: μ_1 and μ_2 are singular to each other.
$\mu_n \overset{w}{\to} \mu$: μ_n converges to μ weakly.

$(\Omega, \mathcal{A}, P)$: measure space where P is a probability measure.
E, Var, Cov: expectation, variance, and covariance.
$X \overset{D}{=} Y$: X and Y have the same probability distribution.
$X_n \Rightarrow X$, $X_n \overset{D}{\to} X$, $X_n \overset{w}{\to} X$: X_n converges to X in distribution.
$X_n \xrightarrow{P} X$: X_n converges to X in probability.
$X_n \overset{a.s.}{\to} X$, $X_n \overset{a.e.}{\to} X$: X_n converges to X almost surely/everywhere.
$F_n \Rightarrow F$, $F_n \overset{w}{\to} F$: F_n converges weakly to F.

Equations are numbered within a chapter. Definitions, theorems, examples, remarks, lemmas, corollaries are numbered within a section, within a chapter.

Preface

Measure Theory and Probability have been treated singly or jointly in numerous excellent books, varying in their level of difficulty and exposure. This book is fashioned after the three one semester Masters courses in Measure Theory, Probability Theory, and Measure Theoretic Probability, for Mathematics and Statistics students at Indian Statistical Institute (ISI). It can serve as a first course on these topics anywhere.

While the Statistics students at the ISI would already have had a heavy dose of Probability, Statistics and Analysis, the Mathematics students' exposure to basic Probability would be scant or practically non-existent. At the same time, some of these Mathematics students aspire to gain knowledge in Probability for a future teaching/research career. This book aims to be an easily accessible resource for all these groups.

Quite frequently, students struggle with Measure Theory. Moreover, students with a purely mathematics background would suddenly encounter a lot of unfamiliar concepts from probability in a Measure Theoretic Probability course. To make up for a possible lack of knowledge in basic probability, we intersperse our material with some examples, exercises, and derivations from basic probability. At the same time, we inject some topics that would be of special interest to the Mathematics students, for example, a few denseness results in function spaces and the fundamental theorem of integral calculus. The material on Probability is more or less standard, but we have broadly stuck to the syllabi mentioned above, and have excluded many topics that the reader would find in probability books mentioned later.

Having a background in basic probability would be helpful, but we do not assume that the reader has had an exposure to Probability. A reasonable familiarity with real analysis is needed.

For clarity, the material is divided into 23 short chapters. At the appearance of a new concept, adequate exercises are provided to strengthen the concept. There are additional exercises at the end of almost every chapter, and some of them are used in subsequent material.

A few results have been stated due to their importance, but the proofs do not belong to a first course. Examples of these are Bochner's theorem, law of iterated logarithm, Talagrand's inequality, and Cramèr's and Sanov's results on large deviation.

In Chapter 1, we list our goals, give a small list of books on Measure Theory and Probability, and recall some basic facts from real analysis.

In Chapter 2, after introducing the six classes of sets, semi-field, field, σ-field, monotone class, π-class, and the λ-class, we prove the Monotone Class Theorem, and the $\pi - \lambda$ Theorem of Dynkin.

In Chapter 3 we introduce positive measure spaces and their completions, and prove the First Borel-Cantelli lemma.

In Chapter 4, we state and prove the Carathéodory Extension Theorem via outer measures. Depending on the target audience, the proof of this theorem may be skipped.

Chapter 5 is devoted to a detailed discussion on the family of Lebesgue-Stieltjes measures on $\mathbb{R}$ and $\mathbb{R}^d$. This leads to the concept of a cumulative distribution function for probability measures, and of course, provides us with the construction of the Lebesgue measure on $\mathbb{R}^d$.

Chapter 6 is a quick introduction to measurable functions.

In Chapter 7, we define integrals of measurable functions with respect to general measures, and cover the essential results on integration, such as Fatou's Lemma, and Monotone and Dominated Convergence Theorems.

Chapter 8 is on the basic inequalities of Chebyshev, Hölder, Jensen, Markov and Minkowski .

The basic topological properties of various L^p spaces are discussed in Chapter 9.

Chapter 10 is on product spaces and transition measures. This is done in details for finite products. For countably infinite product spaces, we cover only the case when all measures and transition measures are probability measures. The important Theorems of Fubini and Tonelli on iterated integrals and interchange of the order of integration on product spaces is also covered in this chapter.

Chapter 11 essentially begins the Probability part, by introducing random vectors ($\mathbb{R}^d$-valued measurable functions on probability spaces). Basic concepts such as expectation, and moment generating function are introduced. The Kolmogorov Extension Theorem, guaranteeing the existence of a countable collection of random variables with a given set of consistent finite dimensional distributions, is stated without proof.

Chapter 12 is on moments and cumulants. We establish a one to one correspondence between them via the notion of multiplicative extension. This serves as a crucial tool to establish Isserlis' formula, which is also known as Wick's formula, to calculate the moments of products of Gaussian random variables. Moments and cumulants can also be used to establish convergence in distribution, in particular providing an alternate proof of the central limit theorem later.

Chapter 13 introduces convergence in measure/probability and almost uniform convergence. The concept of uniform integrability, along with its immediate consequences to convergence of integrals, are discussed.

Chapter 14 introduces the key concept in Probability, that of independence of events, σ-fields, and random variables. It also proves the Second Borel-Cantelli lemma. Conditional probability is introduced in its rudimentary form. We also briefly discuss conditional independence.

Chapter 15 is on the $0-1$ laws of Kolmogorov, and Hewitt-Savage.

Chapter 16 is on sums of independent rvs. The strong law of large numbers (SLLN) is stated and proved for pair-wise independent identically distributed rvs. This proof of Etemadi does not use the maximal inequality. A more traditional proof is outlined in the exercises. The maximal inequalities of Kolmogorov and of Lévy are stated and proved. The celebrated three series theorem of Kolmogorov that establishes the equivalence of the a.s. convergence of a series of independent rvs with the convergence of three appropriate series, those of sums of probabilities, of expectations, and of variances of the summands, is stated and proved. The basic law of iterated logarithm is stated without proof. Two exponential inequalities are covered: Hoeffding's inequality with its proof, and Talagrand's inequality without a proof.

Chapter 17 discusses the essentials of weak convergence (only for finite measures on $\mathcal{B}(\mathbb{R}^d)$), including Helly's selection principle, tightness, weak compactness, and Prokhorov's theorem.

Chapter 18 covers the basic properties of characteristic functions in one and several dimensions, including Bochner's theorem (without proof), Lev́y;s inversion formula in $\mathbb{R}^d$ and his continuity theorem.

Chapter 19 is on the celebrated Central limit theorem (CLT) in $\mathbb{R}^d$ for independent and identically distributed (iid) random variables. We present two proofs. One is via characteristic functions. The other uses the method of moments and the Mallows/Wasserstein metric, but for simplicity only for real-valued random variables. Some other easy CLTs (such as the Lindeberg CLT) which drop the iid assumption are also given. The converse of the CLT (Feller's theorem showing the necessity of the Lindeberg condition) is discussed briefly.

Chapter 20 is on signed measures. The Hahn-Jordan decomposition theorem is proved in all details.

Chapter 21 introduces absolute continuity and singularity of measures, and gives a proof of the Radon-Nikodym theorem using the Hahn-Jordan decomposition. It leads to the Lebesgue Decomposition Theorem, which decomposes any σ-finite measure on the Borel σ-field into sum of three component measures: (i) purely discrete, (ii) absolutely continuous with respect to the Lebesgue measure, and (iii) singular with respect to the Lebesgue measure but without any discrete component.

Chapter 22 is devoted to the study of some properties of real functions such as absolute continuity and bounded variation, and proves the fundamental theorem of integral calculus.

Chapter 23 is on the measure theoretic definition and basic properties of conditional expectation and probability. It also introduces the concept of regular conditional probability distribution. For a measure theory course, this chapter might be omitted, while for a probability course it is the stepping stone for study of martingales and general stochastic processes.

For a basic introduction to martingales, the reader may consider the introductory Bose et al. [2024], which is written in a spirit similar to the current book.

Work for this book has been partially supported by the author's J.C. Bose Fellowship, number JBR/2023/000023 from Anusandhan National Research Foundation, Govt. of India.

I thank Arijit Chakrabarty (ISI), and Rajat Subhra Hazra (Leiden University) for sharing their class notes. Chapters 20, 21, and 23 are based on material from our book Bose et al. [2024]

I thank Ayan Bhattacharya (IIT Bombay) for providing valuable comments. A special thank you to B.Stat. (undergraduate) student Saptashwa Baisya, who has been reading my notes diligently as independent study, and has pointed out several errors.

I am grateful to the Referees and the Handling Editor for their insightful comments and suggestions. These have significantly improved the material. I thank Rajendra Bhatia for encouraging me to write in the TRIM series.

Arup Bose March 16, 2025

Chapter 1

Preliminaries

The motivation and goals of our study may be broadly stated as follows:

1. Measure on an abstract space. In particular, the length and volume measures (Lebesgue measure), and probability measures.

2. Integrals of a function with respect to a measure, extending the Riemann integral. This leads to a fundamental concept of "expectation" of a random variable. Conditions under which the order of integration, or of limit and integration can be interchanged for functions.

4. Basic inequalities of Markov, Chebyshev, Hölder, Minkowski, and Jensen. Advanced inequalities: exponential and Talagrand.

5. Notions of convergence of measurable functions and of measures.

6. Characteristic function (Fourier transform) and its applications.

7. Measures that are not necessarily non-negative and their decomposition into difference of non-negative measures.

8. Absolute continuity and singularity of measures–Radon-Nikodym theorem. Fundamental theorem of calculus.

9. Probabilistic independence and behaviour of independent random vectors. Maximal inequalities of Kolmogorov and Lévy. $0-1$ laws, weak and strong laws, law of iterated logarithm, central limit theorem.

10. Conditional expectation given a σ-field (using the Radon-Nikodym derivative). Existence of a regular conditional probability distribution.

A. Bose, *Elements of Measure and Probability*, Texts and Readings in Mathematics 88,
https://doi.org/10.1007/978-981-95-2758-8_1

1.1 Additional books

Here is a partial list of some excellent books in Measure Theory and Probability. Browse through these as you progress.

(i) Probability: Breiman [1992], Chow and Teicher [2003], Chung [2001], Durrett [2019], Feller [1968], Gut [2013], Grimmett and Stirzaker [2002], Khoshnevisan [2007], Loéve [1977], Neveu [1965], Parthasarathy [2005], Resnick [2005], Rosenthal [2006], Williams [1991].

(ii) Measure Theory and Probability: Ash and Doléans-Dade [2000], Athreya and Lahiri [2006], Athreya and Sunder [2019], Billingsley [1995], Dudley [2002], Goswami and Rao [2025], Roussas [2014], Shorack [2017].

(iii) Measure Theory: Bogachev [2007], Halmos [1950] Cohn [2013], Royden and Fitzpatrick [2015], Stein and Shakarchi [2005], Tao [2011].

1.2 Arithmetic of $\pm\infty$

$\infty - \infty$, $-\infty + \infty$, $\pm\infty / \pm\infty$, are all undefined.

$\infty + \infty = \infty, \quad -\infty - \infty = -\infty,$

$0 \times \infty = 0 \times (-\infty) = 0$, $\infty \times \infty = -\infty \times (-\infty) = \infty$, $\infty \times -\infty = -\infty$,

For $a \in \mathbb{R}$, $a + \infty = \infty, \quad a - \infty = -\infty, \ \dfrac{a}{\infty} = \dfrac{a}{-\infty} = 0,$

For $0 < b \leq \infty$, $b \times \infty = \infty$. For $-\infty < b < 0$, $b \times \infty = -\infty$.

1.3 Bit of set theory

A set A is called **countable** if it has a $1-1$ correspondence with some subset of $\{1, 2, \ldots\}$. Otherwise it is called **uncountable**.

Example 1.3.1. (**Power set**) For any set Ω, $\mathcal{P}(\Omega)$, the class of all subsets of Ω, is called the *power set of* Ω. Let $\Omega = \{1, 2, \ldots\}$. Then $\mathcal{P}(\Omega)$ is uncountable. ▲

Recall the following operations on sets:

(i) Complement $A^c := A' := \{x : x \notin A\}$.

(ii) Difference $A \setminus B := \{x : x \in A \ \text{ and } \ x \notin B\}$.

(iii) Union $A \cup B = \{x : x \in A \text{ or } B\}$. Write $A_1 \cup \cdots \cup A_k = \cup_{n=1}^{k} A_n$.

(iv) Intersection $A \cap B = \{x : x \in A \text{ and } B\}$. Intersection of several sets is written as $A_1 \cap \cdots \cap A_k = \cap_{n=1}^{k} A_n$.

(v) $(A \cap B)^c = A^c \cup B^c$.

Disjointification: This is an extremely useful idea. Suppose that we have sets $\{A_n\}, n = 1, 2, \ldots$. Define

$$B_k := A_k - \cup_{j=1}^{k-1} A_j, k \geq 1, \quad A_0 = \emptyset.$$

Then

(i) $B_1, B_2, \ldots$ are disjoint and $B_k \subset A_k$ for all $k = 1, 2, \ldots$.

(ii) $\cup_{k=1}^{n} B_k = \cup_{k=1}^{n} A_k$ for every n, and $\cup_{k=1}^{\infty} B_k = \cup_{k=1}^{\infty} A_k$.

Suppose $\{x_k\}$ is a sequence of real numbers. Then

$$\limsup x_n := \inf_{n \geq 1} \sup_{k \geq n} x_k \quad \text{and} \quad \liminf x_n := \sup_{n \geq 1} \inf_{k \geq n} x_k. \tag{1.1}$$

Exercise 1.3.1 Show that,

(a) $\liminf x_n$ and $\limsup x_n$ always exist but can be $-\infty$ and ∞.

(b) $\limsup x_n \geq \liminf x_n$.

(c) $\lim x_n$ exists iff $\limsup x_n = \liminf x_n =: x$, and $\lim x_n = x$.

(d) For the sequence $1, -1, 1, -1, \ldots$, find $\limsup x_n$ and $\liminf x_n$.

(e) Let $\{x_n\}$ be a sequence such that $\sum_{k=1}^{\infty} |x_k - x_{k+1}| \leq \epsilon_k$ such that $\sum_{k=1}^{\infty} \epsilon_k < \infty$. Show that $\{x_n\}$ is Cauchy and hence converges.

Taking a cue from the above, we give the following definition.

Definition 1.3.1. Suppose $\{A_n\}$ is a sequence of sets. Then define

$$\limsup A_n := \cap_{n=1}(\cup_{k=n}^{\infty} A_k) \text{ and } \liminf A_n := \cup_{n=1}(\cap_{k=n}^{\infty} A_k). \quad \diamond$$

$\limsup A_n$ and $\liminf A_n$ are also written as $\overline{\lim} A_n$ and $\underline{\lim} A_n$.

Exercise 1.3.2 Show that

(a) $x \in \limsup A_n$ if and only if x belongs to A_k for infinitely many k's.

(b) $x \in \liminf A_n$ if and only if for some $k = k(x)$, $x \in \cap_{j \geq k} A_j$.

(c) As a consequence of (i) and (ii), $\limsup A_n \supseteq \liminf A_n$.

If the two sets in (c) are equal, then the common *set* is called $\lim A_n$.

Exercise 1.3.3 Construct sets $\{A_n\}$ so that $\limsup A_n \neq \liminf A_n$.

Exercise 1.3.4 If $\{A_n\}$ is a non-decreasing sequence of sets (that is $A_n \subseteq A_{n+1}$ for all n), then show that $\lim A_n = \cup_{n=1}^{\infty} A_n$.

Exercise 1.3.5 If $\{A_n\}$ is a non-increasing sequence of sets (that is $A_n \supseteq A_{n+1}$ for all n), then show that $\lim A_n = \cap_{n=1}^{\infty} A_n$.

Exercise 1.3.6 Suppose $\{A_n\}, n = 1, 2, \ldots$ is a sequence of sets. Define

$$C_n := \cup_{k=n}^{\infty} A_k, \quad D_n := \cap_{k=n}^{\infty} A_k, n = 1, 2, \ldots.$$

Show that

(a) $C_n \supseteq D_n$ for all n. Further, $\{C_n\}$ and $\{D_n\}$ are respectively, decreasing and increasing sequences of sets.

(c) $\limsup A_n = \cap_{n=1}^{\infty} C_n$, and $\liminf A_n = \cup_{n=1}^{\infty} D_n$.

Exercise 1.3.7 For real-valued functions, lim sup and lim inf are defined point-wise using Eqn. (1.1). So, if $\{f_n(\cdot)\}$ is a sequence of functions on a non-empty set Ω, then their lim sup and lim inf are:

$$f(\omega) := \limsup f_n(\omega) \quad \text{and} \quad g(\omega) := \liminf f_n(\omega).$$

Let $\{A_n\}$ be a sequence of subsets of Ω. Define the functions $\{f_n\}$ as

$$f_n(\omega) := \begin{cases} 0 & \text{if } \omega \notin A_n, \\ 1 & \text{if } \omega \in A_n. \end{cases}$$

How are $\limsup A_n$, $\liminf A_n$, and $\limsup f_n$ and $\liminf f_n$ related?

Exercise 1.3.8 Let $\{A_n\}$ and $\{B_n\}$ be sequences of sets. How are $\overline{\lim} A_n$, $\overline{\lim} B_n$, $\overline{\lim}(A_n \cup B_n)$, $\overline{\lim}(A_n \cap B_n)$ etc. related?

1.4 Topological notions

We review some results for metric spaces, though we shall mostly work with $\mathbb{R}^d$. For details, please refer to any book on real analysis.

Definition 1.4.1. (Metric) Let M be a non-empty set. We say that $d : M \times M \to [0, \infty)$ is a metric and (M, d) is a metric space if

(i) $d(x,y) = d(y,x) \geq 0$ for all $x, y \in M$.

(ii) $d(x,x) = 0$ for all $x \in M$.

(iii) $d(x,y) \leq d(x,z) + d(z,y)$ for all $x, y, z \in M$. ◊

Definition 1.4.2. (**Norm**) Let V be a vector space (over the field of real or complex numbers).

(a) A function $||\cdot|| : V \to \mathbb{R}$ is called a *norm* if

(i) $||v|| \geq 0$ for all $v \in V$.

(ii) $||\alpha v|| = |\alpha|\ ||v||$ for all $v \in V$ and scalar α.

(iii) $||v_1 + v_2|| \leq ||v_1|| + ||v_2||$, for all $v_1, v_2 \in V$.

(iv) $||v|| = 0$ implies $v = 0$. ◊

Exercise 1.4.1 If $||\cdot||$ is a norm, then show that $d(v_1, v_2) := ||v_1 - v_2||$ is a metric.

Exercise 1.4.2 Consider $\mathbb{R}^d$. Show that $||\cdot||$ defined in (1.2) is a norm.

$$||x|| = [\sum_{i=1}^{d} x_i^2]^{1/2}, x = (x_1, \ldots, x_d)' \in \mathbb{R}^d. \tag{1.2}$$

Definition 1.4.3. Let $\{x_n\}$ be a sequence in a metric space $(M,\ d)$.

(a) $\{x_n\}$ is said to converge to x if there exist an $x \in M$ such that given any $\epsilon > 0$, there exists an N such that for all $n \geq N$, $d(x_n, x) \leq \epsilon$.

(b) $\{x_n\}$ is said to be a **Cauchy sequence** if given any $\epsilon > 0$, there exists an N such that for all $n, m \geq N$, $d(x_n, x_m) \leq \epsilon$.

(c) $(M,\ d)$ is called **complete** if every Cauchy sequence converges. ◊

Definition 1.4.4. Let $(M,\ d)$ be a metric space.

(a) For any $x \in M$ and $r > 0$, $B(x,r) = \{y \in M : d(x,y) < r\}$ is called the **open ball** of radius r centered at x.

(b) Let $A \subset M$. Any $x \in M$ is said to be a **limit point** of A if for all $r > 0$, $B(x,r)$ contains at least one element of $A \setminus \{x\}$. If A has no limit points, then it is called **nowhere dense**.

(c) A set $V \subset M$ is said to be **open** if for every element of $x \in V$ there is an $r > 0$ such that $B(x,r) \subset V$.

(d) A set $C \subset M$ is said to be **closed** if its complement is open.

(e) A set $K \subset M$ is called **compact** if for every collection of open sets $\{K_\alpha, \alpha \in I\}$ such that $\cup_{\alpha \in I} K_\alpha \supseteq K$, there is a finite $F \subset I$ such that $\cap_{\alpha \in F} K_\alpha \supseteq K$; that is, every open cover of K has a finite sub-cover. ◊

Definition 1.4.5. (**Boundary of a set**) Let (M, d) be a metric space and $A \subset M$. An $x \in M$ is said to be a boundary point of A if, for all $r > 0$, $B(x, r)$ contains elements from both A and A^c. The set of all boundary points of A is denoted by ∂A. ◊

Definition 1.4.6. (**Closure and interior**) Let (M, d) be a metric space and $A \subset M$. The **closure** of A is the union of A and the set of its limit points. The **interior** of A is the set of all $x \in M$ such that there is an open ball $B(x, r) \subset A$. The closure and interior of A are denoted by $\bar{A}$ and A^o respectively. ◊

Exercise 1.4.3 Show that for $A \subset \mathbb{R}$, $(\bar{A} \setminus A) \subset \partial A$, and $(A \setminus A^o) \subset \partial A$.

Theorem 1.4.1. *(a) Arbitrary union of open sets is open. Intersection of finitely many open sets is open.*

(b) Arbitrary intersection of closed (respectively compact) sets is closed (respectively compact). Union of finitely many closed (respectively compact) sets is closed (respectively compact).

(c) Any open set in $\mathbb{R}$ is a countable union of disjoint open intervals.

(d) Any set K in $\mathbb{R}^d$ is compact if and only if it is closed and bounded.

(e) If $\{K_\alpha, \alpha \in I\}$ are compact sets in $\mathbb{R}^d$ such that $\cap_{\alpha \in I} K_\alpha = \emptyset$, then there is a finite subset F of I such that $\cap_{\alpha \in F} K_\alpha = \emptyset$. ◆

1.5 Functions

Axiom of choice. Let $\mathcal{X}$ be any collection of non-empty sets. Then there is a function $f : \mathcal{X} \to \cup_{A \in \mathcal{X}} A$ so that for every $A \in \mathcal{X}$, $f(A) \in A$.

Loosely speaking this means that given any $\mathcal{X}$, we can construct a set B containing one element from each $A \in \mathcal{X}$. This axiom is used in Section 5.2.3 to construct a "non-Borel measurable" set.

Definition 1.5.1. (**Indicator function**) For $A \subset \Omega$, $\mathbf{1}_A : \Omega \to \mathbb{R}$ defined below is called the *indicator function* of the set A:

$$\mathbf{1}_A(\omega) = \begin{cases} 1 & \text{if } \omega \in A, \\ 0 & \text{if } \omega \notin A. \end{cases} \tag{1.3}$$

◊

Definition 1.5.2. (**Positive and negative parts**) Consider a function $f:\Omega\to\bar{\mathbb{R}}$ and define

$$f^+(\omega)=\begin{cases} f(\omega) & \text{if } f(\omega)\geq 0,\\ 0 & \text{otherwise.}\end{cases} \tag{1.4}$$

$$f^-(\omega)=\begin{cases} -f(\omega) & \text{if } f(\omega)\leq 0\\ 0 & \text{otherwise.}\end{cases} \tag{1.5}$$

Then f^+ and f^- are called the *positive and negative parts* of f. ◇

Exercise 1.5.1 If f is real-valued, then $|f|=f^+ + f^-$ and $f=f^+ - f^-$.

1.5.1 Image and inverse image

Definition 1.5.3. (i) Suppose $f:\Omega_1\to\Omega_2$ is a function. Then the **inverse of** f, denoted by f^{-1} is defined by

$$f^{-1}(A_2)=\{x\in\Omega_1 : f(x)\in A_2\}\ \text{ for all }\ A_2\in\mathcal{P}(\Omega_2).$$

(ii) Let $f:\Omega_1\to\Omega_2$ and $g:\Omega_2\to\Omega_3$ be functions. Their composition $g\circ f:\Omega_1\to\Omega_3$ is defined as $g\circ f(x)=g(f(x))$, $x\in\Omega_1$. ◇

Exercise 1.5.2 Suppose $f:\Omega_1\to\Omega_2$ is any function. Suppose $\{A_\alpha,\alpha\in I\}$ and $\{B_\alpha,\alpha\in I\}$ are subsets of Ω_1 and Ω_2 respectively. Show that

$$\begin{aligned} f(\cup_{\alpha\in I}A_\alpha) &= \cup_{\alpha\in I} f(A_\alpha),\\ f(\cap_{\alpha\in I}A_\alpha) &\subseteq \cap_{\alpha\in I} f(A_\alpha),\ \text{(equality may not hold)},\\ (f(A_\alpha))^c &= (f(A_\alpha^c))\ \text{does } \textbf{not} \text{ hold in general,}\end{aligned}$$

$$\begin{aligned} f^{-1}(B_\alpha^c) &= (f^{-1}(B_\alpha))^c,\\ f^{-1}(\cup_{\alpha\in I}B_\alpha) &= \cup_{\alpha\in I} f^{-1}(B_\alpha),\\ f^{-1}(\cap_{\alpha\in I}B_\alpha) &= \cap_{\alpha\in I} f^{-1}(B_\alpha).\end{aligned}$$

1.5.2 Semi-continuity and approximation

Recall that a function $f:(M,\ d)\to\mathbb{R}^d$ is continuous if for every open set $B\subseteq\mathbb{R}^d$, $f^{-1}(B)$ is an open set of M.

Lemma 1.5.1. (**Urysohn**) Let $(M,\ d)$ be a metric space, K be a closed set and $U \supseteq K$ be an open set. Then there exists a continuous function $f : M \to [0,\ 1]$ such that $\mathbf{1}_K(\omega) \leq f(\omega) \leq \mathbf{1}_U(\omega)$ for all $\omega \in M$. ♦

Theorem 1.5.1. *If $F \subset \mathbb{R}$ is closed, then $\mathbf{1}_F$ is the limit of some non-increasing uniformly continuous functions f_n, $0 \leq f_n \leq 1$.* ♦

Definition 1.5.4. (**Semi-continuous function**) A function $f : \mathbb{R} \to \mathbb{R}$ is **lower semi-continuous** (lsc) at x if $\liminf_{y \to x} f(y) \geq f(x)$. If f is lsc at every x, then we say that it is a lower semi-continuous function. A function f is **upper semi-continuous** (usc) if $-f$ is lsc. ◊

Exercise 1.5.3 Suppose $A \subset \mathbb{R}$. Show the following:

(a) If A is open then $\mathbf{1}_A$ is a lower semi-continuous function.

(b) If A is closed then $\mathbf{1}_A$ is an upper semi-continuous function.

Exercise 1.5.4 Suppose that f is lower or upper semi-continuous such that $|f| \leq M < \infty$. Show that there exists continuous functions $\{g_n\}$ such that $|g_n| \leq M$, and $g_n \to f$ point-wise.

Definition 1.5.5. For any bounded function f, define the **upper and lower envelope** functions of f respectively as,

$$\overline{f} = \inf\{g : g \text{ is usc},\ g \geq f\},\ \underline{f} = \sup\{g : g \text{ is lsc},\ g \leq f\}.$$ ◊

Exercise 1.5.5 (a) Show that $\underline{f}$ and $\overline{f}$ are respectively lsc and usc.

(b) Further,

$$\overline{f}(x) = \limsup_{y \to x} f(y) \quad \text{and} \quad \underline{f}(x) = \liminf_{y \to x} f(y) \quad \text{for all} \quad x,$$
$$f(x) = \underline{f}(x) = \overline{f}(x) \quad \text{for all continuity points } x \text{ of } f.$$

1.6 Riemann integral

Consider any partition of $[a,\ b]$, a, b finite, and its norm as:

$$\pi := \{a = x_0 < x_1 < \cdots < x_n = b\},\ |\pi| := \max_{1 \leq i \leq n} (x_i - x_{i-1}).$$

Let f be a bounded real valued function on $[a,\ b]$. Define,

$$\begin{aligned} m_i &:= \inf\{f(x) : x_{i-1} < x \leq x_i\},\ i = 1, \ldots, n, \\ M_i &:= \sup\{f(x) : x_{i-1} < x \leq x_i\},\ i = 1, \ldots, n, \\ M &:= \sup_{a \leq x \leq b} |f(x)| < \infty\,. \end{aligned}$$

We have suppressed the dependence of m_i and M_i on π in our notation. Define the *upper and lower sums* of f on $[a,\ b]$ by,

$$U_{\pi,n} := \sum_{i=1}^{n} M_i(x_i - x_{i-1}), \quad L_{\pi,n} := \sum_{i=1}^{n} m_i(x_i - x_{i-1}).$$

Definition 1.6.1. Let $f : [a,\ b] \to \mathbb{R}$ be bounded. Then f is said to be Riemann integrable if $\lim_{|\pi|\to 0} L_{\pi,n} = \lim_{|\pi|\to 0} U_{\pi,n}$, is finite and is independent of the sequence of partitions. The common value is called the *Riemann integral* of f on $[a,\ b]$, written as $R(f) = \int_a^b f(x)dx$. ◇

Exercise 1.6.1 Show that if f is a bounded continuous function on $[a,\ b]$, then it is Riemann integrable.

Exercise 1.6.2 Give an example of a sequence of functions $\{f_n\}$ on $[a,\ b]$ such that they are bounded by 1, each f_n is Riemann integrable, $f_n(x) \to f(x)$ for all $x \in [a,\ b]$ but f is not Riemann integrable.

Chapter 2

Classes of sets

As we shall see later, measures can be defined only for classes of sets with certain properties. This chapter provides the definition and properties of several relevant classes of sets. In particular, we introduce the notions of semi-field, field, monotone class, π- and the λ-systems, σ-field and the Borel σ-field. We also prove the monotone class and the $\pi - \lambda$ theorems.

2.1 Sample space and events

A **sample space** is a non-empty set, usually denoted by Ω. As an example, the four outcomes of two tosses of a coin can be listed as $\Omega = \{HH, TT, TH, HT\}$. More generally, this set Ω can be *countably infinite* or *uncountably infinite*. We shall see plenty of examples of all kinds later.

Then we have a collection of *appropriate* subsets of Ω which we shall call **events**. This collection is contextual, and, even with the same Ω, we may have more than one relevant collection. This choice may vary, depending on our goal. Natural conditions on this collection are imposed, which can be more, or less strict, again depending on our goals.

Calculation of measure, or probability, is always done for **events**. For example, probability of "one heads in two tosses", is the probability of the event $\{HT, TH\}$. Events can become complicated (for example by combination of "simple" events), and soon we face the question "*what constitutes the collection of all events in a given scenario?*"

This collection of events shall vary depending on the problem at hand. But we may turn around this question and stipulate some "natu-

A. Bose, *Elements of Measure and Probability*, Texts and Readings in Mathematics 88,
https://doi.org/10.1007/978-981-95-2758-8_2

ral" conditions that events should satisfy, irrespective of what question on measure or probability we are trying to answer. As an example, for which types of sets in $\mathbb{R}$ can we talk about their "lengths"? These sets would of course include all intervals, and their finite unions. We shall see later that many other sets would be included (all open and closed sets, for example), but many more will be forced to be left out.

In the next sections, we define *six* different types of subsets of Ω, and investigate their properties. These are going to be useful to us later, especially in the construction and extension of measures. en

2.2 Semi-field

While this is not a full class of events, it serves as a very convenient stepping stone to larger classes of events. It will also play a major role in the development of the length measure.

Definition 2.2.1. (**Semi-field**) Let Ω be a non-empty set. A (non-empty) collection $\mathcal{S}$ of subsets of Ω is called a *semi-field* if,

(i) it is closed under finite intersection, and

(ii) for any $A \in \mathcal{S}$, A^c is a finite disjoint union of sets from $\mathcal{S}$. ◇

Exercise 2.2.1 Let Ω be the interval $(0,\ 1]$. Let $\mathcal{S}$ be the collection of all sub-intervals of Ω of the form $(a,\ b]$. Show that $\mathcal{S}$ is a semi-field.

For any collection $\mathcal{C}$ of subsets of Ω, and any subset A of Ω, define,

$$\mathcal{C} \cap A := \{B \cap A : B \in \mathcal{C}\}. \tag{2.1}$$

Exercise 2.2.2 Let $\mathcal{S}$ be a semi-field and let E be any subset of Ω. Show that $\mathcal{S} \cap E$ is a semi-field (of subsets of E).

2.3 Field

Definition 2.3.1. (**Field**) Let Ω be a non-empty set. Any collection $\mathcal{F}$ of subsets of Ω is said to be a *field* (or an **algebra**) of subsets of Ω, if **all** the following three conditions are satisfied:

(i) $\Omega \in \mathcal{F}$,

(ii) if $A \in \mathcal{F}$, then $A^c \in \mathcal{F}$,

(iii) if $\{A_i\}_{1 \leq i \leq n}$ belong to $\mathcal{F}$, then $\cup_{i=1}^{n} A_i \in \mathcal{F}$. ◇

Exercise 2.3.1 Suppose $\mathcal{F}$ is a collection of subsets of Ω such that $\Omega \in \mathcal{F}$, and $A, B \in \mathcal{F}$ implies $A \setminus B \in \mathcal{F}$. Show that $\mathcal{F}$ is a field.

Exercise 2.3.2 Show that any field is closed under **finite intersection** and **finite differencing**, and in (iii) of Definition 2.3.1, union could be replaced by intersection.

The empty set is a member of $\mathcal{F}$, and any field is a semi-field. A semi-field or a field is always defined with respect to a sample space Ω. Later, we may not always mention this space explicitly, since after all, our main focus will be on events. This definition of field/algebra is in no way connected to the definition of field/algebra in the theory of algebra.

Exercise 2.3.3 Show that a semi-field which is closed under complementation is a field. Caution! A field must satisfy three conditions.

Exercise 2.3.4 Show that the $\mathcal{S}$ of Exercise 2.2.1 is not a field.

Example 2.3.1. Let $\Omega := (0, 1]$, and

$$\mathcal{F} := \left\{ \cup_{j=1}^{n} (a_j, b_j] : a_1, \ldots, a_n, b_1, \ldots, b_n \in [0, 1],\, n \geq 1 \right\}.$$

Then $\mathcal{F}$ is a field. That $\mathcal{F}$ is closed under finite unions is obvious. To see that $\mathcal{F}$ is closed under complement, observe that

$$\begin{aligned} \left(\cup_{j=1}^{n} (a_j, b_j] \right)^c &= \cap_{j=1}^{n} (a_j, b_j]^c \\ &= \cap_{j=1}^{n} \Big((0, a_j] \cup (b_j, 1] \Big) =: \cap_{j=1}^{n} (A_{1j} \cup A_{2j}) \\ &= \cup_{(j_1, \ldots, j_n) \in \{1,2\}^n} \left(A_{j_1 1} \cap \ldots \cap A_{j_n n} \right), \end{aligned}$$

for some appropriate $A_{j_t, t}$, $1 \leq t \leq n$ each of which is a left-open, right-closed interval. ▲

Exercise 2.3.5 Suppose $\mathcal{S}$ is a semi-field. Show that the collection of all finite disjoint union of sets from $\mathcal{S}$ is a field.

Exercise 2.3.6 Let Ω be a non-empty set, and $A \subset \Omega$ also be non-empty.

(a) What is the smallest field containing Ω?

(b) What is the smallest field containing A and Ω?

Exercise 2.3.7 Let Ω be any set, and

$$\mathcal{F}_0 := \{ A \subset \Omega : \text{either } A \text{ is finite or } A^c \text{ is finite} \}.$$

(a) Show that $\mathcal{F}_0$ is a field, called the "**co-finite**" field.

(b) If $\Omega := \mathbb{N} := \{1, 2, \ldots\}$, does the set $\{2, 4, 6, , \ldots\}$ belong to $\mathcal{F}_0$?

Exercise 2.3.8 Show that arbitrary intersection of fields is a field.

Definition 2.3.2. (**Minimal field**) Let $\mathcal{C}$ be a collection of subsets of Ω. The smallest field containing all sets of $\mathcal{C}$ is called the *minimal field* containing $\mathcal{C}$ or the field generated by $\mathcal{C}$, and is written as $\mathcal{F}(\mathcal{C})$. ◇

Exercise 2.3.9 Let $\mathcal{C}$ be a class of subsets of Ω. Verify that we can describe the smallest field $\mathcal{F}(\mathcal{C})$ containing $\mathcal{C}$ in the following way. Let

$$\begin{aligned} \mathcal{D} &:= \{A : A = \cap_{j=1}^{n} B_j, n \geq 1, \text{ each } B_j, \text{or } B_j^c \text{ is in } \mathcal{C}\}, \\ \mathcal{G}_1 &:= \{\cup_{i=1}^{n} A_i : n \geq 1, \text{each } A_i \in \mathcal{D}\}, \\ \mathcal{G}_2 &:= \{\cup_{i=1}^{n} A_i : n \geq 1, \text{each } A_i \in \mathcal{D}, \text{ and they are disjoint}\}. \end{aligned}$$

Show that $\mathcal{G}_1 = \mathcal{G}_2 = \mathcal{F}(\mathcal{C})$.

Exercise 2.3.10 Suppose $\mathcal{F}$ is a field. Let B be a subset of Ω. Consider the collection of sets $\mathcal{F} \cap B$ as defined in (2.1). Show that this is a field of subsets of B (hence complement is taken with respect to B).

2.4 σ-field

Suppose we have an infinite sequence of coin tossing games, and we are interested in the event A that one head occurs, somewhere. Let A_i be the event that the ith toss is head. Then $A = \cup_{i=1}^{\infty} A_i$. Thus, in general, we would like countable union of events to also be events. This leads to the following definition of the most widely accepted class of events.

Definition 2.4.1. (**σ-field**) Let $\Omega \neq \emptyset$ and $\mathcal{A}$ be a collection of subsets of Ω. Then $\mathcal{A}$ is said to be a *σ-field* if, it is a field, and also satisfies: if $\{A_i\}_{1 \leq i < \infty}$ belong to $\mathcal{A}$, then $\cup_{i=1}^{\infty} A_i \in \mathcal{A}$. That is $\mathcal{A}$ is a field and is closed under countable unions. The pair $(\Omega, \mathcal{A})$ is called a **measurable space** and sets in $\mathcal{A}$ are called measurable sets. ◇

Example 2.4.1. Suppose we toss a coin infinite times. Let Ω be the set of all infinite sequences of H and T, or be the set $\{0, 1, \ldots\}$, which indicates how many tosses were required before we got the first H. In both cases we can take the class of all subsets of Ω, $\mathcal{P}(\Omega)$ as the σ-field. The sizes of the two σ-fields are of different orders of ∞. ▲

Exercise 2.4.1 Give an example of a field which is not a σ-field.

Example 2.4.2. For a given non-empty set Ω, $\mathcal{A}$ defined below is a σ-field, called the "**co-countable**" σ-field.

$$\mathcal{A} := \{A \subset \Omega : \text{either } A \text{ is countable, or } A^c \text{ is countable}\} . \qquad \blacktriangle$$

Exercise 2.4.2 Show that σ-fields are closed under *countable intersection.*

Exercise 2.4.3 Show that a field which is closed under increasing limits is a σ-field.

Exercise 2.4.4 Suppose $\{\mathcal{A}_n\}$ is a sequence of σ-fields of subsets of Ω. Then is their union a σ-field? What if the sequence is increasing?

Exercise 2.4.5 Show that intersection of σ-fields is again a σ-field.

Definition 2.4.2. (**Minimal σ-field**) Let $\mathcal{C}$ be a collection of subsets of Ω. The smallest σ-field containing all sets of $\mathcal{C}$ is the *minimal σ-field* containing $\mathcal{C}$, or the **σ-field generated by** $\mathcal{C}$, and is written as $\sigma(\mathcal{C})$. ◇

Exercise 2.4.6 Let Ω be a non-empty set. Let $\mathcal{F}$ and $\mathcal{A}$ be the co-finite field, and the co-countable σ-field respectively. Show that $\mathcal{A} = \sigma(\mathcal{F})$.

Definition 2.4.3. (**Restricted σ-field**) Let $(\Omega, \mathcal{A})$ be a measurable space and $\emptyset \neq B \in \mathcal{A}$. Let $\mathcal{A}_B := \mathcal{A} \cap B$. Then the σ-field $\mathcal{A}_B$ of subsets of B is called the *restriction of $\mathcal{A}$ on B.* ◇

Exercise 2.4.7 Verify that $\mathcal{A} \cap B$ is indeed a σ-field of subsets of B.

Exercise 2.4.8 Let $\mathcal{C} \subset \mathcal{P}(\Omega)$ and $\emptyset \neq A \subset \Omega$. Show that

$$\sigma(\mathcal{C} \cap A) = \sigma(\mathcal{C}) \cap A, \quad \text{a } \sigma\text{-field of subsets of } A. \tag{2.2}$$

Exercise 2.4.9 Let $\Omega = \mathbb{R}$ and $\mathcal{C} = \{\{x\} : x \in \mathbb{R}\}$. Describe $\sigma(\mathcal{C})$.

Definition 2.4.4. (**Countably generated field and σ-field**) A field or a σ-field $\mathcal{A}$ of subsets of Ω, is said to be *countably generated,* if there exists a countable collection $\{C_i\}$ such that $\mathcal{A} = \sigma(C_1, C_2, \ldots)$. ◇

2.5 Borel σ-field

When the sample space is $\mathbb{R}$, $\mathbb{R}^d$, or more generally a metric space M, the most natural σ-field to work with is the so-called Borel σ-field on it.

Definition 2.5.1. (**Borel σ-field on** $\mathbb{R}$) The minimal σ-field containing all intervals of $\mathbb{R}$ is called the *Borel σ-field* of $\mathbb{R}$, denoted by $\mathcal{B}(\mathbb{R})$. ◇

Exercise 2.5.1 Show that $\mathcal{B}(\mathbb{R})$ is countably generated.

Exercise 2.5.2 Consider $\Omega = \mathbb{R}$. Show that the smallest σ-field that contains each of the following collections equals $\mathcal{B}(\mathbb{R})$.

(i) all intervals of finite length;

(ii) all closed intervals;

(iii) all open intervals;

(iv) all left-open right-closed intervals (including $-\infty$ as a left limit);

(v) all left-closed right-open intervals (including ∞ as a right limit).

Exercise 2.5.3 Show that $\mathcal{B}(\mathbb{R})$ contains the following collection of sets. For which of these collections is the smallest σ-field equal to $\mathcal{B}(\mathbb{R})$?

(i) all finite sets;

(ii) all compact sets;

(iii) all closed sets.

$\mathcal{B}(\mathbb{R})$ contains many more sets. In Section 5.2.3, we shall see that it does not equal $\mathcal{P}(\mathbb{R})$, the collection of all subsets of $\mathbb{R}$.

The Borel σ-field on $\bar{\mathbb{R}}$, written $\mathcal{B}(\bar{\mathbb{R}})$ is the smallest σ-field generated by all intervals of the form $(a, b]$, $a, b \in \bar{\mathbb{R}}$.

Exercise 2.5.4 Let $f : [a,\ b] \to \mathbb{R}$ be a bounded function. Show that the set of continuity points of f is a Borel set.

Definition 2.5.2. (**Borel σ-field**) Let $(M,\ d)$ be a metric space. The smallest σ-field $\mathcal{B}(M)$ containing all open sets is called the *Borel σ-field of M*. The elements of $\mathcal{B}(M)$ are called Borel sets. ◇

Exercise 2.5.5 Let A be a Borel set in $\mathbb{R}^d$. Show that $\bar{A}$, A^o, and ∂A (see Definitions 1.4.5 and 1.4.6) are also Borel sets.

2.6 Monotone class

Definition 2.6.1. (**Monotone class**) Let Ω be a non-empty set. Any class of subsets $\mathcal{M}$ of Ω, is said to be a *monotone class* if it is closed under increasing and decreasing limits of sets. That is,

(i) if $\{A_n\}$ is a non-decreasing sequence from $\mathcal{M}$, then $\lim A_n \in \mathcal{M}$, and

(ii) if $\{A_n\}$ is a non-increasing sequence from $\mathcal{M}$, then $\lim A_n \in \mathcal{M}$. ◇

Clearly any σ-field is a monotone class. A field which has only a finite number of sets is automatically a monotone class.

Exercise 2.6.1 Construct examples of fields that are not monotone classes and vice-versa.

Exercise 2.6.2 Show that if $\mathcal{M}$ is a field and is also a monotone class, then it is a σ-field.

Exercise 2.6.3 Show that arbitrary intersection of monotone classes is a monotone class. Hence, given a class of sets $\mathcal{C}$, there is the **smallest monotone class** that contains $\mathcal{C}$, usually denoted by $\mathcal{M}(\mathcal{C})$.

The following result links the notions of field, monotone class and σ-field, is going to be a crucial tool and shall be used innumerable times.

Theorem 2.6.1 (**Monotone class theorem**). *Let $\mathcal{F}$ be a field. Then*

(a) $\mathcal{M}(\mathcal{F}) = \sigma(\mathcal{F})$;

(b) if $\mathcal{M}$ is a monotone class such that $\mathcal{M} \supseteq \mathcal{F}$, then $\mathcal{M} \supseteq \sigma(\mathcal{F})$. ◆

Proof. (a) We immediately have $\mathcal{M}(\mathcal{F}) \subseteq \sigma(\mathcal{F})$ since both are monotone classes. We need to show the other inclusion.

(i) Fix $A \in \mathcal{M}(\mathcal{F})$. Let

$$\mathcal{M}_A := \{B \in \mathcal{M}(\mathcal{F}) : A \cap B, A \cap B^c, \text{ and } A^c \cap B \in \mathcal{M}(\mathcal{F})\}. \tag{2.3}$$

It is easy to check that $\mathcal{M}_A$ is a monotone class.

(ii) Now suppose $A \in \mathcal{F}$. Then every $B \in \mathcal{F}$ satisfies the condition in (2.3). Thus $\mathcal{F} \subseteq \mathcal{M}_A$. This implies

$$\mathcal{M}(\mathcal{F}) \subseteq \mathcal{M}_A, \text{ since both are monotone classes containing } \mathcal{F}.$$

On the other hand, by definition, $\mathcal{M}_A \subseteq \mathcal{M}(\mathcal{F})$. Hence

$$\mathcal{M}_A = \mathcal{M}(\mathcal{F}). \tag{2.4}$$

(iii) But (2.4) means that for any $B \in \mathcal{M}(\mathcal{F})$, and any $A \in \mathcal{F}$,

$$A \cap B,\ A \cap B^c \text{ and } A^c \cap B \in \mathcal{M}(\mathcal{F})\,. \tag{2.5}$$

Hence, $\mathcal{M}_B \supseteq \mathcal{F}$. This in turn implies that $\mathcal{M}_B \supseteq \mathcal{M}(\mathcal{F})$. But we already know $\mathcal{M}_B \subseteq \mathcal{M}(\mathcal{F})$. Hence

$$\mathcal{M}_B = \mathcal{M}(\mathcal{F}). \tag{2.6}$$

We now claim that $\mathcal{M}(\mathcal{F})$ is a field. This follows immediately since (2.4), (2.5) and (2.6) hold whenever $A, B \in \mathcal{M}(\mathcal{F}) = \mathcal{M}_A$.

By Exercise 2.6.2, $\mathcal{M}(\mathcal{F})$ is a σ-field. But then $\mathcal{M}(\mathcal{F})$ and $\sigma(\mathcal{F})$ must be equal.

Part (b) follows easily from Part (a). ■

2.7 π- and λ-classes

The so-called π- and λ- classes serve to disentangle the defining properties of a σ-field. They lead to a result which, as a technical tool, is an alternative to the monotone class Theorem 2.6.1. In this book, we shall hardly ever use this result and instead rely on Theorem 2.6.1.

Definition 2.7.1. (**π-class**) A class of subsets $\mathcal{P}$ of Ω is a *π-class*, if it is closed under finite intersections. ◇

Definition 2.7.2. A class of subsets $\mathcal{D}$ of Ω is a *λ-class*, if the following three conditions hold.

(i) $\Omega \in \mathcal{D}$.

(ii) If $E, F \in \mathcal{D}$ and $E \subset F$, then $F \setminus E \in \mathcal{D}$.

(iii) For all $n \in \mathbb{N}$, if $E_n \in \mathcal{D}$ and $E_n \subset E_{n+1}$, then $\cup_{n\in\mathbb{N}} E_n \in \mathcal{D}$. ◇

Exercise 2.7.1 Show that arbitrary intersection of λ-classes is a λ-class.

Any σ-field is a λ-class. If $\mathcal{C} \subset \mathcal{P}(\Omega)$, then $\lambda(\mathcal{C})$ denotes the smallest λ-class that contains $\mathcal{C}$. Since any σ-field is a λ-class, $\lambda(\mathcal{C}) \subset \sigma(\mathcal{C})$.

The next lemma ties the three notions of π-class, λ-class and σ-field.

Lemma 2.7.1. $\mathcal{A}$ is a σ-field if and only if, it is a π- and a λ-class. ♦

Proof. Clearly, any σ-field is a π- and a λ-class. So, suppose that $\mathcal{A}$ is a π-class and a λ-class. Then we have the following observations.

(i) Since $\mathcal{A}$ is a λ-class, $\Omega \in \mathcal{A}$.

(ii) Let $A \in \mathcal{A}$. Then $A^c = \Omega \setminus A \in \mathcal{A}$ since $\mathcal{A}$ is a λ-class. So $\mathcal{A}$ is closed under complementation.

(iii) Let $E, F \in \mathcal{A}$. By (ii), E^c, $F^c \in \mathcal{A}$. Now $E \cup F = (E^c \cap F^c)^c \in \mathcal{A}$, since $\mathcal{A}$ is a π-class and (ii) holds. So, in view of (i) and (ii), $\mathcal{A}$ is a field.

Now let $E_1, E_2, \dots$ be a sequence in $\mathcal{A}$. Then by (ii), $F_n = \cup_{i=1}^n E_i \in \mathcal{A}$ for every n. But then F_n forms an increasing sequence, and since $\mathcal{A}$ is a λ-class, $\cup_n E_n = \cup_n F_n \in \mathcal{A}$. This proves the lemma completely. ■

The following useful lemma gives a condition on $\mathcal{C}$ so that $\lambda(\mathcal{C}) = \sigma(\mathcal{C})$. Its proof is similar to the proof of the monotone class Theorem 2.6.1.

Lemma 2.7.2 (Dynkin). If $\mathcal{P}$ is a π-class, then $\lambda(\mathcal{P}) = \sigma(\mathcal{P})$. ♦

Proof. Since any σ-field is a λ-class, $\lambda(\mathcal{P}) \subset \sigma(\mathcal{P})$. By Lemma 2.7.1, it is enough to show that $\lambda(\mathcal{P})$ is a π-class. We establish this in two steps.

Step 1. Let

$$\mathcal{D}_1 = \{B \in \lambda(\mathcal{P}) : B \cap C \in \lambda(\mathcal{P}) \text{ for all } C \in \mathcal{P}\}.$$

We claim that $\lambda(\mathcal{P}) = \mathcal{D}_1$. To see this, observe the following:

(i) Evidently, $\Omega \in \mathcal{D}_1$.

(ii) Let $B_1, B_2 \in \mathcal{D}_1$ with $B_1 \subset B_2$. We claim that $B_2 \setminus B_1 \in \mathcal{D}_1$. To see this, let $C \in \mathcal{P}$. Then

$$(B_2 \setminus B_1) \cap C = (B_2 \cap C) \setminus (B_1 \cap C) \in \lambda(\mathcal{P}),$$

since the two intersections on the right side belong to $\lambda(\mathcal{P})$ by definition of $\mathcal{D}_1$. Hence their difference belongs to $\lambda(\mathcal{P})$, it being a λ-class.

(iii) Now let $B_n \uparrow B, B_n \in \mathcal{D}_1$. To show that $B \in \mathcal{D}_1$. Let $C \in \mathcal{P}$. Then $(B_n \cap C) \in \lambda(\mathcal{P})$. This sequence converges to $B \cap C$, which then belongs to $\lambda(\mathcal{P})$ since it is a λ-class. So $B \in \mathcal{D}_1$.

Hence $\mathcal{D}_1$ is a λ-class. Now, since $\mathcal{P}$ is a π-class, it is closed under intersection. So, observing the definition of $\mathcal{D}_1$, clearly $\mathcal{P} \subset \mathcal{D}_1$. Hence $\lambda(\mathcal{P}) \subset \mathcal{D}_1$. But by definition $\mathcal{D}_1 \subset \lambda(\mathcal{P})$. Hence $\lambda(\mathcal{P}) = \mathcal{D}_1$.

Step 2. Let

$$\mathcal{D}_2 = \{C \in \lambda(\mathcal{P}) : B \cap C \in \lambda(\mathcal{P}), \text{ for all } B \in \lambda(\mathcal{P})\}.$$

We claim that $\mathcal{D}_2$ is a λ-class. We omit its proof which would be similar to the proof of Step 1.

Now let $C \in \mathcal{P}$ and $B \in \lambda(\mathcal{P}) = \mathcal{D}_1$. Then $B \cap C \in \lambda(\mathcal{P})$ and hence $C \in \mathcal{D}_2$. Hence $\mathcal{P} \subset \mathcal{D}_2$. This in turn implies that $\lambda(\mathcal{P}) \subset \mathcal{D}_2$, but by definition, $\mathcal{D}_2 \subset \lambda(\mathcal{P})$. Hence $\mathcal{D}_2 = \lambda(\mathcal{P})$. Observing the definition of $\mathcal{D}_2$, we conclude that $\lambda(\mathcal{P})$ is a π-class, as desired. ■

Exercise 2.7.2 Show that Lemma 2.7.2 is equivalent to the statement: Let $\mathcal{P}$ be a π-class and $\mathcal{D}$ be a λ-class. If $\mathcal{P} \subset \mathcal{D}$, then $\sigma(\mathcal{P}) \subset \mathcal{D}$.

2.8 Exercises

Exercise 2.8.1 (a) Let $\mathcal{S}_i, i = 1, 2$ be semi-fields of subsets of $\Omega_i, i = 1, 2$ respectively. Show that $\mathcal{S}_1 \times \mathcal{S}_2$ is a semi-field of subsets of $\Omega_1 \times \Omega_2$.

(b) Show by an example, that the claim in (a) is not true if semi-fields are replaced by either fields or σ-fields.

Exercise 2.8.2 (a) Let $\emptyset \neq \Omega$, and $A_1, A_2 \subset \Omega$. Describe $\sigma(A_1, A_2, \Omega)$.

(b) Suppose $\mathcal{C} := \{A_i\}, 1 \leq i \leq n$ is a collection of *non-empty disjoint* subsets of Ω such that their union is Ω. That is, it is a **partition** of Ω. Describe $\mathcal{F}(\mathcal{C})$. How many sets are in it?

(c) A set A in a field $\mathcal{F}$ is said to be an **atom**, if there is no non-empty proper subset of A which is in $\mathcal{F}$.

(i) If $\mathcal{F}$ is finite, describe its atoms in terms of intersection and complements of sets from $\mathcal{F}$.

(ii) Show that any field which is finite, has exactly 2^n sets for some $n \geq 1$.

Exercise 2.8.3 Let $\mathcal{C} := \{A_i\}$ be a countable partition of Ω.

(a) Describe $\mathcal{F}(\mathcal{C})$. How many sets are there in this field? What will happen if the partition is uncountable?

(b) Describe $\sigma(\mathcal{C})$. How many sets are there in this σ-field?

Exercise 2.8.4 Show that there is no countably infinite σ-field.

Exercise 2.8.5 Let $\mathcal{C}$ be any class of subsets of Ω which contains the empty set. Show that $E \in \sigma(\mathcal{C})$ iff there exists $C_1, C_2, \ldots \in \mathcal{C}$ (this collection is allowed to depend on E) such that $E \in \sigma(C_1, C_2, \ldots)$.

Exercise 2.8.6 Let $E \in \mathcal{B}(\mathbb{R})$. Show that $\{(x,y) : x+y \in E\}$ is a Borel set in $\mathbb{R}^2$. Can you generalise to other functions of (x,y)?

Exercise 2.8.7 Let $\mathcal{F} \subset \mathcal{P}(\Omega)$, such that $\Omega \in \mathcal{F}$ and $\mathcal{F}$ is closed under complementation and finite disjoint union. Prove or give a counter-example to the claim "then $\mathcal{F}$ a field".

Exercise 2.8.8 Can you prove Dynkin's Lemma 2.7.2 using Monotone Class Theorem 2.6.1 and vice-versa?

Chapter 3

Introduction to measures

When we calculate weighted average, length, area of a surface, volume of a solid, area under a curve, and the probability of an event, there is an underlying scale to compute the *measure*—the length measure (Lebesgue measure on $\mathbb{R}$), the surface measure, the volume measure and a probability measure. We will now give formal definition for measures on fields and σ-fields and look at their basic properties.

3.1 Countably additive measure

Definition 3.1.1. (**Set function**) Let $\mathcal{C} \subset \mathcal{P}(\Omega)$. Any function μ from $\mathcal{C}$ to $\bar{\mathbb{R}}$ will be called a *set function* on $\mathcal{C}$. $\diamond$

Definition 3.1.2. (**Measure**) (a) Suppose $\mu : \mathcal{A} \to \mathbb{R}^+ \cup \{\infty\}$ is a set function on a σ-field $\mathcal{A}$. Then μ is called a **measure** on $\mathcal{A}$ if,

$$\mu(\cup_{n=1}^{\infty} A_n) = \sum_{n=1}^{\infty} \mu(A_n) \text{ for all disjoint } \{A_i\} \text{ from } \mathcal{A}. \tag{3.1}$$

(b) Suppose $\mathcal{F}$ is a field. Then a set function $\mu : \mathcal{F} \to \mathbb{R}^+ \cup \{\infty\}$ is called a **measure** on $\mathcal{F}$ if

$$\mu(\cup_{n=1}^{\infty} A_n) = \sum_{n=1}^{\infty} \mu(A_n) \text{ for all disjoint } \{A_i\} \text{ with } \cup_{n=1}^{\infty} A_n \in \mathcal{F}. \tag{3.2}$$

The triplets $(\Omega, \mathcal{F}, \mu)$ and $(\Omega, \mathcal{A}, \mu)$ are called **measure spaces**. $\diamond$

A. Bose, *Elements of Measure and Probability*, Texts and Readings in Mathematics 88,
https://doi.org/10.1007/978-981-95-2758-8_3

Exercise 3.1.1 The sum in (3.2) does not depend on the order in which the sequence $\{A_i\}$ is written. This is due to the following fact: suppose $\{x_i\}, i \geq 1$ is any sequence of non-negative numbers. Let $\pi(i), i \geq 1$ be any permutation of $i, i \geq 1$. Then, $\sum_{i=1}^{\infty} x_i = \sum_{i=1}^{\infty} x_{\pi(i)}$.

Remark 3.1.1. (a) Every event must have a measure. That is, $\mu(A)$ must be defined for all $A \in \mathcal{A}$ (or $\mathcal{F}$). Further, $\mu(A)$ can be ∞. To avoid triviality, we shall assume that, there is at least one A in $\mathcal{A}$ or $\mathcal{F}$ with $\mu(A) < \infty$. In that case, it follows easily $\mu(\emptyset) = 0$. The alternative to all this is to add the condition $\mu(\emptyset) = 0$ to Definition 3.1.2.

(b) Properties (3.1) and (3.2) are called **countable additivity** of μ on $\mathcal{A}$ and $\mathcal{F}$ respectively. One may demand only the weaker *finite additivity.* Section 3.5 has a brief discussion on finitely additive measures. In any case, countable additivity is a generally accepted required property.

(c) Why do not we always take $\mathcal{P}(\Omega)$ as the underlying σ-field, and declare every subset of Ω as an event? Unfortunately, many reasonable measures do NOT satisfy (3.1) on $\mathcal{P}(\Omega)$, but only on a smaller σ-field. The "length measure" is a prime example. We shall see this later.

(iv) The important notion of *signed measures* that are not necessarily non-negative shall be discussed in Chapter 20. ●

Exercise 3.1.2 In each case below, verify that μ is a measure:

(i) Let $\Omega \neq \emptyset$. Then the **counting measure** is defined as:

$$\mu(A) := \#A \text{ for all } A \in \mathcal{P}(\Omega).$$

(ii) Let $\Omega = \mathbb{N}$, and $\{x_i, i \in \Omega\}$ be non-negative real numbers. Define

$$\mu(A) := \sum_{i \in A} x_i, \ A \in \mathcal{P}(\Omega).$$

Definition 3.1.3. (**Probability, finite and σ-finite measure**) Let μ be a measure on $\mathcal{A}$ (or $\mathcal{F}$). Then it is called

(a) a **probability** if $\mu(\Omega) = 1$, and in that case $(\Omega, \mathcal{A}, \mu)$ or $(\Omega, \mathcal{F}, \mu)$ is called a **probability space**;

(b) **finite** if $\mu(\Omega) < \infty$;

(c) **σ-finite** if there is $\{A_i\}$, $\mu(A_i) < \infty$ for all i, and $\cup_{n=1}^{\infty} A_n = \Omega$. ◇

Check that in (c) above, $\{A_i\}$ can be taken to be disjoint without any loss. Such a collection is called a **measurable partition**. The counting measure is σ-finite if Ω is countable.

Exercise 3.1.3 If $(\Omega, \mathcal{A}, P)$ is a probability space, then show that,

$$P(\emptyset) = 0, \quad \text{and} \quad P(A^c) = 1 - P(A) \quad \text{for all} \quad A \in \mathcal{A}.$$

Exercise 3.1.4 Let $\Omega = \{0, 1, \ldots\}$, and $\mathcal{A} = \mathcal{P}(\Omega)$.

(a) Fix $0 < p < 1$. Let $q = 1 - p$. Define μ for singleton sets as

$$\mu(\{i\}) := pq^i, \ i = 0, 1, \ldots.$$

Show that μ can be extended to a probability measure on $\mathcal{A}$ in a natural way. It is called the **geometric measure** with parameter p.

(b) Show that μ defined below is a probability measure. It is called the **Poisson measure** with parameter $\lambda > 0$.

$$\mu(A) := \exp\{-\lambda\} \sum_{i \in A} \frac{\lambda^i}{i!}, \ A \subseteq \Omega. \tag{3.3}$$

Exercise 3.1.5 Let Ω be a finite set. Define μ on the σ-field $\mathcal{P}(\Omega)$ as,

$$\mu(A) := \frac{\#(A)}{\#(\Omega)}, \ A \subseteq \Omega.$$

Show that μ is a probability measure. It is called the **uniform measure** on Ω. There cannot be a uniform measure on any Ω which is infinite.

3.2 Sub-additivity and continuity

Here are some easy consequences of Definition 3.1.2.

Theorem 3.2.1. *(a) If μ is a measure on $\mathcal{F}$ or $\mathcal{A}$, and $B \subseteq A$ (from $\mathcal{F}$ or $\mathcal{A}$ resp.), then $\mu(B) \leq \mu(A)$.*

(b) Let μ be a measure on $\mathcal{A}$, and $\{A_i\}$ be from $\mathcal{A}$. Then

$$\mu(\cup_{n=1}^{\infty} A_n) \leq \sum_{n=1}^{\infty} \mu(A_n). \tag{3.4}$$

(c) Let μ be a measure on $\mathcal{F}$ and $\{A_i\}$ be from $\mathcal{F}$. Then

$$\mu(\cup_{n=1}^{\infty} A_n) \leq \sum_{n=1}^{\infty} \mu(A_n) \quad \text{whenever} \quad \cup_{n=1}^{\infty} A_n \in \mathcal{F}\,. \tag{3.5}$$

(d) Suppose that μ is a measure on $\mathcal{A}$ and $\{A_i\}$ are from $\mathcal{A}$. Then

$$\lim \mu(A_n) = \mu(\cup_{n=1}^{\infty} A_n) \quad \textit{if} \quad \{A_i\} \quad \textit{is a non-decreasing sequence.} \tag{3.6}$$

The result continues to hold if μ is a measure on $\mathcal{F}$ and $\cup_{n=1}^{\infty} A_n \in \mathcal{F}$.
(e) Let μ be a measure on $\mathcal{A}$ and $\{A_i\}$ be non-increasing from $\mathcal{A}$. Then

$$\lim \mu(A_n) = \mu(\cap_{n=1}^{\infty} A_n) \quad \textit{if} \quad \mu(A_k) < \infty \textit{ for some } k\,.$$ ◆

Property (a) is known as **monotonicity**. Relations (3.4) and (3.5) are known as **countable sub-additivity**. Relation (3.6) is called **continuity from below**.

Proof of Theorem 3.2.1. Part (a) is easy, since $A = B \cup (A \setminus B)$. Further, (c) implies (b). We prove (c) by disjointification. Define

$$B_k := A_k - \cup_{j=1}^{k-1} A_j, \;\; k \geq 1\,. \tag{3.7}$$

Then $\{B_k\}$ are disjoint, $B_k \in \mathcal{F}$, $B_k \subset A_k$ and $\cup_{j=1}^{k} A_j = \cup_{j=1}^{k} B_j$, for all k. Hence

$$\begin{aligned} \mu(\cup_{k=1}^{\infty} A_k) &= \sum_{k=1}^{\infty} \mu(B_k), \text{ by countable additivity} \\ &\leq \sum_{k=1}^{\infty} \mu(A_k), \text{ by (a)}\,. \end{aligned}$$

(d) Let $\{B_k\}$ be as in (3.7). Since $\{A_k\}$ is non-decreasing, for all $n \geq 2$, $\cup_{k=1}^{n} B_k = \cup_{k=1}^{n} A_k = A_n$, $A = \cup_{n=1}^{\infty} A_n = \cup_{n=1}^{\infty} B_n \in \mathcal{F}$ or $\mathcal{A}$. Hence,

$$\begin{aligned} \mu(A) &= \sum_{k=1}^{\infty} \mu(B_k) \text{ (countable additivity on } \mathcal{F} \text{ or } \mathcal{A}) \\ &= \lim_{n\to\infty} \sum_{k=1}^{n} \mu(B_k) \\ &= \lim_{n\to\infty} \mu(\cup_{k=1}^{n} B_k) = \lim_{n\to\infty} \mu(A_n)\,. \end{aligned}$$

(e) Let $C_{n,k} := A_k \setminus A_n$, $n \geq k$. Then $C_{n,k} \uparrow A_k \setminus \cap_{n=1}^{\infty} A_n$. By Part (d),

$$\mu(C_{n,k}) \uparrow \mu(A_k \setminus \cap_{n=1}^{\infty} A_n) = \mu(A_k) - \mu(\cap_{n=1}^{\infty} A_n) \text{ as } n \to \infty .$$

But then we can write $\mu(A_k) = \mu(C_{n,k}) + \mu(A_n)$, and all these quantities being finite, $\mu(A_n) = \mu(A_k) - \mu(C_{n,k}) \to \mu(\cap_{n=1}^{\infty} A_n)$. ∎

3.3 First Borel-Cantelli lemma

There is more than one lemma that goes by the name of Borel and Cantelli. Here is the first and the simplest of these. It has numerous uses, especially when μ is a probability measure. The Second Borel-Cantelli lemma will appear in Chapter 14 when we discuss the concept of "independence" of events in probability spaces.

Lemma 3.3.1. Let $\{A_n\}$ be a sequence of sets in $(\Omega, \mathcal{A}, \mu)$. Then

$$\sum_{n=1}^{\infty} \mu(A_n) < \infty \text{ implies } \mu(\limsup A_n) = 0. \tag{3.8}$$

♦

Proof. Recall that

$$\limsup A_n = \cap_{n=1}(\cup_{k=n}^{\infty} A_k).$$

By Theorem 3.2.1(b), and the summability condition in (3.8),

$$\mu(\cup_{k=n}^{\infty} A_k) \leq \sum_{k=n}^{\infty} \mu(A_k) < \infty \text{ for all } n.$$

Hence by Theorem 3.2.1(e),

$$\begin{aligned}
\mu(\limsup A_n) &= \mu(\cap_{n=1}(\cup_{k=n}^{\infty} A_k)) \\
&= \lim_{n\to\infty} \mu(\cup_{k=n}^{\infty} A_k) \\
&\leq \lim_{n\to\infty} \sum_{k=n}^{\infty} \mu(A_k) = 0, \text{ as } \sum_{k=1}^{\infty} \mu(A_k) < \infty .
\end{aligned}$$

∎

3.4 Null set and completion

Definition 3.4.1. (μ**-Null set**) Let $(\Omega, \mathcal{F}, \mu)$ be a measure space where $\mathcal{F}$ is a field. Then $A \in \mathcal{F}$ is said to be a *μ-null set* (or simply a null set) if $\mu(A) = 0$. The class of null sets is denoted by $\mathcal{N}_\mu$ or $\mathcal{N}$. ◇

The only μ-null set in a measure space could be the empty set.

Exercise 3.4.1 Show that in $(\Omega, \mathcal{A}, \mu)$ where $\mathcal{A}$ is a σ-field, a countable union of μ-null sets is again a μ-null set. Give an example to show that an uncountable union of μ-null sets need not be a μ-null set.

Definition 3.4.2. (**Almost surely/everywhere**) A property $Q(\cdot)$ on $(\Omega, \mathcal{A}, \mu)$ is said to hold *almost everywhere*-$[\mu]$ or *almost surely*-$[\mu]$, if there exists a μ-null set A such that it holds for all $\omega \in A^c$. We often write μ-almost surely, μ-almost everywhere, **a.s.** μ, or **a.e** μ. ◇

For example, suppose that $(\Omega, \mathcal{A}, \mu)$ is a measure space. Let $\{f_n\}$ and f be real-valued functions on Ω. Then we would say f_n converges to f a.s. μ, if there is a null set A such that $f_n(\omega) \to f(\omega)$ for all $\omega \notin A$.

Remark 3.4.1 Definition 3.4.2 implies that Q holds *outside some null set*, say A. So, $Q(\omega)$ is allowed to hold for some $\omega \in A$. The set $\{\omega \in \Omega : Q(\omega) \text{ holds}\}$ may not be measurable. Also, the "a.s."' or "a.e." is with respect to a given measure μ. ●

Definition 3.4.3. (**Complete measure**) Let $(\Omega, \mathcal{F}, \mu)$ be a measure space. Then μ is called *complete* if $B \subseteq A$ for some μ-null set A, implies $B \in \mathcal{F}$. We also say that $(\Omega, \mathcal{F}, \mu)$ is a **complete measure space.** ◇

Exercise 3.4.2 Give examples of both, measure spaces that are complete, and not complete.

Theorem 3.4.1 (**Completion of a measure space**). *Let $(\Omega, \mathcal{A}, \mu)$ be a measure space where $\mathcal{A}$ is a σ-field. Let*

$$\mathcal{A}_\mu := \{A \cup B : A \in \mathcal{A},\ B \subseteq N,\ \text{for some } N \in \mathcal{N}_\mu\},$$

where $\mathcal{N}_\mu$ is the collection of all μ-null sets from $\mathcal{A}$.
(a) Then $\mathcal{A}_\mu$ is the smallest σ-field containing $\mathcal{A}$ and all subsets of $\mathcal{N}_\mu$.
(b) Extend μ from $\mathcal{A}$ to $\mathcal{A}_\mu$ as follows:

$$\mu_c(A \cup B) := \mu(A),\ \ A \in \mathcal{A}, B \subseteq N \in \mathcal{N}_\mu. \tag{3.9}$$

Then $(\Omega, \mathcal{A}_\mu, \mu_c)$ is a complete measure space and $\mu_c \equiv \mu$ on $\mathcal{A}$. ◆

The measure space $(\Omega, \mathcal{A}_\mu, \mu_c)$ is said to be the **completion** of $(\Omega, \mathcal{A}, \mu)$, and we often continue to denote μ_c by μ.

Proof. (a) To show that $\mathcal{A}_\mu$ is closed under countable unions, it suffices to observe that,

$$\cup_{n=1}^{\infty}(A_n \cup B_n) = (\cup_{n=1}^{\infty} A_n) \cup (\cup_{n=1}^{\infty} B_n),$$

and invoke Exercise 3.4.1. To show $\mathcal{A}_\mu$ is closed under complementation, consider $A \cup B \in \mathcal{A}_\mu$ where $A \in \mathcal{A}$ and $B \subseteq N \in \mathcal{N}_\mu$. Then

$$\begin{aligned} (A \cup B)^c &= A^c \cap B^c \\ &= (A^c \cap N^c) \cup (A^c \cap (B^c \setminus N^c)) \text{ as } N^c \subset B^c \\ &= (A^c \cap N^c) \cup (A^c \cap (N \setminus B). \end{aligned}$$

Since $A^c \cap N^c \in \mathcal{A}$, and $A^c \cap (N \setminus B) \subseteq N \in \mathcal{N}_\mu$, we get $(A \cup B)^c \in \mathcal{A}_\mu$. This proves that $\mathcal{A}_\mu$ is closed under complementation, and is a field.

(b) To first verify that (3.9) is a valid definition, let $A_1 \cup B_1 = A_2 \cup B_2$, where $A_i \in \mathcal{A}$ and $B_i \subset N_i$ where N_i are μ-null sets. Then we must show that $\mu(A_1) = \mu(A_2)$. Note that

$$\begin{aligned} \mu(A_1) &= \mu(A_1 \cap A_2) + \mu(A_1 \setminus A_2) \\ &= \mu(A_1 \cap A_2) + 0, \text{ since } A_1 \setminus A_2 \subseteq N_2, \text{ and } A_1 \setminus A_2 \in \mathcal{A} \\ &\leq \mu(A_2). \end{aligned}$$

Similarly, $\mu(A_2) \leq \mu(A_1)$, and hence $\mu(A_1) = \mu(A_2)$.

The proof of countably additivity of μ_c on $\mathcal{A}_\mu$ is left as an exercise.

We now show that $(\Omega, \mathcal{A}_\mu, \mu_c)$ is a complete measure space. Take any arbitrary μ_c-null set from $\mathcal{A}_\mu$. Then it can be written as $A \cup B \in \mathcal{A}_\mu$ where $A \in \mathcal{A}$, $B \subseteq N \in \mathcal{A}$, such that $\mu(A) = 0$, **and** $\mu(N) = 0$.

Let $M \subseteq A \cup B$. Now, $M \subseteq A \cup N$ where $A \cup N \in \mathcal{A}$, and moreover, $\mu(A \cup N) = 0$. Hence, by the definition of $\mathcal{A}_\mu$, $M \in \mathcal{A}_\mu$. ■

3.5 Finitely additive measure

We restrict ourselves only to a very brief discussion of finitely additive measures. For more information on such measures, see Bhaskara Rao and Bhaskara Rao [1983].

Definition 3.5.1. (**Finitely additive measure**) Suppose $\mathcal{F}$ is a field and $\mu : \mathcal{F} \to \mathbb{R}^+ \cup \{\infty\}$. Then μ is called *finitely additive* if, for all choices of integers n and disjoint sets $\{A_i\}, 1 \le i \le n$ from $\mathcal{F}$,

$$\mu(\cup_{k=1}^n A_k) = \sum_{k=1}^n \mu(A_k)\,. \tag{3.10}$$

◇

Definition 3.5.2. A set function μ on $\mathcal{F}$ is said to be **continuous from above at** $\emptyset$ if, whenever $\{A_i\}$ is a non-increasing sequence of sets from $\mathcal{F}$ *and* $\cap_{n=1}^\infty A_n = \emptyset$, we have $\lim \mu(A_n) = 0$. ◇

Exercise 3.5.1 Suppose μ is finitely additive on a field $\mathcal{F}$.

(a) Let μ be continuous from below: that is, (3.6) holds if $\{A_i\}$ is a non-decreasing sequence from $\mathcal{F}$ **and** $\cup_{n=1}^\infty A_n \in \mathcal{F}$. Show that then μ is a measure on $\mathcal{F}$.

(b) Suppose μ is **continuous from above** at $\emptyset$. Show that then μ is countably additive on $\mathcal{F}$.

Exercise 3.5.2 Suppose Ω is countably infinite. Define $\mu(\cdot)$ on $\mathcal{P}(\Omega)$ as

$$\mu(A) = \begin{cases} 0 & \text{if } A \text{ is finite,} \\ \infty & \text{if } A \text{ is infinite.} \end{cases}$$

Show that

(a) μ is finitely additive but not countably additive.

(b) μ is not continuous from below at Ω.

(c) μ is not continuous from above at $\emptyset$.

Exercise 3.5.3 Suppose Ω is countably infinite. Let $\mathcal{F}$ be the co-finite field. Define $\mu(\cdot)$ on $\mathcal{F}$ as

$$\mu(A) = \begin{cases} 0 & \text{if } A \text{ is finite,} \\ 1 & \text{if } A^c \text{ is finite.} \end{cases}$$

Show that

(a) μ is finitely additive but not countably additive.

(b) μ is not continuous from below at Ω.

The proof of the following theorem is left as an exercise.

Theorem 3.5.1. *If μ is a* ***finitely additive*** *measure on a field $\mathcal{F}$, then,*

(a) $\mu(\emptyset) = 0$ (it is assumed that there is some $A \in \mathcal{F}$ with $\mu(A) < \infty$);

(b) for all $A, B \in \mathcal{F}$, $\mu(A \cup B) + \mu(A \cap B) = \mu(A) + \mu(B)$;

(c) if $A, B \in \mathcal{F}, B \subseteq A$, then $\mu(A) = \mu(B) + \mu(A - B) \geq \mu(B)$.

(d) For any sequence of sets $\{A_n\}$ from $\mathcal{F}$,

(i) $\mu(\cup_{k=1}^{n} A_k) \leq \sum_{k=1}^{n} \mu(A_k)$. This is called ***finite sub-additivity.***

(ii) If $\{A_n\}$ are disjoint and $\cup_{k=1}^{\infty} A_k \in \mathcal{F}$, then

$$\mu(\cup_{k=1}^{\infty} A_k) \geq \sum_{k=1}^{\infty} \mu(A_k) \quad \textbf{\textit{(countable super-additivity)}}. \tag{3.11}$$ ◆

Exercise 3.5.4 Let μ be a finitely additive and countably sub-additive measure on a field or a σ-field. Show that μ is countably additive.

3.6 Towards Lebesgue-Stieltjes measures

The few examples of countably additive measures that we have seen so far have been very simple. How do we build a collection of interesting non-trivial measures? Maybe we can start by checking out the "length measure" on the real-line. Call such a "target" measure λ. Note that :

1. For $a < b$, λ must satisfy $\lambda(a, b) = \lambda(a, b] = \lambda[a, b) = \lambda[a, b] = b - a$.

2. Since $\mathcal{B}(\mathbb{R})$ is the smallest σ-field containing all intervals, we should extend λ in a countably additive way at least to $\mathcal{B}(\mathbb{R})$.

3. We shall see in Section 5.2.3 that λ is not extendable to $\mathcal{P}(\mathbb{R})$.

4. For every $\omega \in \mathbb{R}$, $\{\omega\}$ is a Borel set and $\lambda\{\omega\} = 0$.

5. The measure λ is **σ-finite** since $\mathbb{R} = \cup_{n=-\infty}^{\infty}(n,\ n+1]$.

6. This λ must be **translation invariant**. That is $\lambda(A+x) = \lambda(A)$ for every $A \in \mathcal{B}(\mathbb{R})$ and every $x \in \mathbb{R}$.

7. Thus we can restrict our attention to the interval $(0,\ 1]$, complete our definition of λ on $\mathcal{B}(0,\ 1]$, and then use it to define λ on $\mathcal{B}(\mathbb{R})$.

8. A general programme would be to define a measure on "simpler sets" (semi-field maybe) and then extend the definition a σ-field. For example, suppose that F is a non-decreasing continuous function F. Define $\mu(a,\ b] = F(b) - F(a)$. Is this extendable to a countably additive measure on $\mathcal{B}(\mathbb{R})$?

9. There is another interesting question. Once we are successful in defining λ on $\mathcal{B}(\mathbb{R})$, can we use it to construct the area measure or the volume measures on $\mathcal{B}(\mathbb{R} \times \mathbb{R})$ and $\mathcal{B}(\mathbb{R}^3)$?

10. Lebesgue measure and more are discussed in Chapter 5.

3.7 Exercises

Exercise 3.7.1 Suppose μ is a probability measure on a σ-field and $\{A_n\}$ is a sequence of measurable sets. Show that

(a) for every n,

$$\mu(\cup_{i=1}^n A_i) = \sum_{i-1}^{n} \mu(A_i) - \sum_{1 \leq i_1 < i_2 \leq n} \mu(A_{i_1} \cap A_{i_2})$$
$$+ \sum_{1 \leq i_1 < i_2 < i_3 \leq n} \mu(A_{i_1} \cap A_{i_2} \cap A_{i_3}) - \cdots (-1)^n \mu(A_1 \cdots \cap A_n).$$

(b) if $\mu(A_n \cap A_m) = 0$ for any $n \neq m$, then $\mu(\cup_{n=1}^{\infty} A_n) = \sum_{n=1}^{\infty} \mu(A_n)$.

Exercise 3.7.2 Define μ on the co-countable σ-field of an uncountable set Ω as

$$\mu(A) = \begin{cases} 0 & \text{if } A \text{ is countable,} \\ 1 & \text{if } A^c \text{ is countable.} \end{cases}$$

(a) Show that μ is a countably additive measure.

(b) Show that μ is not σ-finite.

Exercise 3.7.3 Let μ_1, μ_2 be finite measures on $\sigma(\mathcal{F})$ where $\mathcal{F}$ is a field.

(a) If $\mu_1 \equiv \mu_2$ on $\mathcal{F}$, then show that $\mu_1 \equiv \mu_2$ on $\sigma(\mathcal{F})$.

(b) Is the conclusion in (a) valid if the measures are not finite?

Exercise 3.7.4 Let $(\Omega, \mathcal{A}, \mu)$ be a measure space. Let $\{A_n\}$ be a sequence of sets in $\mathcal{A}$. Show that $\mu(\liminf A_n) \leq \liminf \mu(A_n)$. If further μ is a finite measure, then show that $\mu(\limsup A_n) \geq \limsup \mu(A_n)$.

Exercise 3.7.5 Let $(\Omega, \mathcal{A}, \mu_n)$ be a sequence of measure spaces where $\{\mu_n\}$ is non-decreasing. That is, for any set A, $\mu_n(A)$ is a non-decreasing sequence. Define $\mu(A) := \lim \mu_n(A)$, $A \in \mathcal{A}$. Is μ a measure?

Exercise 3.7.6 If μ_1 and μ_2 are measures on $(\Omega, \mathcal{A})$, show that $\mu_1 + \mu_2$ is also a measure. When is the set function $\mu_1 - \mu_2$ defined? If it is defined, is it countably additive?

Exercise 3.7.7 Let $(\Omega, \mathcal{F}, \mu)$ br a measure space, and $A_1, A_2 \in \mathcal{F}$. When can we conclude that

$$\mu(A_1 \cup A_2) = \mu(A_1) + \mu(A_2) - \mu(A_1 \cap A_2)\,?$$

Exercise 3.7.8 Suppose $(\Omega, \mathcal{A}, \mu)$, is a measure space and $\{A_i\}$ is a sequence of sets from $\mathcal{A}$. When can we conclude that

$$\mu(\cup_{i=1}^{\infty} A_i \geq \sum_{i=1}^{\infty} \mu(A_i) - \sum_{1 \leq i < j < \infty} \mu(A_i \cap A_j)\,?$$

Exercise 3.7.9 Let P and G be the Poisson and the geometric measures on $(\mathbb{N}, \mathcal{P}(\mathbb{N}))$ where $\mathbb{N} = \{0, 1, \ldots\}$. Identify a function f such that

$$P(A) = \sum_{\omega \in A} f(\omega) G(\{\omega\}), \text{ for all } A \in \mathcal{P}(\mathbb{N})\,.$$

What is the corresponding function to express G in terms of P?

Exercise 3.7.10 Let μ be a σ-finite measure. Show that there is no uncountable collection of disjoint sets each with a positive measure.

Chapter 4

Extension of measures

Before we construct measures that are non-trivial, we explore whether we can extend a measure defined on a field $\mathcal{F}$ to $\sigma(\mathcal{F})$, and in a unique way. As we have seen in Exercise 3.7.3 any extension will be unique if the measure is finite. We shall prove a general result that guarantees a unique extension when the measure is σ-finite. The developments leading to the extension theorem, and its proof may be skipped at the first reading, but the result itself will be vital to us, especially in the construction of Lebesgue and Lebesgue-Stieltjes measures on $\mathbb{R}^d$.

4.1 Extension to the class of increasing limits

Lemma 4.1.1. Let $(\Omega, \mathcal{F}, \mu)$ be a measure space where $\mathcal{F}$ is a field and μ is finite. Suppose $\{A_n\}$ and $\{B_n\}$ are two sequences of non-decreasing sets from $\mathcal{F}$ such that $A_n \uparrow A$ and $B_n \uparrow B$ (A or B need not be in $\mathcal{F}$).
(a) If $A \subseteq B$, then $\lim \mu(A_n) \leq \lim \mu(B_n)$.
(b) If $A = B$, then $\lim \mu(A_n) = \lim \mu(B_n)$. ♦

Proof. Fix an integer m. Then $A_m \cap B_n \in \mathcal{F}$, and $A_m \cap B_n \uparrow A_m \in \mathcal{F}$. Since μ is a measure on $\mathcal{F}$,

$$\begin{aligned} \mu(A_m) &= \lim_{n\to\infty} \mu(A_m \cap B_n), \text{ by Theorem 3.2.1(d)} \\ &\leq \lim_{n\to\infty} \mu(B_n), \text{ by Theorem 3.2.1(a), as } A_m \cap B_n \subseteq B_n\,. \end{aligned}$$

Now let $m \to \infty$ to complete the proof of (a).

(b) follows by reversing the roles of A and B in the above argument. ■

A. Bose, *Elements of Measure and Probability*, Texts and Readings in Mathematics 88,
https://doi.org/10.1007/978-981-95-2758-8_4

Lemma 4.1.1 implies that if we add the limits of increasing sets to the collection $\mathcal{F}$, then there is only one way of extending μ to such sets. Now suppose $(\Omega, \mathcal{F}, \mu)$ is a finite measure space where $\mathcal{F}$ is a field. Define

$$\mathcal{I}_{\mathcal{F}} := \{A: \ A_n \uparrow A, \text{ for some } A_n \in \mathcal{F}\}. \tag{4.1}$$

Clearly, $\mathcal{F} \subset \mathcal{I}_{\mathcal{F}}$. Extend μ to sets in $\mathcal{I}_{\mathcal{F}}$ by

$$\mu_I(A) = \lim \mu(A_n), \text{ where } A_n \uparrow A, \ A_n \in \mathcal{F}. \tag{4.2}$$

Lemma 4.1.2. Let μ be a finite measure on $(\Omega, \mathcal{F}, \mu)$ where $\mathcal{F}$ is a field, and let $\mathcal{I}_{\mathcal{F}}$ and μ_I be as in (4.1) and (4.2). Then we have the following.

(a) The class $\mathcal{I}_{\mathcal{F}}$ is closed under finite union and intersection.

(b) Definition (4.2) is unambiguous. That is, it does not depend on the specific sequence $\{A_n\}$.

(c) For all sets $A \in \mathcal{F}$, $\mu(A) = \mu_I(A)$.

(d) If $G_1, G_2 \in \mathcal{I}_{\mathcal{F}}$ and $G_1 \subseteq G_2$, then $\mu_I(G_1) \leq \mu_I(G_2)$.

(e) $\mu_I(G_1 \cup G_2) + \mu_I(G_1 \cap G_2) = \mu_I(G_1) + \mu_I(G_2)$ for all $G_1, G_2 \in \mathcal{I}_{\mathcal{F}}$. So μ_I is finitely additive.

(f) If $G_n \in \mathcal{I}_{\mathcal{F}}$ and $G_n \uparrow G$, then $G \in \mathcal{I}_{\mathcal{F}}$, and $\mu_I(G_n) \uparrow \mu_I(G)$. ♦

Proof. (a) Let $G_1, G_2 \in \mathcal{I}_{\mathcal{F}}$. Let $A_n, B_n \in \mathcal{F}$ be such that $A_n \uparrow G_1$ and $B_n \uparrow G_2$. Then $A_n \cup B_n, A_n \cap B_n \in \mathcal{F}$. Moreover, $A_n \cup B_n \uparrow G_1 \cup G_2$ and $A_n \cap B_n \uparrow G_1 \cap G_2$. Hence $G_1 \cup G_2$ and $G_1 \cap G_2$ are in $\mathcal{I}_{\mathcal{F}}$.

Parts (b), (c) and (d) follow from Lemma 4.1.1.

(e) Let $G_1, G_2 \in \mathcal{I}_{\mathcal{F}}$. Then there exists non-decreasing $A_n, B_n \in \mathcal{F}$ such that $A_n \uparrow G_1$ and $B_n \uparrow G_2$. By additivity of μ on $\mathcal{F}$,

$$\mu(A_n \cup B_n) + \mu(A_n \cap B_n) = \mu(A_n) + \mu(B_n) \text{ for all } n.$$

The sets involved increase to $G_1 \cup G_2$, $G_1 \cap G_2$, G_1 and G_2 respectively. By Lemma 4.1.1, their μ-measures increase to $\mu_I(G_1 \cup G_2)$, $\mu_I(G_1 \cap G_2)$, $\mu_I(G_1)$ and $\mu_I(G_2)$ respectively. This proves Part (e).

(f) This is proved by a *diagonalisation* argument. Let $A_{nm} \in \mathcal{F}$ be such that $A_{nm} \uparrow G_n$ as $m \to \infty$. Define

$$D_m = A_{1m} \cup A_{2m} \cup \cdots \cup A_{mm}.$$

That is, D_m is the union of the sets in the mth column of the matrix $A_{ij}, 1 \le i \le m, 1 \le j$. Now, $D_m \in \mathcal{F}$ and $D_m \uparrow \cup_{k=1}^{\infty} D_k$. Further,

$$A_{nm} \subseteq D_m \subseteq G_m \text{ for all } n \le m. \tag{4.3}$$

Hence

$$\mu(A_{nm}) \le \mu(D_m) \le \mu_I(G_m). \tag{4.4}$$

If we let $m \to \infty$ in (4.3), we obtain

$$G_n \subseteq \cup_{m=1}^{\infty} D_m \subseteq G.$$

Let $n \to \infty$ to conclude that D_m increases to G. Hence $G \in \mathcal{I}_{\mathcal{F}}$, and

$$\lim \mu(D_m) = \mu_I(G).$$

If we now let $m \to \infty$ in (4.4), we obtain

$$\mu_I(G_n) \le \lim_{m\to\infty} \mu(D_m) \le \lim_{m\to\infty} \mu_I(G_m).$$

Now let $n \to \infty$ to obtain

$$\lim_{n\to\infty} \mu_I(G_n) = \lim_{m\to\infty} \mu(D_m) = \mu_I(G). \qquad \blacksquare$$

4.2 Outer measure

Recall that we always assume, there is a set A such that $0 < \mu(A) < \infty$ for any measure μ under our consideration. Lemma 4.1.2(d–f) prompts us to consider a general class of sets $\mathcal{G}$ as described in the next lemma.

Lemma 4.2.1. Let $\mathcal{G} \subset \mathcal{P}(\Omega)$ and $\mu : \mathcal{G} \to [0, \infty)$ be such that

(i) Ω, $\emptyset \in \mathcal{G}$, and $\mathcal{G}$ is closed under finite union, finite intersection, and increasing limits.

(ii) $\mu(\emptyset) = 0$, $\mu(\Omega) < \infty$.

(iii) If $G_1, G_2 \in \mathcal{G}$, $G_1 \subseteq G_2$, then $\mu(G_1) \le \mu(G_2)$. That is, μ is monotone.

(iv) $\mu(G_1 \cup G_2) + \mu(G_1 \cap G_2) = \mu(G_1) + \mu(G_2)$ for all $G_1, G_2 \in \mathcal{G}$.

(v) If $G_n \in \mathcal{G}$ and $G_n \uparrow G$, then $\mu(G_n) \uparrow \mu(G)$.

Define

$$\mu^*(A) = \inf\{\mu(G) : G \in \mathcal{G}, G \supseteq A\}, \ A \subseteq \Omega. \tag{4.5}$$

Then

(a) $\mu^*(\emptyset) = 0$, and $\mu^*(\Omega) = \mu(\Omega)$;

(b) for all $A \in \mathcal{G}$, $\mu^*(A) = \mu(A)$;

(c) if $A \subseteq B$, then $\mu^*(A) \le \mu^*(B)$;

(d) for all $A \subseteq \Omega$, $\mu^*(A) \le \mu(\Omega) < \infty$;

(e) for all $A, B \subseteq \Omega$,

$$\mu^*(A \cup B) + \mu^*(A \cap B) \le \mu^*(A) + \mu^*(B).$$

Hence, for any $A \subset \Omega$,

$$\mu^*(A) + \mu^*(A^c) \ge \mu^*(\Omega) + \mu^*(\emptyset) = \mu(\Omega) + \mu(\emptyset) = \mu(\Omega); \tag{4.6}$$

(f) If $A_n \uparrow A$, then $\mu^*(A_n) \uparrow \mu^*(A)$. ♦

Now, $\mu^*(G) = \mu(G)$ for all $G \in \mathcal{G}$. It is important to keep in mind that in Lemma 4.2.1, μ need not be a measure, and μ^* depends on both $\mathcal{G}$ and μ. The set function μ^* also need not be a measure.

Proof Lemma 4.2.1. We skip the proof of Parts (a)–(d).

(e) Fix $\epsilon > 0$. Choose $G_1, G_2 \in \mathcal{G}$ such that $G_1 \supseteq A$, $G_2 \supseteq B$ and

$$\mu(G_1) \le \mu^*(A) + \epsilon, \ \mu(G_2) \le \mu^*(B) + \epsilon.$$

Then since $A \cup B \subset G_1 \cup G_2$ and $A \cap B \subset G_1 \cap G_2$,

$$\begin{aligned} \mu^*(A) + \mu^*(B) + 2\epsilon &\ge \mu(G_1) + \mu(G_2) \\ &= \mu(G_1 \cup G_2) + \mu(G_1 \cap G_2) \\ &\ge \mu^*(A \cup B) + \mu^*(A \cap B). \end{aligned}$$

Since ϵ was arbitrary, proof of Part (e) is complete.

(f) As μ^* is non-decreasing and $A_n \uparrow A$, $\mu^*(A) \ge \lim \mu^*(A_n)$. To prove the reverse inequality, as in the proof of (e), get $G_n \in \mathcal{G}$ such that

$$G_n \supseteq A_n \text{ and } \mu(G_n) \le \mu^*(A_n) + \epsilon 2^{-n}, \text{ for all } n \ge 1.$$

We now claim that

$$\mu(\cup_{k=1}^{m} G_k) \leq \mu^*(A_m) + \epsilon \sum_{k=1}^{m} 2^{-k}, \text{ for all } m \geq 1. \tag{4.7}$$

We prove this by induction. Clearly, this holds for $m = 1$. Let this hold for all $m \leq n$. We shall prove it for $m = n + 1$. First note that

$$\mu\big((\cup_{k=1}^{n} G_k) \cap G_{n+1}\big) \geq \mu(G_n \cap G_{n+1}) \geq \mu^*(A_n \cap A_{n+1}) = \mu^*(A_n). \tag{4.8}$$

Hence (since $\mu^*(A) < \infty$ for every A, subtraction is allowed),

$$\begin{aligned}
\mu(\cup_{k=1}^{n+1} G_i) &= \mu(\cup_{k=1}^{n} G_k) + \mu(G_{n+1}) - \mu\big((\cup_{k=1}^{n} G_k) \cap G_{n+1}\big), \text{ using (iv)} \\
&\leq \mu^*(A_n) + \sum_{k=1}^{n} \frac{\epsilon}{2^k} + \mu^*(A_{n+1}) + \frac{\epsilon}{2^{n+1}} - \mu^*(A_n), \\
&\qquad \text{(by (4.7) for } m = n, \text{ and (4.8))} \\
&\leq \mu^*(A_{n+1}) + \epsilon \sum_{k=1}^{n+1} 2^{-k}.
\end{aligned}$$

This establishes (4.7). Note that $A = \cup_{k=1}^{\infty} A_k \subseteq \cup_{k=1}^{\infty} G_k$. Hence,

$$\begin{aligned}
\mu^*(A) &\leq \mu^*(\cup_{k=1}^{\infty} G_k), \text{ by (c)} \\
&= \mu(\cup_{k=1}^{\infty} G_k), \text{ by Condition (i)} \\
&= \lim_{n\to\infty} \mu(\cup_{k=1}^{n} G_k), \text{ using Condition (v)} \\
&\leq \lim_{n\to\infty} [\mu^*(A_n) + \epsilon \sum_{k-1}^{n} 2^{-k}], \text{ by (4.7)} \\
&= \lim_{n\to\infty} \mu^*(A_n) + \epsilon\,.
\end{aligned}$$

Since ϵ is arbitrary, (f) is proved, and proof of the lemma is complete. ■

Exercise 4.2.1 Show that conditions of Lemma 4.2.1 hold for $(\mathcal{I}_{\mathcal{F}}, \mu_I)$ given in (4.1)–(4.2). So its conclusions hold for the corresponding μ_I^*.

In general μ^*, and in particular μ_I^*, are defined on $\mathcal{P}(\Omega)$. They have some measure-like properties. This motivates the following:

Definition 4.2.1. (**Outer measure**) A set function ν^* on $\mathcal{P}(\Omega)$ is said to be an *outer measure* if the following three conditions are satisfied.

(i) $\nu^*(\emptyset) = 0$.

(ii) ν^* is monotone on $\mathcal{P}(\Omega)$.

(iii) ν^* is countably sub-additive on $\mathcal{P}(\Omega)$. ◇

Exercise 4.2.2 Exhibit an outer measure which is not a measure.

Exercise 4.2.3 Verify that the set function μ^* defined in (4.5) via $(\mathcal{G}, \mu)$ is a finite outer measure. As a consequence, due to Exercise 4.2.1, the set function μ_I^* is also a finite outer measure.

4.3 Extension of finite measures

Though an outer measure is not a measure in general, we know that our outer measure μ_I^* agrees with μ on $\mathcal{F}$. We have the following theorem for an outer measure of this type and this is motivated by Eqn. (4.6).

Theorem 4.3.1. *Let $(\Omega, \mathcal{F}, \mu)$ be a finite measure space where $\mathcal{F}$ is a field. Let μ_I^* be the outer measure constructed from μ. Define*

$$\mathcal{H}_{\mathcal{F}} := \{H \subseteq \Omega : \mu_I^*(H) + \mu_I^*(H^c) = \mu(\Omega)\}.$$

Then $\mathcal{H}_{\mathcal{F}} \supseteq \mathcal{F}$ is a σ-field and μ_I^* is a (finite) measure on $\mathcal{H}_{\mathcal{F}}$. In particular, μ_I^* is the unique extension of μ from $\mathcal{F}$ to $\sigma(\mathcal{F})$. ◆

Remark 4.3.1. (a) An observation that often turns out to be useful is that, due to Lemma 4.2.1(e), $\mathcal{H}_{\mathcal{F}}$ can also be described as:

$$\mathcal{H}_{\mathcal{F}} = \{H \subseteq \Omega : \mu_I^*(H) + \mu_I^*(H^c) \leq \mu(\Omega)\}. \tag{4.9}$$

(b) Note that in general $\mathcal{H}_{\mathcal{F}}$ can be strictly larger than $\sigma(\mathcal{F})$. We shall explore possible special features of this σ-field later. ●

Proof of Theorem 4.3.1. Without loss we assume that $\mu(\Omega) = 1$. We prove the theorem in four steps.

(i) Recall the definition of $\mathcal{I}_{\mathcal{F}}$ from Eqn. (4.1). We claim that $\mathcal{I}_{\mathcal{F}} \subseteq \mathcal{H}_{\mathcal{F}}$. To see this, suppose

$$A_n \in \mathcal{F} \text{ such that } A_n \uparrow G \in \mathcal{I}_{\mathcal{F}}.$$

Then

$$\begin{aligned}\mu_I^*(G) + \mu_I^*(G^c) &= \lim[\mu(A_n) + \mu_I^*(G^c)], \text{ since } A_n \uparrow G \\ &\le \limsup[\mu(A_n) + \mu_I^*(A_n^c)], \text{ since } G^c \subseteq A_n^c \\ &= \lim[\mu(A_n) + \mu(A_n^c)], \text{ since } A_n^c \in \mathcal{F} \\ &= 1.\end{aligned}$$

Hence by (4.9), $G \in \mathcal{H}_{\mathcal{F}}$.

(ii) From the definition of $\mathcal{H}_{\mathcal{F}}$, it follows that $\Omega \in \mathcal{H}_{\mathcal{F}}$, and $\mathcal{H}_{\mathcal{F}}$ is closed under complementation.

(iii) $\mathcal{H}_{\mathcal{F}}$ is a field and μ_I^* is finitely additive on $\mathcal{H}_{\mathcal{F}}$. To show this, let $H_1, H_2 \in \mathcal{H}_{\mathcal{F}}$. Let,

$$\begin{aligned}A &= H_1 \cup H_2, \quad x = \mu_I^*(A) + \mu_I^*(A^c), \\ B &= H_1 \cap H_2, \quad y = \mu_I^*(B) + \mu_I^*(B^c).\end{aligned}$$

The following two inequalities follow from Lemma 4.2.1(e).

$$\mu_I^*(A) + \mu_I^*(B) \le \mu_I^*(H_1) + \mu_I^*(H_2), \tag{4.10}$$

$$\mu_I^*(A^c) + \mu_I^*(B^c) \le \mu_I^*(H_1^c) + \mu_I^*(H_2^c). \tag{4.11}$$

If we now add (4.10) and (4.11), and use the fact that $H_1, H_2 \in \mathcal{H}_{\mathcal{F}}$, and also Lemma 4.2.1(e), then we arrive at

$$2 \le x + y \le \mu_I^*(H_1) + \mu_I^*(H_2) + \mu_I^*(H_1^c) + \mu_I^*(H_2^c) = 2.$$

By Lemma 4.2.1(e), $x \ge 1$ and $y \ge 1$. Thus, $x = y = 1$. In other words $A = H_1 \cup H_2 \in \mathcal{H}_{\mathcal{F}}$ and $B = H_1 \cap H_2 \in \mathcal{H}_{\mathcal{F}}$. That is, $\mathcal{H}_{\mathcal{F}}$ is a field.

Moreover, equality holds in (4.10) and (4.11), and that proves finite additivity of μ_I^* on $\mathcal{H}_{\mathcal{F}}$. This establishes (iii).

(iv) $\mathcal{H}_{\mathcal{F}}$ is a σ-field and μ_I^* is countably additive on $\mathcal{H}_{\mathcal{F}}$. To show this, observe that we have already shown in (ii) above that $\mathcal{H}_{\mathcal{F}}$ is a field. Hence by Exercise 2.4.3, it would be a σ-field, if it is closed under increasing limits. So suppose $H_n \in \mathcal{H}_{\mathcal{F}}$ and $H_n \uparrow H$. By Lemma 4.2.1(e), $\mu_I^*(H) + \mu_I^*(H^c) \ge 1$. On the other hand, by Lemma 4.2.1(f), $\mu_I^*(H_n) \uparrow \mu_I^*(H)$. Fix $\epsilon > 0$. Then for all large n, $\mu_I^*(H) \le \mu_I^*(H_n) + \epsilon$.

Hence, for all large n, since $H^c \subseteq H_n^c$,

$$\begin{aligned} 1 &\le \mu_I^*(H) + \mu_I^*(H^c) \\ &\le \mu_I^*(H_n) + \epsilon + \mu_I^*(H_n^c) \text{ for all large } n \\ &= 1 + \epsilon. \end{aligned}$$

As ϵ is arbitrary, $\mu_I^*(H)+\mu_I^*(H^c) = 1$, or $H \in \mathcal{H}_\mathcal{F}$. Thus $\mathcal{H}_\mathcal{F}$ is a σ-field.

On $\mathcal{H}_\mathcal{F}$, μ_I^* is finitely additive, and is continuous from below. By Exercise 3.5.1(a), it is countably additive. This proves (iv), completing the proof of the first part. Uniqueness follows from Exercise 3.7.3. ■

From the constructive proof that we have given, (recall (4.6) and (4.9)), μ is *not* extendable to a σ-field larger than $\mathcal{H}_\mathcal{F}$. How large is $\mathcal{H}_\mathcal{F}$? This has a "complete" answer.

Theorem 4.3.2. *Suppose $(\Omega, \mathcal{F}, \mu)$ is a finite measure space, where $\mathcal{F}$ is a field. Then $(\Omega, \mathcal{H}_\mathcal{F}, \mu_I^*)$ is the completion of $(\Omega, \sigma(\mathcal{F}), \mu_I^*)$.* ◆

Proof. We first show that $(\Omega, \mathcal{H}_\mathcal{F}, \mu_I^*)$ is a complete measure space. To see this, suppose $A \in \mathcal{H}_\mathcal{F}$ is a μ_I^*-null set, and let $B \subseteq A$. Then

$$\begin{aligned} \mu(\Omega) &\le \mu_I^*(B) + \mu_I^*(B^c) \\ &\le \mu_I^*(A) + \mu_I^*(B^c), \text{ since } B \subseteq A \\ &= \mu_I^*(B^c), \quad \text{since } \mu_I^*(A) = 0 \\ &\le \mu_I^*(\Omega) = \mu(\Omega). \end{aligned}$$

Hence $B \in \mathcal{H}_\mathcal{F}$, proving that $(\Omega, \mathcal{H}_\mathcal{F}, \mu_I^*)$ is complete.

We now show that $(\Omega, \mathcal{H}_\mathcal{F}, \mu_I^*)$ is the completion of the measure space $(\Omega, \sigma(\mathcal{F}), \mu_I^*)$. For brevity, let us write $\mathcal{A} := \sigma(\mathcal{F})$. Let $\mathcal{A}_{\mu_I^*}$ be the completion of $\mathcal{A}$ with respect to μ_I^*. We have to show that $\mathcal{A}_{\mu_I^*} = \mathcal{H}_\mathcal{F}$.

We first show that $\mathcal{A}_{\mu_I^*} \subset \mathcal{H}_\mathcal{F}$. Consider any $A \cup B \in \mathcal{A}_{\mu_I^*}$ where $A \in \mathcal{A}$, $B \subseteq N$, and $N \in \mathcal{N}_{\mu_I^*}$ is a μ_I^* null set from $\mathcal{A}$. Since $\mathcal{A} \subseteq \mathcal{H}_\mathcal{F}$, we have $A, N \in \mathcal{H}_\mathcal{F}$. On the other hand, $(\Omega, \mathcal{H}_\mathcal{F}, \mu_I^*)$ is a complete measure space. Hence $B \in \mathcal{H}_\mathcal{F}$. This implies $A \cup B \in \mathcal{H}_\mathcal{F}$, and so $\mathcal{A}_{\mu_I^*} \subseteq \mathcal{H}_\mathcal{F}$.

Now let $A \in \mathcal{H}_\mathcal{F}$. Recall the outer measure μ_I^* from (4.5). That implies, there exist sequences of sets $\{C_n, D_n\}$ from $\mathcal{A}$ such that

$$C_n \subseteq A \subseteq D_n, \quad \mu_I^*(C_n) \to \mu_I^*(A), \mu_I^*(D_n) \to \mu_I^*(A). \tag{4.12}$$

Let

$$C = \cup_{n=1}^{\infty} C_n, \quad D = \cap_{n=1}^{\infty} D_n \,.$$

Then

$$A = C \cup (A \setminus C) \text{ and } C \in \mathcal{A}. \tag{4.13}$$

But

$$(A \setminus C) \subseteq (D \setminus C) \in \mathcal{A}, \tag{4.14}$$

and

$$\mu_I^*(D \setminus C) \le \mu_I^*(D_n \setminus C_n) = \mu_I^*(D_n) - \mu_I^*(C_n) \to 0. \tag{4.15}$$

Thus $\mu_I^*(D \setminus C) = 0$. Now using (4.13), (4.14), and (4.15), $A \in \mathcal{A}_{\mu_I^*}$. Hence $\mathcal{H}_{\mathcal{F}} \subset \mathcal{A}_{\mu_I^*}$, and this proves the result completely. ■

4.4 Extension of σ-finite measures

Theorem 4.4.1 (Carathéodory Extension). *Suppose $(\Omega, \mathcal{F}, \mu)$ is a countably additive measure space, where $\mathcal{F}$ is a field and μ is σ-finite. Then it can be uniquely extended to the measure space $(\Omega, \sigma(\mathcal{F}), \mu)$.* ◆

Proof. We use Theorem 4.3.1 in a predictable way. Write

$$\Omega = \cup_{n=1}^{\infty} A_n, \text{ where } A_n \text{ are disjoint}, A_n \in \mathcal{F}, \mu(A_n) < \infty \text{ for all } n \ge 1. \tag{4.16}$$

Define the measures μ_n on $\mathcal{F}$ as

$$\mu_n(A) = \mu(A \cap A_n), \;\; A \in \mathcal{F}. \tag{4.17}$$

Then it is easy to check that for each $n \ge 1$, $(\Omega, \mathcal{F}, \mu_n)$ is a countably additive finite measure space. Then by Theorem 4.3.1 there is a unique extension $(\Omega, \sigma(\mathcal{F}), \mu_n^*)$. Now it is obvious what we should do: define

$$\mu^*(A) = \sum_{n=1}^{\infty} \mu_n^*(A), \;\; A \in \sigma(\mathcal{F}).$$

Then μ^* is countably additive since each μ_n^* is so (by Exercise 3.1.1, non-negative numbers can be added in any order).

It is easy to establish that $(\Omega, \sigma(\mathcal{F}), \mu^*)$ is an extension of $(\Omega, \mathcal{F}, \mu)$.

It remains to prove uniqueness. Suppose ν is a measure on $\sigma(\mathcal{F})$ which agrees with μ on $\mathcal{F}$. Then we have to prove that $\nu = \mu^*$ on $\sigma(\mathcal{F})$. Define

$$\nu_n(A) := \nu(A \cap A_n), \; A \in \sigma(\mathcal{F}) .$$

Then $(\Omega, \sigma(\mathcal{F}), \nu_n)$ is a finite measure space. Moreover, if $A \in \mathcal{F}$,

$$\begin{aligned} \nu_n(A) &= \mu(A \cap A_n), \text{ since } \nu = \mu \text{ on } \mathcal{F} \\ &= \mu_n(A), \text{ by (4.17)} \\ &= \mu_n^*(A), \text{ since } \mu_n^* = \mu_n \text{ on } \mathcal{F}. \end{aligned}$$

Hence by Theorem 4.3.1, $\nu_n = \mu_n^*$ on $\sigma(\mathcal{F})$. But then for $A \in \sigma(\mathcal{F})$,

$$\nu(A) = \sum_{n=1}^{\infty} \nu_n(A) = \sum_{n=1}^{\infty} \mu_n^*(A) = \mu^*(A) .$$ ■

4.5 An approximation result

Measures of sets in $\sigma(\mathcal{F})$ can be approximated by those in $\mathcal{F}$ in the σ-finite case. This result will be handy in future.

Theorem 4.5.1 (Approximation by sets in the field). *Let μ be a σ-finite countably additive measure on $(\Omega, \mathcal{F})$ where $\mathcal{F}$ is a field. Then for $\epsilon > 0$, and $A \in \sigma(\mathcal{F})$ with $\mu(A) < \infty$, there is an $F \in \mathcal{F}$ so that,*

$$\mu(A \Delta F) < \epsilon . \tag{4.18}$$

Note that the unique extension to $\sigma(\mathcal{F})$ here is also denoted by μ. ◆

Proof. We prove it in three steps.

(i) Recall the class of increasing limits of sets (4.1) from $\mathcal{F}$:

$$\mathcal{I}_{\mathcal{F}} = \{A : A = \cup_{n=1}^{\infty} A_n, \; A_n \in \mathcal{F}\}.$$

By Theorem 3.2.1(d), (4.18) holds for any $A \in \mathcal{I}_{\mathcal{F}}$ with $\mu(A) < \infty$.

(ii) Now suppose μ is finite. By Eqn. (4.5), the outer measure of any set A can be approximated by the measure of sets G from $\mathcal{I}_{\mathcal{F}}$, which in turn can be approximated by sets from $F \in \mathcal{F}$ by (i). This proves the theorem for the case where μ is a finite measure.

(iii) Now suppose μ is σ-finite on $\mathcal{F}$. Fix $A \in \sigma(\mathcal{F})$. Suppose $\{A_n\} \subset \mathcal{F}$ satisfies (4.16). Define μ_n on $\sigma(\mathcal{F})$ by

$$\mu_n(G) := \mu_n(G \cap A_n), \ G \in \sigma(\mathcal{F}).$$

Each μ_n is a finite measure. By Part (ii), there exists $B_n \in \mathcal{F}$ such that

$$\mu_n(A \Delta B_n) < \frac{\epsilon}{2^n}. \tag{4.19}$$

Note that

$$\begin{aligned}
\frac{\epsilon}{2^n} &\geq \mu_n(A\Delta B_n) \\
&= \mu((A\Delta B_n) \cap A_n) \\
&= \mu\left(A\Delta(B_n \cap A_n) \cap A_n\right) \\
&= \mu_n(A\Delta(B_n \cap A_n)) = \mu_n(A\Delta \tilde{B}_n),
\end{aligned}$$

where $\tilde{B}_n = B_n \cap A_n$. The extra property we have is that $\tilde{B}_n \subseteq A_n$ and is in $\mathcal{F}$. Thus in (4.19) we may, and do, assume that $B_n \subseteq A_n$.

Let $C = \cup_{n=1}^{\infty} \tilde{B}_n$. Then $C \in \sigma(\mathcal{F})$ and $C \cap A_n = \tilde{B}_n$. Hence

$$\begin{aligned}
\mu_n(A\Delta C) &= \mu((A\Delta C) \cap A_n) \\
&= \mu\left((A\Delta \tilde{B}_n) \cap A_n\right) \\
&= \mu_n(A\Delta \tilde{B}_n) < \epsilon/2^n.
\end{aligned}$$

This implies that

$$\mu(A\Delta C) = \sum_{n=1}^{\infty} \mu_n(A\Delta C) < \epsilon.$$

It remains to address the fact that C may not belong to $\mathcal{F}$. Note that

$$(\cup_{k=1}^{n} \tilde{B}_k \setminus A) \uparrow (C \setminus A), \quad (A \setminus \cup_{k=1}^{n} \tilde{B}_k) \downarrow (A \setminus C).$$

Now observe the following two facts:

$$\mu(\cup_{k=1}^{n} \tilde{B}_k \setminus A) \uparrow \mu(C \setminus A), \text{ by Theorem 3.2.1(d)},$$

$$\mu(A \setminus \cup_{k=1}^{n} \tilde{B}_k) \downarrow \mu(A \setminus C), \text{ by Theorem 3.2.1(e) using } \mu(A) < \infty.$$

Hence $\mu(A\Delta \cup_{k=1}^n \tilde{B}_k) \to \mu(A\Delta C)$ as $n \to \infty$. This proves the result by choosing $F = \cup_{k=1}^n \tilde{B}_k \in \mathcal{F}$, for large enough n. ■

4.6 Exercises

Exercise 4.6.1 Justify the existence of $\{C_n\}$ and $\{D_n\}$ in (4.12).

Exercise 4.6.2 Let μ be a measure on a field $\mathcal{F}$ and let μ^* be the corresponding outer measure. Show that for any $\epsilon > 0$ and any E such that $\mu^*(E) < \infty$, there exists a set $F \in \mathcal{F}$ such that $\mu^*(E\Delta F) < \epsilon$.

Exercise 4.6.3 Consider $\Omega = (0,\ 1]$. Let $\mathcal{F}$ be the set of all finite disjoint unions of sets of the form $(a,\ b]$ from $(0,\ 1]$. Recall that $\mathcal{F}$ is a field and generates the Borel σ-field $\mathcal{B}(\Omega)$. Let $\mathbb{Q}$ be the set of rational numbers. Consider the two measure spaces $(\Omega, \mathcal{B}(\Omega), \mu_i)$ where

$$\mu_i(A) = i\#(A \cup \mathbb{Q}),\ \ A \in \mathcal{B}(\Omega),\ \ i = 1, 2\,.$$

Note that both measures are σ-finite on $\mathcal{B}(\Omega)$. However, for every non-empty set $A \in \mathcal{F}$, $\mu_i(A) = \infty$, and they are identical on $\mathcal{F}$. Why does this not contradict the extension theorem?

Chapter 5

Lebesgue-Stieltjes measures

By the help of the extension theorem, we shall now exhibit an important general class of countably additive measures on $\mathcal{B}(\mathbb{R}^d)$, called the Lebesgue-Stieltjes measures. The length, area, volume measures and probability measures on $\mathbb{R}^d$ will be special cases.

5.1 Distribution function

Definition 5.1.1. (Distribution function on $\mathbb{R}$) Any function $F : \mathbb{R} \to \mathbb{R}$ is called a *distribution function*(df) if, (i) and (ii) below hold.

(i) F is non-decreasing. That is, $F(a) \leq F(b)$ for all $a \leq b$, $a, b \in \mathbb{R}$, and

(ii) F is right continuous. That is, $\lim_{y \downarrow x} F(y) = F(x)$ for every $x \in \mathbb{R}$.

F is called a **probability distribution function** if, in addition,

(iii) $\lim_{x \to -\infty} F(x) = 0$, and $\lim_{x \to \infty} F(x) = 1$.

In the probability literature, a probability distribution function goes by the name **cumulative distribution function (cdf)**. ◇

Remark 5.1.1. (a) Distribution functions can take negative values, and be unbounded. For example $F(x) := x$, $x \in \mathbb{R}$ is a distribution function.
(b) If F is a cdf, then $0 \leq F(x) \leq 1$ for all x.
(c) A distribution function need not be left continuous. However, the left limit $F(x-) := \lim_{y \to x, y < x} F(y)$ exists for each $x \in \mathbb{R}$. A distribution function F is continuous at x if and only if $F(x-) = F(x)$. ●

A. Bose, *Elements of Measure and Probability*, Texts and Readings in Mathematics 88,
https://doi.org/10.1007/978-981-95-2758-8_5

Example 5.1.1. The *standard Gaussian/normal cdf* $\Phi(\cdot)$ is given by:

$$\Phi(x) := \frac{1}{\sqrt{2\pi}} \int_{-\infty}^{x} \exp(-y^2/2)dy, \ x \in \mathbb{R}. \qquad \blacktriangle$$

Example 5.1.2. The *gamma cdf* with parameters $\lambda, p > 0$ is given by:

$$F_{\lambda,p}(x) := \begin{cases} 0 \ \text{ if } \ x \leq 0, \\ \displaystyle\int_0^x \frac{\lambda^p}{\Gamma(p)} e^{-\lambda y} y^{p-1} dy \ \text{ if } \ x > 0. \end{cases} \tag{5.1}$$ ▲

Example 5.1.3. The *standard exponential cdf* is a special case of the gamma cdf with $\alpha = 1$ and $p = 1$, and takes the form

$$F(x) := \begin{cases} 0 \ \text{ if } \ x \leq 0 \\ 1 - e^{-x} \ \text{ if } \ x > 0. \end{cases} \tag{5.2}$$ ▲

5.2 Lebesgue-Stieltjes measures

We now delve into a very important class of measures.

Definition 5.2.1. (Lebesgue-Stieltjes measure on $\mathcal{B}(\mathbb{R})$) A countably additive measure μ on $\mathcal{B}(\mathbb{R})$ is said to be *Lebesgue-Stieltjes* if $\mu(I) < \infty$ for every bounded interval I of $\mathbb{R}$. ◈

Exercise 5.2.1 (a) Verify that any Lebesgue-Stieltjes measure is σ-finite, and any finite measure is Lebesgue-Stieltjes.

(b) Give an example of a σ-finite measure which is not Lebesgue-Stieltjes.

Theorem 5.2.1 (From Lebesgue-Stieltjes measure to distribution function). *Let μ be a Lebesgue-Stieltjes measure on $\mathcal{B}(\mathbb{R})$.*

(a) Fix $F_\mu(0) \in \mathbb{R}$ arbitrarily, and define

$$F_\mu(x) := \begin{cases} F_\mu(0) + \mu(0, \ x] \ \text{ if } \ x > 0, \\ F_\mu(0) - \mu(x, \ 0] \ \text{ if } \ x < 0. \end{cases} \tag{5.3}$$

Then F_μ is a distribution function.

(b) If μ is a finite measure, then we may choose $F_\mu(0) = \mu(-\infty, \ 0]$. With this choice (5.3) reduces to

$$F_\mu(x) = \mu(-\infty, \ x], \ \textit{for all } x \in \mathbb{R}. \tag{5.4}$$

This F_μ is a cdf if $\mu(\mathbb{R}) = 1$. ◆

Proof. To establish non-decreasingness, let $a < b$. Then from (5.3),

$$F_\mu(b) - F_\mu(a) = \mu(a,\ b] \geq 0\,. \tag{5.5}$$

To establish right continuity, fix $x \in \mathbb{R}$ and $x_n \downarrow x$. Since μ is finite on every sub-interval, using continuity from above,

$$F_\mu(x_n) - F_\mu(x) = \mu(x,\ x_n] \downarrow \mu(\emptyset) = 0\,,$$

Other parts are trivial to prove. ■

Remark 5.2.1 Whenever μ is a finite measure, we shall choose the distribution function F_μ as described in (5.4). ●

Exercise 5.2.2 Find F_P and identify its nature in each case below.

(a) P is the probability measure where $P\{0\} = P\{1\} = 1/2$.

(b) P is the probability measure where $P\{i\} = 2^{-i},\ i = 1, 2, \ldots$.

Exercise 5.2.3 Suppose μ is a finite measure, and $\mu\{x\} > 0$ for some $x \in \mathbb{R}$. Show that F_μ is not left continuous at x.

Definition 5.2.2. A measure μ is said to be **concentrated** on B if $\mu(\Omega \setminus B) = 0$. ◇

Remark 5.2.1. Consider the countable set $S = \{x_1, x_2, \ldots\} \subset \mathbb{R}$. Let μ be a measure concentrated on S. Let $\mu\{x_i\} = a_i > 0, i \geq 1$. Then

(i) μ is Lebesgue-Stieltjes if for every bounded interval I, $\sum_{x_i \in I} a_i < \infty$.

(iii) The distribution function F_μ is continuous at x if and only if $x \notin S$.

(iv) At every $x_i \in S$, $F_\mu(x_i)$ has a jump of a_i.

(v) If $x < y, x, y \in S$ and there is no point between x and y which is in S, then F_μ is constant on $[x, y)$. ●

5.2.1 Lebesgue-Stieltjes measures on $\mathcal{B}(\mathbb{R})$

We shall now explore how to obtain a countably additive measure from a distribution function F. Note that its definition for finite left open right closed intervals is already set by Eqn. (5.5). The point is to show that this leads to a countably additive unique measure on $\mathcal{B}(\mathbb{R})$. Towards this, we make a few observations.

(i) Define the set of all **right semi-closed intervals** of $\bar{\mathbb{R}}$ as

$$(5.6)\quad \bar{\mathcal{S}} := \{A : A = (a,\ b] \text{ or } [-\infty,\ b], \text{ or } (-\infty,\ b],\ a, b \in \bar{\mathbb{R}},\ a \le b\}.$$

Then $\bar{\mathcal{S}}$ is a semi-field of subsets of $\bar{\mathbb{R}}$.

(ii) Let

$$(5.7)\quad \bar{\mathcal{F}} := \{A : A = \cup_{k=1}^{n} I_k : n \ge 1, I_k \in \bar{\mathcal{S}} \text{ for all } k, \text{ and they are disjoint}\}.$$

Then by Exercise 2.3.5, $\bar{\mathcal{F}}$ is a field of subsets of $\bar{\mathbb{R}}$.

(iii) Extend the function $F : \mathbb{R} \to \mathbb{R}$ to $F : \bar{\mathbb{R}} \to \bar{\mathbb{R}}$ by defining

$$F(\infty) := \lim_{x \to \infty} F(x), \quad F(-\infty) := \lim_{x \to -\infty} F(x)\,.$$

The above limits exist but they may equal ∞ and $-\infty$ respectively.

(iv) Define μ_F on $\bar{\mathcal{S}}$ by

$$\begin{aligned} \mu_F(a,\ b] &:= F(b) - F(a),\ a, b \in \bar{\mathbb{R}},\ a < b, \\ (5.8)\qquad \mu_F[-\infty,\ b] &:= F(b) - F(-\infty) = \mu_F(-\infty,\ b],\ b \in \bar{\mathbb{R}}. \end{aligned}$$

Then μ_F is non-negative.

(v) Define μ_F on $\bar{\mathcal{F}}$ by $\mu_F(\emptyset) = 0$ and

$$(5.9)\quad \mu_F(\cup_{k=1}^{n} I_k) := \sum_{k=1}^{n} \mu_F(I_k),\ I_1, \ldots, I_n \in \bar{\mathcal{S}} \text{ and they are disjoint.}$$

Note that $\cup_{k=1}^{n} I_k$ has alternate descriptions as $\cup_{k=1}^{t} J_k$ where J_k are disjoint elements of $\bar{\mathcal{S}}$. Thus we must show that (5.9) does not depend on the specific representation. This is easily done by using (iv), and is left as an exercise.

(vi) μ_F defined in (5.9) is finitely additive on $\bar{\mathcal{F}}$. This is easily proved by using (iv) and (v) and is also left as an exercise.

Lemma 5.2.1. Let F be a distribution function, and μ_F be as in (5.9) defined on $\bar{\mathcal{F}}$ as in (5.7). Then μ_F is countably additive. ♦

Proof. First assume that $F(\infty) - F(-\infty) < \infty$. Then μ_F is finite. Since μ_F is finitely additive, to show that it is countable additive, by Exercise 3.5.1(b), it is enough to show that $\mu_F(A_n) \downarrow 0$ for every sequence $\{A_n\}$ from $\bar{\mathcal{F}}$ that decreases to $\emptyset$.

Recall that each A_n is a finite disjoint union of intervals of the form $(a,\ b]$. Enough to consider only $a \in \mathbb{R}$. Suppose $a_n \downarrow a$. Then for every fixed b, by right continuity,

$$\mu_F(a_n,\ b] = F(b) - F(a_n) \to F(b) - F(a) = \mu_F(a,\ b]\,.$$

Thus for every A_n, we can find sets $B_n \in \bar{\mathcal{F}}$ such that

(i) closures $\bar{B}_n$ (in $\bar{\mathbb{R}}$) of B_n are contained in A_n, and

(ii) $\mu_F(B_n)$ is as close to $\mu_F(A_n)$ as we please.

So fix $\epsilon > 0$ and choose B_n such that

$$0 \le \mu_F(A_n) - \mu_F(B_n) \le \frac{\epsilon}{2^n}. \tag{5.10}$$

Observe that

$$\cap_{k=1}^{\infty} \bar{B}_k \subseteq \cap_{k=1}^{\infty} A_k = \emptyset\,.$$

Since each $\bar{B}_n$ is *compact*, by Theorem 1.4.1(e) there exists a finite m such that $\cap_{k=1}^{m} \bar{B}_k = \emptyset$. Then for all $n \ge m$,

$$\begin{aligned}
\mu_F(A_n) &= \mu_F(A_n \setminus \cap_{k=1}^{m} B_k) + \mu_F(\cap_{k=1}^{m} B_k), \text{as } B_k \subseteq A_k \text{ for } k \ge 1 \\
&= \mu_F(A_n \setminus \cap_{k=1}^{m} B_k) \\
&\le \mu_F(\cup_{k=1}^{m} (A_k \setminus B_k)) \quad \text{as} \quad A_k \downarrow \\
&\le \sum_{k=1}^{n} \mu_F(A_k \setminus B_k) \quad \text{since} \quad \mu_F \quad \text{is finitely additive} \\
&\le \epsilon \text{ by (5.10)}.
\end{aligned}$$

This shows that $\mu_F(A_n) \downarrow 0$ if $F(\infty) - F(-\infty) < \infty$. Hence μ_F is countably additive in this case.

Now suppose $F(\infty) - F(-\infty) = \infty$. Define

$$F_n(x) = \begin{cases} F(x) & \text{if } |x| \le n, \\ F(n) & \text{if } x \ge n, \\ F(-n) & \text{if } x \le -n. \end{cases} \tag{5.11}$$

Each F_n is a *bounded* distribution function. Define the set function μ_n on $\bar{\mathcal{F}}$ corresponding to F_n. These are all finite, and hence countably additive by Step 1. Moreover, it is easily seen that for all $A \in \bar{\mathcal{F}}$,

$$\mu_n(A) \leq \mu_F(A) \text{ and } \mu_n(A) \to \mu_F(A) \text{ as } n \to \infty. \tag{5.12}$$

For countable additivity of μ_F, let $\cup_{n=1}^{\infty} A_n = A \in \bar{\mathcal{F}}$, $A_n \in \bar{\mathcal{F}}$ disjoint. Then, finite additivity of μ_F, by Theorem 3.5.1(d)(ii), implies,

$$\mu_F(A) \geq \sum_{k=1}^{\infty} \mu_F(A_k). \tag{5.13}$$

So if $\sum_{k=1}^{\infty} \mu_F(A_k) = \infty$, the proof would be complete.

If instead, $\sum_{k=1}^{\infty} \mu_F(A_k) < \infty$, then

$$\begin{aligned} \mu_F(A) &= \lim_{n\to\infty} \mu_n(A) \\ &= \lim_{n\to\infty} \sum_{k=1}^{\infty} \mu_n(A_k) \quad \text{since each } \mu_n \text{ is countably additive.} \end{aligned}$$

But then

$$\begin{aligned} 0 &\leq \mu_F(A) - \sum_{k=1}^{\infty} \mu_F(A_k), \text{ by (5.13)} \\ &= \lim_{n\to\infty} \sum_{k=1}^{\infty} [\mu_n(A_k) - \mu_F(A_k)] \text{ since } \sum_{k=1}^{\infty} \mu_F(A_k) < \infty \\ &\leq 0 \text{ since } \mu_n \leq \mu_F \text{ (from (5.12)).} \end{aligned}$$

This completes the proof of the lemma. ■

Theorem 5.2.2 (From Distribution function to Lebesgue-Stieltjes measure). *Let F be a distribution function on $\mathbb{R}$ and let for all $a < b$, $\mu_F(a, b] = F(b) - F(a)$. Then there is a unique extension μ_F which is a Lebesgue-Stieltjes measure on $\mathcal{B}(\mathbb{R})$.* ◆

Proof. By the previous lemma, get countably additive measure μ_F on the field $\bar{\mathcal{F}}$ (as given in (5.7)), of subsets of $\bar{\mathbb{R}}$.

Now consider the class $\mathcal{F}$ of all finite disjoint unions of right-semi-closed intervals of $\mathbb{R}$ (treating (a, ∞) as right-semi-closed). It is easily checked that $\mathcal{F}$ is a field. Define μ_F (abuse of notation) on $\mathcal{F}$ in the natural way.

For example (there is no other choice for μ_F on these sets),

$$\begin{aligned}\mu_F(a,\ \infty) &= F(\infty) - F(a),\ a, b \in \mathbb{R}\\ \mu_F(-\infty,\ b] &= F(b) - F(-\infty),\ a, b \in \mathbb{R}\\ \mu_F(\mathbb{R}) &= F(\infty) - F(-\infty).\end{aligned}$$

Clearly this μ_F is countably additive on $\mathcal{F}$. Further μ_F is σ-finite on $\mathcal{F}$ (but not necessarily on $\bar{\mathcal{F}}$). Now apply Carathéodory Extension Theorem 4.4.1 to complete the construction of μ_F on $\mathcal{B}(\mathbb{R})$. Clearly this measure is Lebesgue-Stieltjes. We omit the details. ■

Exercise 5.2.4 Suppose F is a distribution function and μ_F is the corresponding Lebesgue-Stieltjes measure. Show that for $a < b$, $a, b \in \mathbb{R}$,

$$\begin{aligned}\mu_F(a,\ b] &= F(b) - F(a)\\ \mu_F[a,\ b] &= F(b) - F(a-)\\ \mu_F(a,\ b) &= F(b-) - F(a)\\ \mu_F[a,\ b) &= F(b-) - F(a-).\end{aligned}$$

If F is continuous at a and b, then all four expressions above are equal.

Exercise 5.2.5 Suppose F is a distribution function and μ_F is the corresponding Lebesgue-Stieltjes measure. Show that

$$\begin{aligned}\mu_F(-\infty,\ b] &= F(b) - F(-\infty)\\ \mu_F[a,\ \infty) &= F(\infty) - F(a-)\\ \mu_F(-\infty,\ b) &= F(b-) - F(-\infty)\\ \mu_F(a,\ \infty) &= F(\infty) - F(a)\\ \mu_F(\mathbb{R}) &= F(\infty) - F(-\infty).\end{aligned}$$

5.2.2 Lebesgue measure on $\mathcal{B}(\mathbb{R})$

Definition 5.2.3. (**Lebesgue measure**) The Lebesgue-Stieltjes measure μ_F, given by the distribution function $F(x) = x, x \in \mathbb{R}$, is called the *Lebesgue measure* on $\mathcal{B}(\mathbb{R})$, or simply on $\mathbb{R}$, and is denoted by λ. ◇

This measure satisfies $\lambda(a,\ b] = b - a$ for all $a, b \in \mathbb{R}$, $a < b$. The distribution function $F(x) = x + c$, $x \in \mathbb{R}$, for any c gives rise to the same measure λ on $\mathcal{B}(\mathbb{R})$.

Definition 5.2.4. (**Lebesgue σ-field**) The *Lebesgue σ-field* $\mathcal{L}(\mathbb{R})$ (in short $\mathcal{L}$) is the completion of the $\mathcal{B}(\mathbb{R})$ with respect to the Lebesgue measure λ. Any set in $\mathcal{L}$ is called a *Lebesgue set.* ◇

The Lebesgue measure and the Lebesgue sets on any interval are defined in the natural way. We again emphasize that $\mathcal{B}(\mathbb{R}) \subset \mathcal{L}(\mathbb{R})$.

Exercise 5.2.6 Show that Lebesgue measure λ is translation invariant. That is, for every Borel set B and every real number c, $\lambda(B) = \lambda(B+c)$ where $B+c = \{x+c : x \in B\}$. Can you formulate a "converse"?

Exercise 5.2.7 Let $A \subseteq B$ be Lebesgue sets in $\mathbb{R}$ and $\lambda(A) = \lambda(B)$. Then show that any set C, where $A \subseteq C \subseteq B$ is also Lebesgue measurable, and $\lambda(C) = \lambda(A)$.

5.2.3 Non-Lebesgue set

We now show that $\mathcal{L}(\mathbb{R}) \neq \mathcal{P}(\mathbb{R})$ by an explicit construction that uses the *Axiom of Choice* stated in Section 1.5. For $x, y \in (0,1]$, define

$$x \oplus y := \begin{cases} x+y & \text{if } x+y \leq 1\,, \\ x+y-1 & \text{if } x+y > 1\,. \end{cases}$$

Thus, $x \oplus y \in (0,\ 1]$ for all $x, y \in (0,\ 1]$.

For all $A \subset (0,\ 1]$, and $x \in (0,\ 1]$, define

$$A \oplus x := \{a \oplus x : a \in A\} \subset (0,\ 1]\,.$$

It is left as an exercise to show the following three claims:

(i) For any $y \in (0,\ 1]$, $B \in \mathcal{B}(\mathbb{R})$, $\{x \in (0,1] : x \oplus y \in B\}$ is a Borel set.

(ii) Let $A \in \mathcal{B}(0,\ 1]$, $x \in (0,\ 1]$. Then $\lambda\,(A \oplus x) = \lambda(A)$.

(iii) For $x, y \in (0,\ 1]$, say that $x \sim y$ if $x - y \in \mathbb{Q}$ (the set of rational numbers). Then "$\sim$" is an equivalence relation, and we can write the interval $(0,1] = \cup_{I \in \mathcal{I}} I$ as the union of a disjoint collection, say of some equivalence classes.

Now by Axiom of Choice (see Section 1.5), construct a set H which contains exactly one element from each member of $\mathcal{I}$, that is,

$$\#(H \cap I) = 1 \text{ for all } I \in \mathcal{I}\,.$$

Then we claim that

$$(0,\ 1] = \cup_{r\in\mathbb{Q}\cap(0,\ 1]} H \oplus r\,. \tag{5.14}$$

To see this, fix $x \in (0,\ 1]$. Then there exists $I \in \mathcal{I}$ such that $x \in I$. Let $y \in H \cap I$. Therefore, $x, y \in I$ and hence $x \sim y$. Therefore,

$$x = y \oplus r_0, \text{ for some } r_0 \in (0,\ 1] \cap \mathbb{Q}.$$

Thus, $x \in H \oplus r_0 \subset \cup_{r\in\mathbb{Q}\cap(0,\ 1]} H \oplus r$ and $(0,\ 1] \subset \cup_{r\in\mathbb{Q}\cap(0,\ 1]} H \oplus r$. The reverse inclusion is trivial, and so the proof of (5.14) is complete.

It is easy to see that for distinct $r_1, r_2 \in \mathbb{Q} \cap (0,\ 1]$,

$$(H \oplus r_1) \cap (H \oplus r_2) = \emptyset\,. \tag{5.15}$$

We are now ready to show that $H \notin \mathcal{L}(0,\ 1]$. Suppose if possible, $H \in \mathcal{L}(0,\ 1]$. Using (5.14) and (5.15),

$$\begin{aligned} 1 = \lambda(0,1] &= \sum_{r\in\mathbb{Q}\cap(0,1]} \lambda(H \oplus r) \\ &= \sum_{r\in\mathbb{Q}\cap(0,\ 1]} \lambda(H) \text{ by Step (ii)} \\ &= \begin{cases} 0 & \text{if } \lambda(H) = 0\,, \\ \infty & \text{if } \lambda(H) > 0\,. \end{cases} \end{aligned}$$

This is clearly a contradiction. Thus we have proved that $H \notin \mathcal{L}(\mathbb{R})$.

Exercise 5.2.8 Let λ^* be the outer measure corresponding to λ. Show that for H defined above, $\lambda^*(H) > 0$.

5.2.4 Lebesgue-Stieltjes measures on $\mathcal{B}(\mathbb{R}^d)$

The concepts of Lebesgue-Stieltjes measures and distribution functions on $\mathbb{R}$ can be extended to $\mathbb{R}^d$. We first define such measures.

Definition 5.2.5. Any measure μ on $\mathcal{B}(\mathbb{R}^d)$ is said to be Lebesgue-Stieltjes if, for every *bounded open* set G in $\mathbb{R}^d$, $\mu(G) < \infty$. ◇

These measures are constructed by starting with distribution functions F on $\mathbb{R}^d$, whose definition requires more attention. First, consider the non-decreasing property of distribution functions on $\mathbb{R}$. This now has to be replaced by the non-negativity of appropriate "increments".

We explain this first for $d = 2$. Consider the left-open right-closed *rectangle* $(a_1, b_1] \times (a_2, b_2]$ in $\mathbb{R}^2$. Let $F : \mathbb{R}^2 \to \mathbb{R}$. Then it is called non-decreasing if, for every $a_i < b_i, i = 1, 2$ in $\mathbb{R}$,

$$\mu_F((a_1, b_1] \times (a_2, b_2]) := F(b_1, b_2) - F(a_1, b_2) - F(b_1, a_2) + F(a_1, a_2) \geq 0\,. \tag{5.16}$$

Exercise 5.2.9 One class of *bivariate Gumbel cdf* is defined as

$$F_\theta(x, y) = \begin{cases} 0 & \text{if either } x < 0 \text{ or } y < 0, \\ 1 - e^{-x} - e^{-y} + e^{-(x+y+\theta xy)} & \text{if } 0 < x, y < \infty\,. \end{cases}$$

Show that for every $0 \leq \theta \leq 1$, F_θ is right continuous in every coordinate, and is non-decreasing in the sense of (5.16).

Exercise 5.2.10 Suppose $F(a_1, a_2) = F_1(a_1)F_2(a_2)$, $a_i \in \mathbb{R}$, $i = 1, 2$. where F_1 and F_2 are distribution functions on $\mathbb{R}$. Show that F is non-decreasing on $\mathbb{R}^2$ in the sense of (5.16).

For general d, we shall need a notation for an "increment operator". Let $F : \mathbb{R}^d \to \mathbb{R}$ be any function. Define Δ_{b_i, a_i}, $1 \leq i \leq d$, by

$$\Delta_{b_i, a_i} F(x_1, \ldots, x_d) = F(x_1, \ldots, x_{i-1}, b_i, \ldots x_d) - F(x_1, \ldots, x_{i-1}, a_i, \ldots x_d).$$

Suppose $a_i < b_i, 1 \leq i \leq d$. Then the *right semi-closed interval* $(a,\ b]$ in $\mathbb{R}^d$ is defined as $(a,\ b] = (a_1,\ b_1] \times \cdots \times (a_d,\ b_d]$.

Definition 5.2.6. Any function $F : \mathbb{R}^d \to \mathbb{R}$ is said to be a distribution function if the following two conditions are satisfied.

(i) F is right continuous in every co-ordinate, when the other co-ordinates are held fixed.

(ii) For any $a_i \leq b_i, 1 \leq i \leq d$, consider the left open right closed interval $(a,\ b]$ where $a = (a_1, \ldots a_d)$ and $b = (b_1, \ldots b_d)$ (we say $a \leq b$). Then

$$\mu(a,\ b] := \Delta_{b_1, a_1} \Delta_{b_2, a_2} \cdots \Delta_{b_d, a_d} F(x_1, \ldots, x_d) \geq 0\,. \tag{5.17}$$

(The right side does not depend on the dummy variables $\{x_i\}$). ◇

Given a distribution function F on $\mathbb{R}^d$, the measure of any *rectangle* $(a,\ b]$ in $\mathbb{R}^d$ is defined as in (5.17). Then we proceed as before to extend μ to $\mathcal{B}(\mathbb{R}^d)$. We state this as a theorem and leave its proof as an exercise.

Theorem 5.2.3. *(a) Let F be a distribution function on $\mathbb{R}^d$. Then there is a unique Lebesgue-Stieltjes measure μ on $\mathcal{B}(\mathbb{R}^d)$ which satisfies (5.17) for all $a \leq b$.*

(b) If μ is a Lebesgue-Stieltjes measure on $\mathcal{B}(\mathbb{R}^d)$, then there is a distribution function F which satisfies (5.17) for all $a \leq b$. ◆

Exercise 5.2.11 Let $\{F_i\}, 1 \leq i \leq d$ be distribution functions on $\mathbb{R}$.

(i) Show that F defined below yields a distribution function on $\mathbb{R}^d$:

$$F(a_1, a_2, \ldots, a_d) = F_1(a_1) \cdots F_d(a_d), \ a_i \in \mathbb{R}, \ i = 1, 2, \ldots, d,$$

(ii) Show that the Lebesgue-Stieltjes measure μ_F corresponding to F satisfies: for any left-open right-closed *rectangle* in $\mathbb{R}^d$,

$$\mu_F((a_1, b_1] \times \cdots \times (a_d, b_d]) = \prod_{i=1}^{n} [F_i(b_i) - F_i(a_i)], \text{ for all } a_i < b_i, 1 \leq i \leq, d.$$

Taking $F(x_1, \ldots, x_d) := x_1 \cdots x_d$, yields the Lebesgue measure on $\mathcal{B}(\mathbb{R}^d)$.

Definition 5.2.7. **(Lebesgue measure on $\mathcal{B}(\mathbb{R}^d)$)** The Lebesgue measure λ_d on $\mathcal{B}(\mathbb{R}^d)$, or simply on $\mathbb{R}^d$, is the unique measure for which

$$\lambda_d((a_1, b_1] \times \cdots \times (a_1, b_d]) = \prod_{i=1}^{d} (b_i - a_i), a_i < b_i, a_i, b_i \in \mathbb{R}, i = 1, \ldots, d. \quad \diamond$$

Remark 5.2.2. (a) The completion $\mathcal{L}(\mathbb{R}^d)$ of $\mathcal{B}(\mathbb{R}^d)$ with respect to the Lebesgue measure λ_d is called the Lebesgue σ-field (of $\mathbb{R}^d$). Elements of $\mathcal{L}(\mathbb{R}^d)$ are called Lebesgue sets (in $\mathbb{R}^d$). We shall not have much to do with this σ-field in the rest of the book.

(b) λ_d can be viewed as the "product" of d copies of λ on the d-fold product of $\mathcal{B}(\mathbb{R})$. Chapter 10 discusses product spaces and measures. ●

Exercise 5.2.12 Show that

(a) convex sets in $\mathbb{R}^d$ for $d > 1$ need not be Borel sets.

(b) convex sets in $\mathbb{R}^d$ for any d are Lebesgue sets.

5.2.5 Approximation by open and compact sets

The following result on approximating the measure of a set by the measure of an open or a compact set will be used later while approximating integrals of functions by suitable sub-classes of functions.

Theorem 5.2.4 (Approximation from above and below). *Let μ be a measure on $\mathcal{B}(\mathbb{R}^d)$.*

*(a) (Approximation from below by compact sets) If μ is σ-**finite**, then*

$$\mu(B) = \sup\{\mu(K) : K \subseteq B,\ K \text{ compact}\}, \text{ for every } B \in \mathcal{B}(\mathbb{R}^d). \tag{5.18}$$

*(b) (Approximation from above by open sets) If μ is **finite**, then*

$$\mu(B) = \inf\{\mu(V) : V \supseteq B,\ V \text{ open}\}, \text{ for every } B \in \mathcal{B}(\mathbb{R}^d). \tag{5.19}$$

(c) If μ is Lebesgue-Stieltjes, then (5.19) continues to hold.

(d) Approximation (5.18) continues to hold whenever $\mu(B) < \infty$ even when μ is not σ-finite. ◆

Example 5.2.1. Let $S = \{1, 1/2, 1/3, \ldots\}$. Let μ be the σ-finite measure concentrated on S with $\mu\{1/n\} = n$ for all n. Consider the set $B = \{0\}$. Any open set containing B has infinite measure. Hence μ is not Lebesgue-Stieltjes, and (5.19) does not hold. ▲

Remark 5.2.3. (**Radon measure**) Radon measures are like Lebesgue-Stieltjes measures, on more general spaces. Let $(\Omega, \mathcal{B}(\Omega), \mu)$ be a measure space where Ω is a metric space (more generality is possible) $\mathcal{B}(\Omega)$ is the Borel σ-field. Then μ is said to be a Radon measure if:

(i) For every open set U, $\mu(U) = \sup\{\mu(K) : K \subseteq U,\ K \text{ compact}\}$.

(ii) Every $\omega \in \Omega$ has a neighborhood U such that $\mu(U) < \infty$.

For information on Radon measures, refer to Folland [1999]. ●

Proof of Theorem 5.2.4. (a) First suppose μ is finite. Let

$$\mathcal{M} = \{B \in \mathcal{B}(\mathbb{R}^d) : (5.18) \text{ holds for } B\}.$$

We claim that $\mathcal{M}$ is a monotone class. Let $B_n \in \mathcal{M}, B_n \uparrow B$. Fix $\epsilon > 0$. Let $D_n \subseteq B_n$ be compact so that

$$\mu(B_n) \leq \mu(D_n) + \epsilon.$$

So, $K_n := \cup_{k=1}^n D_k$ is a non-decreasing sequence of compact sets, and,

$$K_n \subseteq B_n \text{ and } (B_n \setminus K_n) \subseteq (B_n \setminus D_n) \text{ for all } n \geq 1.$$

Hence $\mu(B_n \setminus K_n) \le \mu(B_n \setminus D_n) \le \epsilon$ and this implies,

$$\lim_{n\to\infty} \mu(K_n) \le \mu(B) = \lim_{n\to\infty} \mu(B_n) \le \lim_{n\to\infty} \mu(K_n) + \epsilon,$$

so that (5.18) holds for B, and hence $B \in \mathcal{M}$.

Now let $B_n \in \mathcal{M}$, $B_n \downarrow B$. Let $K_n \subset B_n$ be compact so that

$$\mu(B_n) \le \mu(K_n) + \frac{\epsilon}{2^n}.$$

Let $K = \cap_{n=1}^{\infty} K_n$. By Theorem 1.4.1(b), K is compact.

Then

$$\begin{aligned} \mu(B) - \mu(K) &= \mu(B \setminus K), \text{ since } K \subseteq B \\ &\le \mu(\cup_{n=1}^{\infty} (B_n \setminus K_n)), \text{ since } B \subseteq \cup_{n=1}^{\infty} B_n \\ &\le \sum_{n=1}^{\infty} \mu(B_n \setminus K_n) \text{ by countable sub-additivity} \\ &\le \epsilon. \end{aligned}$$

Hence $B \in \mathcal{M}$. Thus $\mathcal{M}$ is a monotone class.

It is easily checked that all sets $(a_1, b_1] \times \cdots \times (a_d, b_d]$ and their finite disjoint unions are contained in $\mathcal{M}$. This class, say $\mathcal{F}$, is a field and $\sigma(\mathcal{F}) = \mathcal{B}(\mathbb{R}^d)$. By the monotone class theorem, $\mathcal{M} = \mathcal{B}(\mathbb{R}^d)$, and hence (a) holds when μ is finite.

Now let μ be σ-finite. Get $B_n \in \mathcal{B}(\mathbb{R}^d)$, $\mu(B_n) < \infty$, and $B_n \uparrow B$. Each B_n can be approximated from within by compact sets (use the finite measures $\mu_n(A) = \mu(A \cap B_n)$, $A \in \mathcal{B}(\mathbb{R}^d)$ and proceed as in the finite case). We also observe that by Theorem 1.4.1(b) finite unions of compact sets are compact. This completely proves (a).

(b) We have

$$\begin{aligned} \mu(B) &\le \inf\{\mu(V) : V \supseteq B,\ V \text{ open}\} \\ &\le \inf\{\mu(W) : W \supseteq B,\ W = K^c, K \text{ compact}\} \\ &= \inf\{\mu(\mathbb{R}) - \mu(W^c) : W^c \subseteq B^c,\ W^c \text{ compact}\} \\ &= \mu(\mathbb{R}) - \sup\{\mu(K) : K \subseteq B^c,\ K \text{ compact}\} \\ &= \mu(\mathbb{R}) - \mu(B^c), \text{ by (a), since } \mu \text{ is finite} \\ &= \mu(B). \end{aligned}$$

(c) Write $\mathbb{R}^d = \cup_{n=1}^{\infty} B_n$ where $\{B_n\}$ are disjoint bounded sets in $\mathcal{B}(\mathbb{R}^d)$. Then for each n, $B_n \subseteq C_n$ for some bounded *open* set C_n. Define

$$\mu_k(A) = \mu(A \cap C_k), \ A \in \mathcal{B}(\mathbb{R}^d).$$

Then each μ_k is a finite measure. Fix $\epsilon > 0$. If B is a Borel subset of B_k, then by (b), there is an open set $W_k \supseteq B$ such that

$$\mu_k(W_k) \le \mu_k(B) + \frac{\epsilon}{2^k}. \tag{5.20}$$

Now note that $V_k = W_k \cap C_k$ is an open set. Moreover, $B \cap C_k = B$ since $B \subseteq B_k \subseteq C_k$. Hence

$$\begin{aligned} \mu(V_k) &= \mu_k(W_k) \text{ by definition of } \mu_k \\ &\le \mu_k(B) + \frac{\epsilon}{2^k} \text{ by (5.20)} \\ &= \mu(B) + \frac{\epsilon}{2^k} \text{ since } B \subseteq C_k. \end{aligned}$$

Fix any $A \in \mathcal{B}(\mathbb{R}^d)$. By the conclusion reached above, let V_k be an open set such that

$$V_k \supseteq A \cap B_k \text{ and } \mu(V_k) \le \mu(A \cap B_k) + \frac{\epsilon}{2^k} \text{ for all } k. \tag{5.21}$$

Let $V = \cup_{n=1}^{\infty} V_n$. Then V is open and $V \supseteq A$. Using (5.21),

$$\mu(V) \le \sum_{n=1}^{\infty} \mu(V_n) \le \mu(A) + \epsilon.$$

This proves (c). Proof of (d) is left as an exercise. ■

5.3 Exercises

Exercise 5.3.1 Show that the set of discontinuity points of any distribution function F on $\mathbb{R}$ is countable. First assume $F(\infty) - F(-\infty) < \infty$.

Exercise 5.3.2 Let F be a continuous probability distribution function. Show that F is uniformly continuous.

Exercise 5.3.3 Suppose F and G are two distribution functions on $\mathbb{R}$ which agree on some countable dense subset of $\mathbb{R}$. Show that $F \equiv G$.

Exercise 5.3.4 Let $D = \{x_1, x_2, \ldots\}$ be a countable set dense in $\mathbb{R}$. Let $F_D : D \to \mathbb{R}$ be a non-decreasing function which is right continuous on D; that is, if $\{x_n\}$ is any sequence from D which decreases to $x \in D$, then $F(x_n) \to F(x)$. Show that F defined below is a distribution function:

$$F(x) = \inf\{F_D(y) : y \in D,\ y > x\},\ x \in \mathbb{R}.$$

Exercise 5.3.5 Show that μ_F as given in (5.9) is well defined, and is finitely additive on $\bar{\mathcal{F}}$.

Exercise 5.3.6 In the proof of Lemma 5.2.1, verify the following:

(i) Claim made in (i) and (ii).

(ii) The step $(A_n \setminus \cap_{k=1}^m B_k) \subseteq \cup_{k=1}^n (A_k \setminus B_k)$.

(iii) Claim made in Eqn. (5.12).

(iv) Eqn. (5.13).

Exercise 5.3.7 Check all the details in the proof of Theorem 5.2.2. Why could not we claim that μ_F is σ-finite on $\bar{\mathbb{R}}$?

Exercise 5.3.8 Suppose f is a non-negative continuous function on $\mathbb{R}$. Fix $F(0)$ arbitrarily and define

$$F(x) := \begin{cases} F(0) + \displaystyle\int_0^x f(t)dt & \text{for } x > 0, \\ \\ F(0) - \displaystyle\int_0^x f(t)dt & \text{for } x < 0. \end{cases}$$

Show that F is a distribution function and is continuous everywhere. The corresponding Lebesgue-Stieltjes measure μ_F satisfies

$$\mu_F(a,\ b] = \int_a^b f(t)dt,\ a, b \in \mathbb{R}.$$

Check that μ_F does not depend on $F(0)$, and $\mu_F\{x\} = 0$ for every $x \in \mathbb{R}$.

Exercise 5.3.9 Let F be a non-decreasing right continuous function on the interval $[a,\ b]$. Show that there is a unique measure μ_F on $\mathcal{B}[a,\ b]$ such that $\mu_F(x, y] = F(y) - F(x)$ for all $a \le x < y \le b$.

Exercise 5.3.10 Consider the measure,

$$\mu(A) = \text{ Number of rational points in } A,\quad A \in \mathcal{B}(\mathbb{R}).$$

Show that μ is a measure but it is not a Lebesgue-Stieltjes measure, and approximation from above fails in Theorem 5.2.4.

Exercise 5.3.11 Verify Steps (i)–(v) in the proof of existence of the non-Borel sets discussed in Section 5.2.3.

Exercise 5.3.12 A measure μ is said to be **non-atomic** if, given any A such that $\mu(A) > 0$, there exists $B \subset A$ such that $0 < \mu(B) < \mu(A)$. Show that Lebesgue measure is non atomic.

Exercise 5.3.13 Consider $(\mathbb{R}^d, \mathcal{B}(\mathbb{R}^d), \lambda_d)$. Show that if $\lambda_d(A) > 0$, and $0 < \alpha < 1$, then there exists $B \subset A$ such that $\lambda_d(B) = \alpha\lambda_d(A)$.

Exercise 5.3.14 Let $A,\ B \in \mathcal{L}(\mathbb{R})$.

(a) Show that $A + B := \{x + y : x \in A, y \in B\} \in \mathcal{L}(\mathbb{R})$.

(b) Show that $\lambda(A + B) \geq \lambda(A) + \lambda(B)$

Exercise 5.3.15 Let λ_d be the Lebesgue measure on $\mathbb{R}^d$. Show that for $E \in \mathcal{L}(\mathbb{R})$, $x \in \mathbb{R}^d$, $\lambda_d(E+x) = \lambda_d(E)$. (Also need to show $E+x \in \mathcal{L}(\mathbb{R})$).

Exercise 5.3.16 (**Cantor set**) Consider the interval $[0,\ 1]$. Remove the middle $1/3$ *open* interval $E_1 = (1/3,\ 2/3)$. From the two disjoint intervals of $[0,\ 1]\setminus E_1$, remove their middle $1/3$ open intervals. Call their union E_2. From the four disjoint closed intervals of $[0,\ 1] \setminus (E_1 \cup E_2)$, remove the *four* middle $1/3$ open intervals. Continue this process. Let $\{E_i\}$ be all the removed intervals. Let $C = [0,\ 1] \setminus \cup_{n=1}^{\infty} E_n$. It is called the *Cantor set.* Show that

(a) C is uncountable and is nowhere dense;

(b) C is a closed set, and every point in C is a limit point;

(c) $\lambda(C) = 0$.

Exercise 5.3.17 (**Cantor distribution function**) In Exercise 5.3.16, $\cup_{i=1}^{n} E_i$ consists of $2^n - 1$ disjoint intervals, say $A_1, \ldots, A_{2^n-1}$, in increasing order of their right end points. Define $F_n : [0,\ 1] \to [0,\ 1]$:

$$F_n(x) = \begin{cases} 0 \ \text{ if } \ x = 0, \\ k/2^n \ \text{ if } \ x \in A_k,\ k = 1, 2, \ldots, 2^n - 1, \\ 1 \ \text{ if } \ x = 1. \end{cases}$$

Complete the definition on $[0,\ 1]$ by linear interpolation.

Show that

(a) each F_n is a non-decreasing continuous function;

(b) $F_n(x)$ converges for every x. Let $F(x) = \lim_{n\to\infty} F_n(x)$. Then $F(0) = 0$ and $F(1) = 1$. Extend F to $\mathbb{R}$ by $F(x) = 1$ for $x \geq 1$, and $F(x) = 0$ for $x \leq 0$. Then F is called the *Cantor (distribution) function*, with $F(-\infty) = 0$ and $F(\infty) = 1$;

(c) F is continuous and non-decreasing;

(d) $F' = 0$ a.e. λ.

(e) Let μ be the measure corresponding to F. Then μ is "orthogonal" to λ; there is a $C \in \mathcal{B}(\mathbb{R})$ such that $\lambda(C) = 0$ and $\mu(C^c) = 0$. Further $\mu\{x\} = 0$ for every $x \in \mathbb{R}$. Measures which satisfy these two conditions are called **singular**. Chapters 20 and 21 have more on these ideas.

Exercise 5.3.18 Consider $([0,\ 1], \mathcal{B}([0,\ 1]))$. Let λ be the Lebesgue measure. Let P be a measure such that whenever $\lambda(A) = 1/2$, we have $P(A) = 1/2$. Show that $\lambda \equiv P$.

Exercise 5.3.19 (Inverse of a probability distribution function) For any probability distribution function F, define

$$F^{-1}(t) = \inf\{x : F(x) \geq t\},\ 0 \leq t \leq 1\,.$$

Show that

(a) F^{-1} is left continuous on $(0,\ 1)$.

(b) $F(F^{-1}(t)) \geq t$ for all $0 \leq t \leq 1$.

(c) $F(F^{-1}(t)) > t$ if and only if t is not in the range of F.

(d) $F^{-1}(F(x)) \leq x$ for all $-\infty < x < \infty$.

(e) $F^{-1}(F(x)) < x$ if and only if $F(y) = F(x)$ for some $y < x$.

Chapter 6

Measurable functions

If f is a continuous function on a closed interval $[a,\ b]$, we understand the area under the curve of f as the **Riemann integral** $\int_a^b f(t)dt$. Note that the "length measure" is involved in the computation of this integral.

Now consider the function f on $[0,\ 1]$ defined as

$$f(x) = \begin{cases} 1 & \text{if } 0 \leq x \leq 1 \text{ is irrational,} \\ 2 & \text{if } 0 \leq x \leq 1 \text{ is rational.} \end{cases} \tag{6.1}$$

Then f is not Riemann integrable in $[0,\ 1]$. On the other hand, f takes the value 2 on the set of all rationals $\mathbb{Q}$ and $\lambda(\mathbb{Q}) = 0$. It has the constant value 1 on $\mathbb{Q}^c$, and $\lambda(\mathbb{Q}^c) = 1$. We are then tempted to conclude that $\int_0^1 f(t)\lambda(dt) = 1$, where $\lambda(dt)$ is some suitable notion of an integral.

Before we develop a notion of a general integral, we need a class of functions that is rich enough, and in particular includes functions such as the one given in (6.1). In this chapter, we introduce these functions, called measurable functions, and their properties. Integration will be developed in the next chapter.

6.1 Measurable functions

Definition 6.1.1. (**Measurable function**) Suppose $(\Omega_i, \mathcal{A}_i)$ $i = 1, 2$ are two measurable spaces. Any function $f : \Omega_1 \to \Omega_2$ is said to be a *measurable* function if $f^{-1}(A_2) \in \mathcal{A}_1$ for every $A_2 \in \mathcal{A}_2$. To indicate that f is measurable, we write $f : (\Omega_1, \mathcal{A}_1) \to (\Omega_2, \mathcal{A}_2)$. ◇

A. Bose, *Elements of Measure and Probability*, Texts and Readings in Mathematics 88,
https://doi.org/10.1007/978-981-95-2758-8_6

Measurability of a function is determined by the underlying σ-fields, and there may not even be any measures defined on these σ-fields. If these σ-fields are clear from the context, we may not mention them. Of course, things become more interesting when we also have measures.

Exercise 6.1.1 Let $f : \Omega_1 \to \Omega_2$ with σ-fields $\mathcal{A}_i,\ i = 1, 2$ respectively. Suppose $\mathcal{C}$ is a class of subsets of Ω_2 such that $\sigma(\mathcal{C}) = \mathcal{A}_2$. Then show that f is measurable if and only if for all $C \in \mathcal{C},\ f^{-1}(C) \in \mathcal{A}_1$.

Exercise 6.1.2 Let $T : (\Omega_1, \mathcal{A}_1) \to (\Omega_2, \mathcal{A}_2)$ be a measurable function, and μ be a measure on $(\Omega_1, \mathcal{A}_1)$. Define the set function μT^{-1} on $\mathcal{A}_2$ by

$$\mu T^{-1}(A_2) = \mu(T^{-1}(A_2)) \quad A_2 \in \mathcal{A}_2. \tag{6.2}$$

Show that μT^{-1} is a measure on $(\Omega_2, \mathcal{A}_2)$.

Definition 6.1.2. (**Induced or push forward measure**) The measure μT^{-1} on $(\Omega_2, \mathcal{A}_2)$ in (6.2) is called the *induced measure* or the **push forward measure** of T. If $(\Omega_2, \mathcal{A}_2) = (\Omega_1, \mathcal{A}_1)$ and $\mu = \mu T^{-1}$, then T is called **measure preserving**. ◇

Later we shall encounter plenty of induced measures.

Exercise 6.1.3 Consider the Lebesgue measure on the interval $[0, 1]$. Show that T defined by $T(x) = 1 - x, 0 \le x \le 1$ is measure preserving.

Exercise 6.1.4 (**Composition of measurable functions**) Suppose that $f_1 : (\Omega_1, \mathcal{A}_1) \to (\Omega_2, \mathcal{A}_2)$, $f_2 : (\Omega_2, \mathcal{A}_2) \to (\Omega_3, \mathcal{A}_3)$ are measurable functions. Show that $f_2 \circ f_1 : (\Omega_1, \mathcal{A}_1) \to (\Omega_3, \mathcal{A}_3)$ is measurable.

6.1.1 Borel measurability

For functions that take values in $\mathbb{R}$, $\bar{\mathbb{R}}$, $\mathbb{R}^d$ etc., the underlying σ-field is usually taken as the Borel σ-fields Any measurable f is then called **Borel measurable** or simply measurable, with the domain space and its σ-field being clear from the context. Borel measurable functions which are defined on a probability space are called **random variables/vectors** and are discussed in details Chapter 11 onwards.

Let f be a **complex-valued** function. Then $f = \mathcal{R}(f) + \iota\mathcal{I}(f)$ where $\iota = \sqrt{-1}$, and $\mathcal{R}(f)$ and $\mathcal{I}(f)$ are real-valued functions. Then, f is said to be (Borel) measurable if both $\mathcal{R}(f)$ and $\mathcal{I}(f)$ are Borel measurable.

Exercise 6.1.5 Suppose $f : \mathbb{R} \to \mathbb{R}$ is continuous. Show that f is measurable with respect to the Borel σ-fields.

Exercise 6.1.6 Let $f : \mathbb{R} \to \mathbb{R}$ be a lower or an upper semi-continuous function, and $|f| \leq M < \infty$. Show that f is Borel measurable. Hence, the upper and lower envelopes of measurable functions are measurable.

Exercise 6.1.7 Suppose $f : \Omega \to \bar{\mathbb{R}}$ is a function and Ω is equipped with the σ-field $\mathcal{A}$. Show that the following statements are equivalent:

(i) The function f is Borel measurable.
(ii) For every $c \in \mathbb{R}$, $\{\omega \in \Omega : f(\omega) > c\} \in \mathcal{A}$.
(iii) For every $c \in \mathbb{R}$, $\{\omega \in \Omega : f(\omega) \leq c\} \in \mathcal{A}$.
(iv) For every $c \in \mathbb{R}$, $\{\omega \in \Omega : f(\omega) \geq c\} \in \mathcal{A}$.
(v) For every $c \in \mathbb{R}$, $\{\omega \in \Omega : f(\omega) < c\} \in \mathcal{A}$.
(vi) For every $a, b \in \mathbb{R}$, $\{\omega \in \Omega : a \leq f(\omega) \leq b\} \in \mathcal{A}$.

Construct other similar statements that are equivalent to the above.

Exercise 6.1.8 If f and g are measurable functions, show that the sets such as $\{\omega : f(\omega) > g(\omega)\}$, $\{\omega : f(\omega) \geq g(\omega)\}$ etc. are measurable.

Exercise 6.1.9 Suppose $f_1, f_2 : \Omega \to \bar{\mathbb{R}}$ are Borel measurable. Show that the sum difference, product and ratio of these functions are also Borel measurable, whenever they are defined.

Exercise 6.1.10 Let $f : (\Omega, \mathcal{A}) \to (\bar{\mathbb{R}}, \mathcal{B}(\bar{\mathbb{R}}))$ be measurable. Recall the functions f^+ and f^- from (1.4) and (1.5). Show that,

(i) both f^+ and f^- are non-negative measurable functions;

(ii) $|f| = f^+ + f^-$, and is measurable;

(ii) $f = f^+ - f^-$.

Definition 6.1.3. (Projection) The mappings $p_i : \mathbb{R}^d \to \mathbb{R}$ defined below are called the *projection maps.*

$$p_i(x) = x_i, \;\; x = (x_1, x_2, \ldots, x_d)' \in \mathbb{R}^d, 1 \leq i \leq d\,. \qquad \diamond$$

Lemma 6.1.1. (Projection maps) (a) Suppose $p_i : \bar{\mathbb{R}}^d \to \bar{\mathbb{R}}$, $1 \leq i \leq d$ are the projection maps. Then each p_i is Borel measurable.

(b) Suppose $f : (\Omega, \mathcal{A}) \to \bar{\mathbb{R}}^d$. Then f is Borel measurable if and only if $f_i = p_i \circ f$ is Borel measurable for all $1 \leq i \leq d$. ♦

The proof of Lemma 6.1.1 is left as an exercise.

Lemma 6.1.2. Let $\{f_n\} : (\Omega, \mathcal{A}) \to (\bar{\mathbb{R}}, \mathcal{B}(\bar{\mathbb{R}}))$ be such that for all $\omega \in \Omega$, $f(\omega) := \lim_{n\to\infty} f_n(\omega)$ exists. Then f is Borel measurable. ♦

Proof. This follows from the string of equalities for any $x \in \mathbb{R}$:

$$\begin{aligned}
\{\omega : f(\omega) > x\} &= \{\omega : \lim_{n\to\infty} f_n(\omega) > x\} \\
&= \cup_{k=1}^{\infty}\{\omega : f_n(\omega) > x + \frac{1}{k} \text{ for all large } n\} \\
&= \cup_{k=1}^{\infty} \liminf_{n\to\infty} \{\omega : f_n(\omega) > x + \frac{1}{k}\} \\
&= \cup_{k=1}^{\infty} \cup_{n=1}^{\infty} \cap_{t=n}^{\infty} \{\omega : f_t(\omega) > x + \frac{1}{k}\}.
\end{aligned}$$ ■

Exercise 6.1.11 Let $\{f_n\}$ be a sequence of Borel measurable functions from Ω to $\bar{\mathbb{R}}$. Show that $\limsup f_n$ and $\liminf f_n$ are measurable.

6.1.2 Simple functions

Recall the definition of the indicator function from (1.3). Let $(\Omega, \mathcal{A})$ be a measurable space. Then for any set $A \in \mathcal{A}$, $\mathbf{1}_A$ is Borel measurable.

Definition 6.1.4. (**Simple function**) A function $s : (\Omega, \mathcal{A}) \to \bar{\mathbb{R}}$ is called *simple* if it is measurable and takes finitely many values. Then

$$s = \sum_{i=1}^{n} x_i \mathbf{1}_{A_i}, \tag{6.3}$$

for some $\{x_i\} \in \bar{\mathbb{R}}$, and a measurable partition $\{A_i\}$ of Ω. ◇

In (6.3), $\{x_i\}$ need not be distinct, though we can always redefine our partition so that they are so. Thus, a simple function can have different representations. Further, *both* ∞ and $-\infty$ are possible values of $\{x_i\}$.

Exercise 6.1.12 Suppose s is a simple function on $(\Omega, \mathcal{A})$. Show that:

(a) $-s$ and s^2 are also simple functions.

(b) For any $A \in \mathcal{A}$, $s\mathbf{1}_A$ is also a simple function.

Exercise 6.1.13 Suppose s_1 and s_2 are simple functions.
(a) When are $s_1 + s_2$ and $s_1 - s_2$ defined? Are they simple whenever they are well-defined?

(b) When is s_1/s_2 defined? Is it simple when it is defined?

(c) Are $s = \max(s_1, s_2)$ and $s = \min(s_1, s_2)$ simple functions?

Exercise 6.1.14 If $f : (\Omega, \mathcal{A}) \to (\bar{\mathbb{R}}, \mathcal{B}(\bar{\mathbb{R}}))$ is measurable, and $A \in \mathcal{A}$, then show that $f\mathbf{1}_A$ is also measurable. When is it a simple function?

Exercise 6.1.15 Let s_n be a sequence of non-negative simple functions, decreasing to f. Show that f is measurable.

6.1.3 Approximation by simple functions

The following result on the approximation of measurable functions by simple functions will be very crucial later.

Lemma 6.1.3 (Approximation by simple functions). Suppose that $f : (\Omega, \mathcal{A}) \to (\bar{\mathbb{R}}, \mathcal{B}(\bar{\mathbb{R}}))$.
(a) If $f \geq 0$, then there exists a sequence of non-negative finite-valued simple functions $\{s_n\}$ such that $s_n(\omega) \uparrow f(\omega)$ for every $\omega \in \Omega$.

(b) There exists a sequence of finite-valued simple functions $\{s_n\}$ such that $|s_n| \leq |f|$, and $s_n(\omega) \to f(\omega)$ for all $\omega \in \Omega$.

(c) If f is **bounded**, then $\{s_n\}$ in (b) can be chosen such that $s_n \to f$ uniformly on Ω. ♦

Proof. (a) Define sets $\{A_k\}$ and B_n, and simple functions $\{s_n\}$ as

$$\begin{aligned} A_k &= f^{-1}[\frac{k-1}{2^n}, \frac{k}{2^n}),\ k = 1, \ldots, n2^n,\ B_n := \Omega - \cup_{k=1}^{n2^n} A_k, \\ s_n &= \sum_{k=1}^{n2^n} \frac{k-1}{2^n} \mathbf{1}_{A_k} + n\mathbf{1}_{B_n}. \end{aligned}$$

Then $A_n, B_n \in \mathcal{A}$, and $B_n \downarrow \emptyset$. By construction, $s_n \uparrow$, $s_n \leq f$, and on B_n^c, $f(\omega) - s_n(\omega) \leq 1/2^n$. Thus $\{s_n\}$ has the desired properties.

(b) Consider f^+ and f^-. Choose $\{s_{1n}\}$ and $\{s_{2n}\}$ respectively for them as in (a). Let $s_n = s_{1n} - s_{2n}$. Then $\{s_n\}$ has the desired properties.

(c) For bounded f, $B_n = \emptyset$ for all large n, and on B_n^c, $|s_n - f| \leq 1/2^n$. ∎

6.2 Exercises

Exercise 6.2.1 Let $f : \Omega_1 \to \Omega_2$ be a function. Let $\mathcal{F}_2$ be a field of subsets of Ω_2. Consider $\mathcal{F}_1 = \{A_1 : A_1 = f^{-1}(A_2),\ A_2 \in \mathcal{F}_2\}$. Is $\mathcal{F}_1$ a field? What can you say about $\mathcal{F}_1$ when $\mathcal{F}_2$ is a σ-field or a semi-field?

Exercise 6.2.2 Give an example of a function $f : (\Omega_1, \mathcal{A}_1) \to (\Omega_2, \mathcal{A}_2)$ (which is measurable) but for some set $A_1 \in \mathcal{A}_1$, $f(A_1) \notin \mathcal{A}_2$.

Exercise 6.2.3 Give an example of a function $f : (\Omega, \mathcal{A}) \to \bar{\mathbb{R}}$ such that $|f|$ is measurable but f is not measurable.

Exercise 6.2.4 If f is a monotone function from $\mathbb{R}$ to $\mathbb{R}$, then show that it is measurable. In particular, any distribution function is measurable.

Exercise 6.2.5 Let $f, g : (\Omega, \mathcal{A}, \mu) \to (\mathbb{R}^d, \mathcal{B}(\mathbb{R}^d))$. Show that

(a) $A := \{\omega : f(\omega) \neq g(\omega)\} \in \mathcal{A}$. Hence $f = g$ a.s. $[\mu]$ iff $\mu(A) = 0$,

(b) equality a.s. $[\mu]$ is an equivalence relation on the set of measurable functions. In everything that we shall encounter later, without any loss, functions in each equivalence class are identified as same.

Exercise 6.2.6 Show that $f : (\Omega, \mathcal{A}) \to \mathbb{R}^+$ is Borel measurable if and only if it is an increasing limit of non-negative simple functions.

Exercise 6.2.7 Show that any $\mathbb{R}^d$-valued function f on $(\Omega, \mathcal{A})$ is Borel measurable if and only if any one of the following holds:

(a) $f^{-1}(U) \in \mathcal{A}$ for every open set U.

(b) $f^{-1}(C) \in \mathcal{A}$ for every closed set C.

Exercise 6.2.8 Suppose $\{f_n\} : (\Omega, \mathcal{A}) \to (\mathbb{R}^d, \mathcal{B}(\mathbb{R}^d))$. Then show that $A := \{\omega : \lim f_n(\omega) \text{ exists}\} \in \mathcal{A}$.

Exercise 6.2.9 Suppose $\{f_n\} : (\Omega, \mathcal{A}) \to (\mathbb{R}^d, \mathcal{B}(\mathbb{R}^d))$. Then show that $A := \{x : \{f_n(x)\} \text{ is a Cauchy sequence}\} \in \mathcal{A}$.

Exercise 6.2.10 Consider a function $f : \mathbb{R} \to \mathbb{R}$. From Exercise 2.5.4, $A = \{x : f \text{ is continuous at } x\} \in \mathcal{B}(\mathbb{R})$. Suppose $\lambda(A^c) = 0$. Show that f is Lebesgue measurable.

Exercise 6.2.11 Let $f : (\Omega, \sigma(\mathcal{C})) \to (\mathbb{R}, \mathcal{B}(\mathbb{R}))$. Show that there is a countable $\mathcal{D} \subset \mathcal{C}$ such that, $f : (\Omega, \sigma(\mathcal{D})) \to (\mathbb{R}, \mathcal{B}(\mathbb{R}))$ is measurable.

Exercise 6.2.12 Suppose $f : \Omega \to \mathbb{R}$ is a function. Describe the smallest σ-field on Ω which makes f Borel measurable.

Exercise 6.2.13 Let $f : (\Omega, \mathcal{A}) \to (\mathbb{R}, \mathcal{B}(\mathbb{R}))$ and define the collection of sets $\mathcal{A}_f = \{f^{-1}(E) : E \in \mathcal{B}(\mathbb{R})\}$.

(a) Show that $\mathcal{A}_f$ is a σ-field.

(b) Let $g : (\Omega, \mathcal{A}_f) \to (\mathbb{R}, \mathcal{B}(\mathbb{R}))$. Show that there exists a function $h : (\mathbb{R}, \mathcal{B}(\mathbb{R})) \to (\mathbb{R}, \mathcal{B}(\mathbb{R}))$ so that, $g = h \circ f$. We can replace $\mathbb{R}$ by $\bar{\mathbb{R}}$.

Chapter 7

Integral

We now develop a notion of integral of suitable functions with respect to arbitrary measures. This integral shall have all the usual properties of the Riemann integral, and for any bounded continuous function on any interval $[a,\ b]$, it will coincide with the Riemann integral.

7.1 Integral of a simple function

Definition 7.1.1. (**Integral of a simple function**) If $s = \sum_{i=1}^{n} x_i \mathbf{1}_{A_i}$ is a simple function on $(\Omega, \mathcal{A}, \mu)$ then we define

$$\int_{\Omega} s(\omega)\mu(d\omega) := \sum_{i=1}^{n} x_i \mu(A_i), \tag{7.1}$$

as long as both ∞ *and* $-\infty$ *do not appear in the above sum.* In that case we say that the integral of s *exists*. It is common and convenient to suppress ω, and write $\int_{\Omega} s d\mu$ for $\int_{\Omega} s(\omega)\mu(d\omega)$. ◇

Note that $\int \mathbf{1}_A d\mu = \mu(A)$ for every $A \in \mathcal{A}$.

Exercise 7.1.1 Show that all representations of s yield the same value of (7.1), and hence the integral is well-defined.

Remark 7.1.1. (a) For Riemann integral (Section 1.6), the domain of f was partitioned into intervals. In Definition 7.1.1, the range of f creates the partition (f is constant on each A_i). This is a very crucial difference.

(b) The integral of a simple function may equal either ∞ or $-\infty$.

A. Bose, *Elements of Measure and Probability*, Texts and Readings in Mathematics 88,
https://doi.org/10.1007/978-981-95-2758-8_7

(c) If s is a non-negative simple function, then its integral exists.

(d) If s is a simple function such that $\int_\Omega s d\mu$ exists, and $a \in \mathbb{R}$, then $\int_\Omega a s d\mu = a \int_\Omega s d\mu$. ●

Exercise 7.1.2 (Monotonicity) If s_1 and s_2 are two simple functions whose integrals exist, and $s_1 \le s_2$, then show that, $\int_\Omega s_1 d\mu \le \int_\Omega s_2 d\mu$. It is possible that, (i) both sides equal $-\infty$, (ii) both sides equal ∞, or (iii) the left side equals $-\infty$ and the right side equals ∞.

Exercise 7.1.3 Let s_1, s_2 be simple functions such that $\int_\Omega s_i d\mu$, $i = 1, 2$ exist. Show that $\int_\Omega (s_1 + s_2) d\mu$ may not exist even when $s_1 + s_2$ is defined. Show that $\int_\Omega (s_1 + s_2) d\mu = \int_\Omega s_1 d\mu + \int_\Omega s_2 d\mu$, when all three integrals exist. (Remember the arithmetic of $\pm\infty$ from Section 1.2).

7.2 Integral of a measurable function

We have seen that a measure is always continuous from below. It is only natural that we define the integral of a non-negative function, by using approximation from below by integrals of simple functions.

Definition 7.2.1. (**Integral of a non-negative function**) Suppose $f : (\Omega, \mathcal{A}) \to (\overline{\mathbb{R}}^+, \mathcal{B}(\overline{\mathbb{R}}^+))$. Define

$$\int_\Omega f d\mu := \sup \Big\{ \int_\Omega s d\mu : s \ \text{ is simple and } \ 0 \le s \le f \Big\}. \qquad \diamond$$

This definition is unambiguous. The value of the integral can be ∞.

Exercise 7.2.1 Let $f : (\Omega, \mathcal{A}) \to (\overline{\mathbb{R}}^+, \mathcal{B}(\overline{\mathbb{R}}^+)$. Let $\{s_n\}$ be non-negative simple functions such that $s_n(\omega) \uparrow f(w)$ for all $\omega \in \Omega$. What happens to $\lim_{n\to\infty} \int_\Omega s_n d\mu$?

The integral of a function f which is not necessarily non-negative, is defined through splitting f into positive and negative parts.

Definition 7.2.2. (**Integral**) For any $f : (\Omega, \mathcal{A}) \to (\bar{\mathbb{R}}, \mathcal{B}(\bar{\mathbb{R}}))$, define

$$\int_\Omega f d\mu = \int_\Omega f^+ d\mu - \int_\Omega f^- d\mu \ \ \textbf{provided not both integrals are } \infty.$$

Then we say that $\int_\Omega f d\mu$ **exists**. Otherwise, we say that $\int_\Omega f d\mu$ **does not exist**. If the integral is finite, then we say that f **is** μ**-integrable**, or simply, integrable We shall often suppress Ω in the notation of the integral. For any set $A \in \mathcal{A}$, define

$$\int_A f d\mu := \int_\Omega f \mathbf{1}_A d\mu. \qquad \diamond$$

The **Lebesgue integral** is nothing but the integral as defined above, when μ is the Lebesgue measure. Section 7.2.8 explores the relation between Lebesgue and Riemann integrals when $\Omega = [a,\ b]$ is any interval.

7.2.1 Properties of the integral

Proof of Theorem 7.2.1 is left as an exercise.

Theorem 7.2.1 (Basic properties). *Suppose that we have functions* $f, f_1, f_2 : (\Omega, \mathcal{A}, \mu) \to (\bar{\mathbb{R}}, \mathcal{B}(\bar{\mathbb{R}}))$. *Then we have the following:*

(a) If $\int f d\mu$ *exists then for every constant* c, $\int c f d\mu = c \int f d\mu$.

(b) If $\int f d\mu$ *exists then* $\left|\int f d\mu\right| \le \int |f| d\mu$.

(c) If f *is non-negative, and* $A \in \mathcal{A}$, *then*

$$\int_A f d\mu = \sup \{ \int_A s d\mu : 0 \le s \le f,\ s \text{ simple}\}.$$

(d) If $\int f d\mu$ *exists, then so does* $\int_A f d\mu$ *for every* $A \in \mathcal{A}$. *If the first integral is finite, then so is the second integral for every* A.

(e) If, $f_1 \ge f_2$ *a.e* μ, *then the following hold.*

(i) If $-\infty < \int f_2 d\mu$ *exists, then* $\int f_1 d\mu$ *exists, and* $\int f_1 d\mu \ge \int f_2 d\mu$.

(ii) If $\int f_1 d\mu < \infty$ *exists then* $\int f_2 d\mu$ *exists, and* $\int f_1 d\mu \ge \int f_2 d\mu$.

(iii) If $\int f_1 d\mu$ *and* $\int f_2 d\mu$ *exist, then* $\int f_1 d\mu \ge \int f_2 d\mu$. ◆

Exercise 7.2.2 Given an example of a sequence of non-negative simple functions $\{s_n\}$ which decreases (to f) but $\lim \int s_n d\mu \neq \int f d\mu$.

7.2.2 Indefinite integral

One way of generating new measures is by using integrals.

Definition 7.2.3. (Indefinite integral). Let $f : (\Omega, \mathcal{A}, \mu) \to (\bar{\mathbb{R}}, \mathcal{B}(\bar{\mathbb{R}}))$ such that $\int_\Omega f d\mu$ exists. Then ν as given in (7.2) is called the *indefinite* *integral* of f with respect to μ.

$$\nu(B) = \int_B f d\mu, \ B \in \mathcal{A}. \tag{7.2}$$

◇

Theorem 7.2.2 (Countable additivity of the integral). *Suppose $f : (\Omega, \mathcal{A}, \mu) \to (\bar{\mathbb{R}}, \mathcal{B}(\bar{\mathbb{R}}))$ and $\int f d\mu$ exists. Then ν, as defined in (7.2), is countably additive. It is a measure if f is non-negative.* ◆

ν takes at most one value out of ∞ and $-\infty$. It is tempting to write $d\nu = f d\mu$. We shall explore this idea in Chapter 21. We can define the two related indefinite integrals:

$$\nu^+(B) := \int_B f^+ d\mu \ \text{ and } \ \nu^-(B) := \int_B f^- d\mu, \ B \in \mathcal{A}.$$

Then $\nu(B) = \nu^+(B) - \nu^-(B)$ for all $B \in \mathcal{A}$. By Theorem 7.2.2, ν^+ and ν^- are measures, and at least one of them is a finite measure.

Proof of Theorem 7.2.2. The proof is trivial when f is simple. Now let f be non-negative. Let s be any simple function, $0 \leq s \leq f$. Let $B = \cup_{n=1}^{\infty} B_n$, $\{B_n\}$ disjoint sets from $\mathcal{A}$. Then

$$\begin{aligned} \int_B s d\mu &= \sum_{n=1}^{\infty} \int_{B_n} s d\mu, \text{ since } s \text{ is simple} \\ &\leq \sum_{n=1}^{\infty} \int_{B_n} f d\mu, \text{ by Theorem 7.2.1(c).} \end{aligned}$$

Now, taking the supremum over all such simple functions s,

$$\nu(B) \leq \sum_{n=1}^{\infty} \nu(B_n) \ \text{(countable sub-additivity).} \tag{7.3}$$

If $\nu(B) = \infty$, then the proof is complete.

So suppose $\nu(B) < \infty$. Since $B_n \subseteq B$, $\mathbf{1}_{B_n} \leq \mathbf{1}_B$ for all n. This implies $\nu(B_n) \leq \nu(B) < \infty$. Fix n and $\epsilon > 0$. Using definition of the integral, and the fact that maximum of finitely many simple functions is simple, get a single simple function s, $0 \leq s \leq f$ so that

$$\int_{B_i} s d\mu \geq \int_{B_i} f d\mu - \frac{\epsilon}{n}, \ i = 1, 2, \ldots, n. \tag{7.4}$$

Then

$$\begin{aligned}\nu(\cup_{i=1}^n B_i) = \int_{\cup_{i=1}^n B_i} f d\mu &\geq \int_{\cup_{i=1}^n B_i} s d\mu, \text{ since } s \leq f \\ &= \sum_{i=1}^n \int_{B_i} s d\mu, \text{ since } s \text{ is simple} \\ &\geq \sum_{i=1}^n \int_{B_i} f d\mu - \epsilon \text{ by (7.4)} \\ &= \sum_{i=1}^n \nu(B_i) - \epsilon.\end{aligned}$$

Since ϵ was arbitrary, we obtain

$$\nu(B_1 \cup \cdots \cup B_n) \geq \sum_{i=1}^n \nu(B_i)\,.$$

This implies that

$$\begin{aligned}\nu(B) &\geq \nu(B_1 \cup \cdots \cup B_n) \\ &\geq \sum_{i=1}^n \nu(B_i) \to \sum_{i=1}^\infty \nu(B_i).\end{aligned}$$

Hence in view of (7.3), ν is countably additive when f is non-negative. For general $f = f^+ - f^-$, use ν^+ and ν^-. Details are omitted. ∎

7.2.3 Monotone convergence theorem (MCT)

This fundamental result is used in the proofs of many results. It reminds us of Theorem 3.2.1(d) on the continuity from below for measures.

Theorem 7.2.3 (Monotone Convergence (MCT)). *Suppose that* $\{f_n\} : (\Omega, \mathcal{A}, \mu) \to (\bar{\mathbb{R}}^+, \mathcal{B}(\bar{\mathbb{R}}^+))$ *is an increasing sequence of functions, and* $f_n(\omega) \uparrow f(\omega)$ *for all* $\omega \in \Omega$. *Then* $\int f_n d\mu \uparrow \int f d\mu$. ◆

Proof. Let $v := \lim_{n\to\infty} \int f_n d\mu$. Then we know from Theorem 7.2.1(e) that $\int f_n d\mu \leq \int f d\mu$ for all n, and hence, $v \leq \int f d\mu$. Fix $0 < b < 1$. Let s be a non-negative simple function, $0 \leq s \leq f$. Let

$$B_n = \{\omega : f_n(\omega) \geq bs(\omega)\}.$$

Then each B_n is measurable, and $B_n \uparrow \Omega$. Now

$$\begin{aligned} v \geq \int f_n d\mu &\geq \int_{B_n} f_n d\mu \\ &\geq b \int_{B_n} s d\mu \\ &\to b \int_{\Omega} s d\mu, \quad \text{as} \quad n \to \infty, \text{ by Theorem 7.2.2} \\ &\to \int_{\Omega} s d\mu, \text{ letting } \; b \to 1\,. \end{aligned}$$

Taking supremum over all possible s, we get $v \geq \int f d\mu$. ■

Exercise 7.2.3 Construct real-valued measurable functions $\{f_n\}$ such that $f_n \uparrow f$, integrals of all these functions exist, but $\lim \int f_n d\mu \neq \int f d\mu$.

Exercise 7.2.4 Construct non-negative measurable functions $\{f_n\}$ such that $f_n \downarrow f$, but $\lim \int f_n d\mu \neq \int f d\mu$. Compare with Exercise 3.7.2(c).

Exercise 7.2.5 Let $\{x_{n,k}\}, n, k \geq 1$ be non-negative reals such that

$$\begin{aligned} x_{n,k} &\leq x_{n+1,k} \text{ for all } \; n, k, \\ x_{n,k} &\uparrow x_k \text{ for each } k = 1, 2, \ldots \end{aligned}$$

Show that $\sum_{k=1}^{\infty} x_{n,k} \uparrow \sum_{k=1}^{\infty} x_k$.

Exercise 7.2.6 If $\{f_n\} : (\Omega, \mathcal{A}, \mu) \to (\bar{\mathbb{R}}^+, \mathcal{B}(\mathbb{R}^+))$, then show that

$$\int (\sum_{i=1}^{\infty} f_i) d\mu = \sum_{i=1}^{\infty} \int f_i d\mu.$$

7.2.4 Linearity and its consequences

Theorem 7.2.4 (Linearity of the integral). *Suppose that we have $f, g : (\Omega, \mathcal{A}, \mu) \to (\bar{\mathbb{R}}, \mathcal{B}(\bar{\mathbb{R}}))$,, Suppose $f + g$ is defined, $\int f d\mu$ and $\int g d\mu$ exist, and $\int f d\mu + \int g d\mu$ is defined. Then $\int (f + g) d\mu$ exists and*

$$\int (f + g) d\mu = \int f d\mu + \int g d\mu.$$

In particular, if f and g are integrable, then $f + g$ is integrable. ◆

Proof. From Exercise 7.1.3 we know that the result is true when f and g are simple functions. Now we proceed in steps.

(i) Suppose f, g are non-negative. Let $\{a_n\}$, $\{b_n\}$ be non-negative simple functions such that $a_n \uparrow f$ and $b_n \uparrow g$. Then $s_n := a_n + b_n \uparrow f+g$. Hence,

$$\begin{aligned} \int s_n d\mu &= \int a_n d\mu + \int b_n d\mu, \text{ by Exercise 7.1.3} \\ &\uparrow \int f d\mu + \int g d\mu, \text{ by MCT.} \end{aligned}$$

By MCT, $\int s_n d\mu \uparrow \int (f+g) d\mu$, proving the result in this special case.

(ii) Now suppose $f \geq 0$, $g \leq 0$, and $h = f + g \geq 0$. This immediately implies that g is finite. Note that $f = h + (-g)$ is the sum of two non-negative measurable functions. Hence

$$\begin{aligned} \int f d\mu &= \int h d\mu + \int (-g) d\mu, \text{ by (i)} \\ &= \int h d\mu - \int g d\mu, \text{ by Theorem 7.2.1(a)} \end{aligned}$$

If $\int g d\mu$ is finite then from the above the result follows.

Let if possible, $\int g d\mu = -\infty$. Then $h \geq 0$, and $\int f d\mu \geq -\int g d\mu = \infty$. This contradicts the hypothesis $\int f d\mu + \int g d\mu$ is defined. Thus this case does not arise.

(iii) If $f \geq 0$ and $g \leq 0$, and $h = f + g \leq 0$ then we can work with $-f$ and $-g$, and apply (ii).

(iv) Now we prove the general case by splitting the range of the functions so as to apply (i), (ii) and (iii). Write $\Omega = \cup_{i=1}^{6} E_i$ where

$$\begin{aligned} E_1 &= \{\omega : f(\omega) \geq 0,\ g(\omega) \geq 0\}, \\ E_2 &= \{\omega : f(\omega) < 0,\ g(\omega) < 0\}, \\ E_3 &= \{\omega : f(\omega) \geq 0,\ g(\omega) < 0,\ h(\omega) \geq 0\}, \\ E_4 &= \{\omega : f(\omega) \geq 0,\ g(\omega) < 0,\ h(\omega) < 0\}, \\ E_5 &= \{\omega : f(\omega) < 0,\ g(\omega) \geq 0,\ h(\omega) \geq 0\}, \\ E_6 &= \{\omega : f(\omega) < 0,\ g(\omega) \geq 0,\ h(\omega) < 0\}. \end{aligned}$$

Now,

$$\begin{aligned}
\int_{E_i} hd\mu &= \int_{E_i} fd\mu + \int_{E_i} gd\mu, \text{ by (i), (ii), and (iii),} \\
\int fd\mu &= \sum_{i=1}^{6} \int_{E_i} fd\mu, \text{ by Theorem 7.2.2} \\
\int gd\mu &= \sum_{i=1}^{6} \int_{E_i} gd\mu, \text{ by Theorem 7.2.2} \\
\int fd\mu + \int gd\mu &= \sum_{i=1}^{6} \int_{E_i} hd\mu \ \text{ using the above three equations} \\
&= \int hd\mu, \text{ by Theorem 7.2.2, } \textbf{provided it exists.}
\end{aligned}$$

To check that $\int hd\mu$ exists, we need to show that *at least one* of $\int h^+ d\mu$ and $\int h^- d\mu$ is finite. Suppose

$$\int h^+ d\mu = \int h^- d\mu = \infty \ \text{ if possible.}$$

In that case there must exist E_i and E_j such that

$$\int_{E_i} h^+ d\mu = \int_{E_j} h^- d\mu = \infty.$$

But then,

$$\int_{E_i} fd\mu = \infty, \text{ or } \int_{E_i} gd\mu = \infty,$$

and hence

$$\int fd\mu = \infty, \text{ or } \int gd\mu = \infty. \tag{7.5}$$

Similarly, working with E_j,

$$\int fd\mu = -\infty, \text{ or} \int gd\mu = -\infty. \tag{7.6}$$

Eqns. (7.5) and (7.6) go against the assumption that $\int fd\mu + \int gd\mu$ is defined. This completes the proof. ■

Exercise 7.2.7 Let $f, g : (\Omega, \mathcal{A}, \mu) \to (\bar{\mathbb{R}}, \mathcal{B}(\bar{\mathbb{R}}))$. Show the following:

(a) $|\int f d\mu| \leq \int |f| d\mu$ whenever $\int f d\mu$ exists. Moreover, $|f|$ is integrable if and only f is integrable.

(b) If $|g| \leq h$ where h is integrable, then g is integrable.

Theorem 7.2.5. *Let $f, g : (\Omega, \mathcal{A}, \mu) \to (\bar{\mathbb{R}}, \mathcal{B}(\bar{\mathbb{R}}))$.*

(a) If $f = 0$ μ-a.s., then $\int f d\mu = 0$.

(b) Suppose $f = g$ a.s. $[\mu]$, and $\int f d\mu$ exists. Then $\int g d\mu$ exists and $\int f d\mu = \int g d\mu$.

(c) If f is μ-integrable, then f is finite a.e. $[\mu]$.

(d) If $f \geq 0$ and $\int f d\mu = 0$, then $f = 0$ a.s. $[\mu]$. ◆

Exercise 7.2.8 (i) In Theorem 7.2.5(c), f need not be bounded. Construct an example where f is a.s. unbounded, but its integral finite.

Proof of Theorem 7.2.5. (a) First suppose $f \geq 0$ everywhere. Fix s, $0 \leq s \leq f$. Then $s = 0$ a.s. So, $\int s d\mu = 0$ and hence $\int f d\mu = 0$. The case $f \geq 0$ a.s. $[\mu]$, is left as an exercise.

For the general case first note that $|f| = 0$ a.s. Hence $\int |f| d\mu = 0$. But then using Exercise 7.2.7, $|\int f d\mu| \leq \int |f| d\mu = 0$.

(b) Let $A = \{\omega : f(\omega) = g(\omega)\}$. Then $A^c \in \mathcal{A}$ and $\mu(A^c) = 0$. Now,

$$\begin{aligned} g &= g\mathbf{1}_A + g\mathbf{1}_{A^c}, \; f = f\mathbf{1}_A + f\mathbf{1}_{A^c}, \\ g\mathbf{1}_{A^c} &= f\mathbf{1}_{A^c} = 0 \text{ a.s. } [\mu]. \end{aligned}$$

Now the result follows from Theorem 7.2.4.

(c) Let $A := \{\omega : |f|(\omega) = \infty\}$. Suppose if possible, $\mu(A) > 0$. Then $\int |f| d\mu \geq \infty \times \mu(A) = \infty$ which is a contradiction. Hence (c) is proved.

(d) Let $B = \{\omega : f(\omega) > 0\}$ and $B_n = \{\omega : f(\omega) > 1/n\}$, $n = 1, 2, \ldots$. Then $B_n \uparrow B$. Now $0 \leq f\mathbf{1}_{B_n} \leq f\mathbf{1}_B$. Hence,

$$0 \leq \int_{B_n} f d\mu \leq \int_B f d\mu \leq \int f d\mu = 0.$$

On the other hand,

$$0 = \int_{B_n} f d\mu \geq \int_{B_n} 1/n d\mu = (1/n)\mu(B_n).$$

Hence $\mu(B_n) = 0$, and this in turn implies that $\mu(B) = 0$. ■

7.2.5 Extended MCT

We now improve the MCT by relaxing the non-negativity assumption. It is pertinent to keep in mind Theorem 3.2.1(e).

Theorem 7.2.6 (Extended MCT). *Suppose that we have functions* $\{f_n\}, g : (\Omega, \mathcal{A}, \mu) \to (\bar{\mathbb{R}}, \mathcal{B}(\bar{\mathbb{R}}))$.
(a) If $f_n \geq g$ *for all* n, $\int g d\mu > -\infty$, *and* $f_n \uparrow f$, *then* $\int f_n d\mu \uparrow \int f d\mu$.
(b) If $f_n \leq g$ *for all* n, $\int g d\mu < \infty$, *and* $f_n \downarrow f$, *then* $\int f_n d\mu \downarrow \int f d\mu$.
(c) Results (a), (b) hold if we add "a. s." at every hypothesis. ◆

Proof. We shall prove (a) and (b) and leave the proof of (c) to the reader.
(a) If $\int g d\mu = \infty$ then by Theorem 7.2.1(e), $\int f_n d\mu = \infty$ and $\int f d\mu = \infty$. So assume that $\int g d\mu < \infty$. Then g is integrable, and by Theorem 7.2.5(c), g is finite almost everywhere.

Redefine g as 0 on $\{\omega : g(w) = \pm\infty\}$. Continue to call this function g. Then $0 \leq f_n - g \uparrow f - g$ a.s. Hence $0 \leq (f_n - g)d\mu \uparrow \int(f - g)d\mu$. Now we can apply additivity Theorem 7.2.4, and the result follows.
(b) Consider the functions $-f_n$ and $-g$, and apply (a). ■

7.2.6 Fatou's lemma

While the MCT is useful, it is inadequate in many applications. We now proceed to derive a stronger result. For any sequence of (extended) real-valued Borel measurable functions $\{f_n\}$, recall the definitions of their lim sup and lim inf:

$$\bar{f}(\omega) := \limsup_{n\to\infty} f_n(\omega) = \inf_{n\geq 1} \sup_{k\geq n} f_k(\omega), \tag{7.7}$$

$$\underline{f}(\omega) := \liminf_{n\to\infty} f_n(\omega) = \sup_{n\geq 1} \inf_{k\geq n} f_k(\omega). \tag{7.8}$$

Exercise 7.2.9 Show that if $\{f_n\} : (\Omega, \mathcal{A}) \to (\bar{\mathbb{R}}, \mathcal{B}(\bar{\mathbb{R}}))$, then $\bar{f}$ and $\underline{f}$ defined in (7.7) and (7.8) are also $\mathcal{B}(\bar{\mathbb{R}})$ measurable.

Lemma 7.2.1 (Fatou). Let $\{f_n\}, f : (\Omega, \mathcal{A}, \mu) \to (\bar{\mathbb{R}}, \mathcal{B}(\bar{\mathbb{R}}))$.
(a) If $f_n \geq f$ for all n and $\int f d\mu > -\infty$, then

$$\liminf_{n\to\infty} \int f_n d\mu \geq \int \liminf_{n\to\infty} f_n d\mu.$$

(b) If $f_n \leq f$ for all n and $\int f d\mu < \infty$, then

$$\limsup_{n\to\infty} \int f_n d\mu \leq \int \limsup_{n\to\infty} f_n d\mu.$$

(c) (a) and (b) remain true if conditions on $\{f_n\}$ and f hold only a.s. ♦

Proof of Lemma 7.2.1. We shall prove only (a) and (b). Part (c) will be left as an easy exercise.

(a) Let $g_n := \inf_{k\geq n} f_k$. Then $f_n \geq g_n \geq f$ for all n, $\int f d\mu > -\infty$, and $g_n \uparrow \underline{f}$. By MCT, Theorem 7.2.6(a), $\int g_n d\mu \uparrow \int \underline{f} d\mu$. Hence

$$\int \liminf f_n d\mu = \int \underline{f} d\mu = \lim \int g_n d\mu = \liminf \int g_n d\mu \leq \liminf \int f_n d\mu\,.$$

(b) Take the negatives, convert limsup to liminf, and apply (a). ■

7.2.7 Dominated convergence theorem (DCT)

This is a major result on pushing a limit inside the integral and is extremely useful. For a useful extension of this result, see Exercise 13.5.2.

Theorem 7.2.7 (Dominated Convergence Theorem (DCT)). *Let $\{f_n\}, h : (\Omega, \mathcal{A}, \mu) \to (\mathbb{R}, \mathcal{B}(\mathbb{R}))$ be such that*

(i) $|f_n| \leq h$ for all n,

(ii) h is integrable,

(iii) $f_n \to f$ a.s. $[\mu]$.

Then f is integrable and $\int f_n d\mu \to \int f d\mu$. The result continues to hold if (i) holds almost surely. ♦

Proof of Theorem 7.2.7. Since (iii) holds, we know that $|f| \leq h$ a.s., and hence f is integrable. Then

$$\begin{aligned}\int \liminf f_n d\mu &\leq \liminf \int f_n d\mu \text{ by Fatou's Lemma 7.2.1(a)}\\ &\leq \limsup \int f_n d\mu\\ &\leq \int \limsup f_n d\mu \text{ by Lemma 7.2.1(b)}\,.\end{aligned}$$

But $\liminf f_n = \limsup f_n = f$ a.s., and the first part of the result follows. The second part is an easy exercise. ■

Exercise 7.2.10 A common mistake while applying DCT is to ignore Condition (ii) in Theorem 7.2.7. The existence of the limit of f_n is not enough for conditions (i) or (ii) to hold. Construct such examples.

7.2.8 Riemann versus Lebesgue integral

Recall Riemann integral from Definition 1.6.1. We now explore the relation between the Riemann and Lebesgue integrals.

Theorem 7.2.8 (Riemann and Lebesgue integral). *Let f be a bounded real-valued function on $[a,\ b]$. Let λ be the Lebesgue measure on $[a,\ b]$. Then the following hold:*

(a) f is Riemann integrable if and only if f is continuous a.e. $[\lambda]$.

(b) If f is Riemann integrable, then f is Lebesgue integrable, and

$$R(f) := \int_a^b f(x)dx = \int_{[a,\ b]} f(x)\lambda(dx)\,. \qquad \blacklozenge$$

Proof. From Exercise 2.5.4 the set $\{\omega : f \text{ is continuous at } \omega\}$ is measurable. Suppose $\sup_{x\in[a,\ b]} |f(x)| \le M < \infty$. Consider any partition $\pi_n := \pi := \{a = x_0 < x_1 < \cdots < x_n = b\}$, and let, $|\pi| := \max_{1\le i\le n}(x_i - x_{i-1})$. Define,

$$\begin{aligned}
M_i &:= \sup\{f(x) : x_{i-1} < x \le x_i\},\ i = 1,\ldots,n,\\
m_i &:= \inf\{f(x) : x_{i-1} < x \le x_i\},\ i = 1,\ldots,n,\\
M &:= \sup_{a\le x\le b} |f(x)| < \infty\,.
\end{aligned}$$

Define the upper and lower *sums* on $[a,\ b]$ by,

$$U_{\pi,n} := \sum_{i=1}^{n} M_i(x_i - x_{i-1}) \ \text{ and } \ L_{\pi,n} := \sum_{i=1}^{n} m_i(x_i - x_{i-1}).$$

Define the upper and lower *functions* $U_{\pi,n}(\cdot)$ and $L_{\pi,n}(\cdot)$ on $[a,\ b]$ by

$$\begin{aligned}
U_{\pi,n}(a) &:= f(a),\ U_{\pi,n}(x) &:= M_i, \text{ if } x_{i-1} < x \le x_i,\ i = 1,\ldots,n,\\
L_{\pi,n}(a) &:= f(a),\ L_{\pi,n}(x) &:= m_i, \text{ if } x_{i-1} < x \le x_i,\ i = 1,\ldots,n.
\end{aligned}$$

Note that

$$L_{\pi,n}(x) \le f(x) \le U_{\pi,n}(x) \ \text{ for all } \ x \in [a,\ b].$$

Moreover these functions are Borel measurable, and bounded by M. Consider the measure space $([a,\ b], \mathcal{L}, \lambda)$. Then $U_{\pi,n}(\cdot)$ and $L_{\pi,n}(\cdot)$ are simple functions, and

$$\int_{[a,\ b]} U_{\pi,n}(x)\lambda(dx) = U_{\pi,n}, \quad \int_{[a,\ b]} L_{\pi,n}(x)\lambda(dx) = L_{\pi,n}\,.$$

Let $\{\pi_n\}$ be any sequence of partitions such that $|\pi_n| \to 0$, and for each n, π_{n+1} is a *refinement* of π_n (that is all points of π_n are also points of π_{n+1}). Then $\{U_{\pi_n,n}(\cdot)\}$ and $\{L_{\pi_n,n}(\cdot)\}$ are respectively non-increasing and non-decreasing sequences of functions, with limits $U(\cdot)$ and $L(\cdot)$ say. These functions depend on $\{\pi_n\}$, but we suppress that dependence in the notation. Irrespective of the choice of $\{\pi_n\}$, we have

$$L(x) \le f(x) \le U(x), \quad \text{for all} \quad x \in [a,\ b]. \tag{7.9}$$

Note that $U(\cdot)$ and $L(\cdot)$, being limits of simple functions, are Borel measurable. Moreover, they are bounded by M on $[a,\ b]$. Hence by DCT Theorem 7.2.7, with the bounding function as the constant function M,

$$\lim_{n\to\infty} U_{\pi_n,n} = \lim_{n\to\infty} \int_{[a,\ b]} U_{\pi_n,n}(x)\lambda(dx) = \int U(x)\lambda(dx), \tag{7.10}$$

$$\lim_{n\to\infty} L_{\pi_n,n} = \lim_{n\to\infty} \int_{[a,\ b]} L_{\pi_n,n}(x)\lambda(dx) = \int L(x)\lambda(dx). \tag{7.11}$$

Now note that for all $a < x < b$,

$$f \text{ is continuous at } x \text{ if and only if } \ U(x) = L(x) = f(x) \tag{7.12}$$

for all choices of the sequence of partitions. Note that in the arguments so far, we have not used any measurability of f.

(a) First suppose f is Riemann integrable. Then by Definition 1.6.1, for *any* sequence of partitions $\{\pi_n\}$ such that $|\pi_n| \to 0$,

$$R(f) = \lim_{n\to\infty} U_{\pi_n,n} = \lim_{n\to\infty} L_{\pi_n,n}.$$

By Eqns. (7.9)–(7.11), for all such sequences of partitions, $U(\cdot) = L(\cdot)$ a.e. $[\lambda]$. The set $A := \{x : U(x) \neq L(x)\}$ is Borel, and $\lambda(A) = 0$. By (7.12) f is continuous on A^c. Hence f is continuous a.e $[\lambda]$.

Now suppose that f is continuous a.e $[\lambda]$. Then from Eqn. (7.12) we get, $U(\cdot) = f(\cdot) = L(\cdot)$ λ-a.e. for any sequence of partitions $\{\pi_n\}$ such that $|\pi_n| \to 0$. Since $U(\cdot)$ and $L(\cdot)$ are Borel measurable, and Lebesgue measure is complete, f is also Lebesgue measurable. Since f is bounded, it is integrable with respect to λ on $[a,\ b]$. Moreover,

$$\int_{[a,\ b]} U(x)\lambda(dx) = \int_{[a,\ b]} f(x)\lambda(dx) = \int_{[a,\ b]} L(x)\lambda(dx), \tag{7.13}$$

irrespective of the sequence $\{\pi_n\}$ *we chose.* Recalling (7.10) and (7.11), it follows that f is Riemann integrable.

(b) Since f is Riemann integrable, by Part (a), it is continuous a.e. $[\lambda]$. But then by (7.10), (7.11) and (7.13) $R(f) = \int_{[a,\ b]} f(x)\lambda(dx)$. ■

Remark 7.2.1. Suppose that $f : (\Omega, \mathcal{A}, \mu) \to (\mathbb{C}, \mathcal{B}(\mathbb{C}))$. Then there are $\mathcal{R}(f), \mathcal{I}(f) : (\Omega, \mathcal{A}, \mu) \to (\mathbb{R}, \mathcal{B}(\mathbb{R}))$ such that $f \equiv \mathcal{R}(f) + \iota\mathcal{I}(f)$. Then $\int f d\mu$ is naturally defined by linearity as $\int \mathcal{R}(f) d\mu + \iota \int \mathcal{I}(f) d\mu$, whenever the relevant integrals exist. We often state a result for real-valued function(s), with the understanding that it carries over to complex-valued functions in a natural way, whenever it makes sense. ●

Exercise 7.2.11 Let f be a complex-valued μ-integrable function.

(a) Show that

$$|\int f d\mu| \leq \int |f| d\mu.$$

Hint: Begin by writing the polar representation $f(\omega) = r(\omega)e^{\iota\theta(\omega)}$.

(b) Show that $|\int_\Omega f d\mu = \int_\Omega |f| d\mu$ if and only if $\arg f$ is a.e. constant on the set $\{\omega : f(\omega) \neq 0\}$.

7.3 Exercises

Exercise 7.3.1 Let $f : (\Omega, \mathcal{A}, \mu) \to (\bar{R}, \mathcal{B}(\bar{\mathbb{R}}))$ be integrable. Show that given $\epsilon > 0$, there is a $\delta > 0$ so that $|\int_E f d\mu| < \epsilon$ whenever $\mu(E) < \delta$. This will be useful in discussing "uniform integrability" in Chapter 13.

Exercise 7.3.2 Consider the measure space $([0,\ 1], \mathcal{B}([0,\ 1], \lambda)$. Define

$$f(x) = \begin{cases} 4 \text{ if } x \text{ is irrational,} \\ 2 \text{ if } x \text{ is rational, } x < 1, \\ \infty \text{ if } x = 1. \end{cases} \tag{7.14}$$

Show that

(i) f is measurable and $f = 4$ almost surely.

(ii) $\int_{[0,\ 1]} f d\lambda = 4$.

(iii) f is not Riemann integrable.

Exercise 7.3.3 (**Change of variable formula**) Suppose that we have $T : (\Omega, \mathcal{A}, \mu) \to (\Omega_1, \mathcal{A}_1, \mu T^{-1})$. Suppose f is any Borel measurable function on Ω_1. Show that, if one of the integrals below exists, then the other exists, and they are equal.

$$\int_{A_1} f(\omega_1)\mu T^{-1}(d\omega_1) = \int_{T^{-1}(A_1)} f(T(\omega))\mu(d\omega), \quad \text{for any} \quad A_1 \in \mathcal{A}_1 .$$

Exercise 7.3.4 Let $T : (a,\ b) \to \mathbb{R}$ be a monotone function with a continuous derivative T'. Let $f : T((a,\ b)), \mathcal{B}(T((a,\ b))) \to (\mathbb{R}, \mathcal{B}(\mathbb{R}))$. Show that f is Lebesgue integrable if and only if $(f \circ T)T'$ is Lebesgue integrable on $(a,\ b)$, and for any $E \in \mathcal{B}((a,\ b))$,

$$\int_E f(T(x))|T'(x)|\lambda(dx) = \int_{T(E)} f(x)\lambda(dx).$$

Hint: For any $A \in \mathcal{B}(a,\ b)$, define $\mu(A) = \int_A |T'(x)|\lambda(dx)$ and use the change of variable formula of Exercise 7.3.3.

Exercise 7.3.5 Let $\{x_{n,k}\}, n, k \geq 1$ be non-negative. Show that

$$\sum_{k=1}^{\infty} (\sum_{n=1}^{\infty} x_{n,k}) = \sum_{n=1}^{\infty} (\sum_{k=1}^{\infty} x_{n,k}).$$

Exercise 7.3.6 Let $\{x_n\}$ be a sequence of real numbers, $x_n \to x$.

(a) If $\{x_n\}$ is non-negative and non-decreasing, using MCT, show that $\left(1 + \frac{x_n}{n}\right)^n \to e^x$. Here x is allowed to equal ∞.

(b) If x is finite, using DCT, show that $\left(1 + \frac{x_n}{n}\right)^n \to e^x$.

Exercise 7.3.7 Suppose $\{p_k, k \geq 0\}$ defines a probability measure. The *generating function* of this distribution is defined as

$$P(s) = \sum_{k=0}^{\infty} p_k s^k, \ 0 \leq s \leq 1 .$$

Show that

$$\frac{d}{ds}P(s) = \sum_{k=1}^{\infty} p_k k s^{k-1} \quad \text{for all} \quad 0 < s < 1\,.$$

Exercise 7.3.8 Let $f, g : (\Omega, \mathcal{A}, \mu) \to (\mathbb{R}, \mathcal{B}(\mathbb{R}))$. Consider the condition

$$\int_A g d\mu \le \int_A f d\mu, \quad \text{for all} \quad A \in \mathcal{A}. \tag{7.15}$$

Show the following:

(a) If f and g are integrable and (7.15) holds, then $g \le f$ a.s.

(b) If $\int f d\mu$, $\int g d\mu$ exist, (7.15) holds, and μ is σ-finite, then $g \le f$ a.s.

Exercise 7.3.9 (**Scheffe's theorem**) Suppose that we have functions $\{f_n\}, f : (\Omega, \mathcal{A}, \mu) \to (\mathbb{R}^+, \mathcal{B}(\mathbb{R}^+))$. Suppose $\int f_n d\mu = \int f d\mu = 1$ for all n, and $f_n \to f$ a.s. Show that $\int |f_n - f| d\mu \to 0$.

Exercise 7.3.10 Suppose $\{p_{n,k}\}$ and $\{p_k\}$ are non-negative sequences such that $\sum_k p_{n,k} = \sum_k p_k = 1$ for all n, and $p_{n,k} \to p_k$ as $n \to \infty$. Show that $\sum_k |p_{n,k} - p_k| \to 0$ as $n \to \infty$.

Exercise 7.3.11 Let $\{f_n\}, \{g_n\} : (\Omega, \mathcal{A}, \mu) \to (\mathbb{R}, \mathcal{B}(\mathbb{R}))$ be such that $|f_n| \le g_n$ for all n, and f_n and g_n converge a.s. to f and g respectively. Further $\int_\Omega g_n d\mu \to \int_\Omega g d\mu < \infty$. Then show that $\int_\Omega f_n d\mu \to \int_\Omega f d\mu$.

Exercise 7.3.12 Suppose $f : \mathbb{R} \to \mathbb{R}$ whose Riemann integral, say $R_{[a,\ b]}(f)$ exists on the interval $[a,\ b]$, for all $a < b$. The **improper integral** of f is defined as

$$R_{(-\infty,\ \infty)}(f) = \lim_{a \downarrow -\infty,\ b \uparrow \infty} R_{[a,\ b]}(f),$$

if the limit exists and is finite.

(a) Show that if f has an improper integral, then it is continuous a.e. λ on $\mathbb{R}$. Show by an example that the converse is not true.

(b) Suppose $f \ge 0$ with $|R_{(-\infty,\ \infty)}(f)| < \infty$. Show that f is integrable on $(\mathbb{R}, \mathcal{L}, \lambda)$, and $R_{(-\infty,\ \infty)}(f) = \int_{(-\infty,\ \infty)} f d\lambda$. Show by example that this may not be true if f is not non-negative.

Exercise 7.3.13 Let $f : \mathbb{R} \to \mathbb{R}$ be Lebesgue integrable. Show that

$$\int_{\mathbb{R}} f(x+t)\lambda(dx) = \int_{\mathbb{R}} f(x)\lambda(dx) \quad \text{for every} \quad t \in \mathbb{R}\,.$$

Exercise 7.3.14 Let $f : \mathbb{R} \to \mathbb{R}$ be Lebesgue integrable. Show that g defined below is continuous at 0:

$$g(t) = \int_{\mathbb{R}} |f(x+t) - f(x)| \lambda(dx).$$

Exercise 7.3.15 Let $f : (\mathbb{R}, \mathcal{B}(\mathbb{R})) \to (\mathbb{R}, \mathcal{B}(\mathbb{R}))$ so that $\int_{\mathbb{R}} |f| d\lambda < \infty$ and $\int_K f d\lambda = 0$ for every compact $K \subset \mathbb{R}$. Show that $f = 0$ a.e. $[\lambda]$.

Exercise 7.3.16 Suppose f is Lebesgue integrable on $[a, b]$. Show that for every $x \in (a, b)$ which is a continuity point of f,

$$\frac{d}{dt} \int_{(a,\ t]} f(y) \lambda(dy) \bigg|_{t=x} = f(x).$$

Exercise 7.3.17 Let $F : \mathbb{R} \to \mathbb{R}$ be such that its derivative $F'(x) = f(x)$ is continuous on (a, b). Then show that

$$\int_{[a,\ b]} f(x) \lambda(dx) = F(b) - F(a).$$

Exercise 7.3.18 Show that strict inequality may hold in Fatou's lemma.

Exercise 7.3.19 Let $f : (\Omega, \mathcal{A}, \mu) \to (\mathbb{R}^+, \mathcal{B}(\mathbb{R}^+))$, and suppose that $0 < \int f d\mu = c < \infty$. Show that

$$\lim_{n\to\infty} \int n \log \left[1 + \left(\frac{f}{n}\right)\right]^{\alpha} d\mu = \begin{cases} \infty & \text{if } 0 < \alpha < 1, \\ c & \text{if } \alpha = 1, \\ 0 & \text{if } 1 < \alpha < \infty. \end{cases} \tag{7.16}$$

Hint: If $\alpha \geq 1$, then the integrand is dominated (by what?). For other values, use Fatou's lemma.

Exercise 7.3.20 (**Beppo Levi theorem**) Suppose we have functions $\{f_n\} : (\Omega, \mathcal{A}, \mu) \to (\mathbb{R}, \mathcal{B}(\mathbb{R}))$ which are integrable, $\sup \int f_n d\mu < \infty$, and $f_n \uparrow f$. Show that f is integrable, and $\int f_n d\mu \to \int f d\mu$.

Exercise 7.3.21 (**Differentiation under the integral**). Suppose $g : (\Omega, \mathcal{A}, \mu) \to (\mathbb{R}^+, \mathcal{B}(\mathbb{R}^+))$. Let $f(t, \omega) : [a,\ b] \times \Omega \to \mathbb{R}$ be such that

(i) $f(t, \omega)$ is measurable for each $t \in [a,\ b]$, $a < b$.

(ii) for a.e. ω, $\frac{\partial}{\partial t} f(t, \omega)$ exists for all t in $(a,\ b)$ and,

(iii) $\sup_t |\frac{\partial}{\partial t} f(t,\omega)| \le g(\omega)$ where $\int g(\omega)\mu(d\omega) < \infty$. Then show that

$$\frac{d}{dt}\int_\Omega f(t,\omega)\mu(d\omega) = \int_\Omega [\frac{\partial}{\partial t} f(t,\omega)]\,\mu(d\omega) \quad \text{for all} \quad a < t < b\,.$$

Exercise 7.3.22 Let $f : (a,\ b) \times \mathbb{R} \to \mathbb{R}$ be such that

(i) The map $\theta \to f(\theta, x)$ is differentiable for each $x \in \mathbb{R}$.

(ii) The map $x \to f(\theta, x)$ is Lebesgue integrable for each $\theta \in (a,\ b)$

(iii) $|f(\theta_1, x) - f(\theta_2, x)| \le g(x)|\theta_1 - \theta_2|$ for all $\theta_1, \theta_2 \in (a,\ b)$ and $x \in \mathbb{R}$,

(iv) g is Lebesgue integrable over $\mathbb{R}$.

Define

$$H(\theta) := \int_{\mathbb{R}} f(\theta, x)\lambda(dx)\,.$$

Show that H is differentiable. Further, $\frac{d}{d\theta} f(\theta, x)$ is integrable over $\mathbb{R}$ for each θ, and

$$\frac{d}{d\theta} H(\theta) = \int_{\mathbb{R}} \frac{d}{d\theta} f(\theta, x)\lambda(dx).$$

Exercise 7.3.23 Show that Theorem 7.2.2 can be proved with the help of MCT and DCT. Note that this is not circuitous.

Exercise 7.3.24 Suppose $f \ge 0$ on $(\Omega, \mathcal{A}, \mu)$ is integrable and let ν be its indefinite integral. Let $g : (\Omega, \mathcal{A}) \to (\mathbb{R}, \mathcal{B}(\mathbb{R}))$. Then show that g is ν-integrable if and only if fg is μ-integrable and in that case,

$$\int_E g d\nu = \int_E f g d\mu, \ \text{for all} \ \ E \in \mathcal{A}\,.$$

Chapter 8

Basic inequalities

Inequalities for integrals, and for measures of sets, are very important. We shall develop a few of the basic inequalities in this chapter. The simpler ones rest on the monotonicity property of the integral. Others use more sophisticated properties, such as convexity.

8.1 Chebyshev's and Markov's inequalities

These inequalities are easy to prove, but are of prime importance in statistics and probability and in other areas of science.

Lemma 8.1.1 (Markov's inequality). Suppose we have a function $f : (\Omega, \mathcal{A}, \mu) \to (\bar{\mathbb{R}}^+, \mathcal{B}(\bar{\mathbb{R}}^+))$. Then for any $\epsilon > 0$, and $p > 0$,

$$\mu\{\omega : f(\omega) \geq \epsilon\} \leq \epsilon^{-p} \int_\Omega f^p d\mu. \qquad \blacklozenge$$

Proof. Define the functions

$$g(\omega) = \begin{cases} 1 & \text{if } \ f(\omega) \geq \epsilon \\ 0 & \text{if } \ f(\omega) < \epsilon, \end{cases}$$
$$h(\omega) = \epsilon^{-p} f^p(\omega).$$

Note that $0 \leq g(\omega) \leq h(\omega)$ for all $\omega \in \Omega$, and the result follows from this by using Theorem 7.2.1(e). ■

Definition 8.1.1. (**Mean** and **Variance**) Let $f : (\Omega, \mathcal{A}, P) \to (\bar{\mathbb{R}}, \mathcal{B}(\bar{\mathbb{R}}))$

A. Bose, *Elements of Measure and Probability*, Texts and Readings in Mathematics 88,
https://doi.org/10.1007/978-981-95-2758-8_8

where P is a probability measure. Then

$$m = \int_\Omega f dP,$$

is called the *mean* of f, whenever the integral exists. If m is finite, then the *variance* of f is defined as

$$\sigma^2 = \int_\Omega (f - m)^2 dP. \tag{8.1}$$ ◊

Note that m can be $\pm\infty$. σ^2 is non-negative and can be ∞.

Exercise 8.1.1 Let $f : (\Omega, \mathcal{A}, P) \to (\bar{\mathbb{R}}, \mathcal{B}(\bar{\mathbb{R}}))$ be integrable. Show that

$$\int f^2 dP \geq \Big(\int f dP\Big)^2. \tag{8.2}$$

Lemma 8.1.2 (Chebyshev's inequality). Suppose we have a function $f : (\Omega, \mathcal{A}, P) \to (\bar{\mathbb{R}}, \mathcal{B}(\bar{\mathbb{R}}))$ where P is a probability measure and f has mean m and finite variance $\sigma^2 > 0$. Then for any $\epsilon > 0$,

$$P\{\omega : |f(\omega) - m| \geq \sigma\epsilon\} \leq \epsilon^{-2}.$$ ♦

This can be proved by defining appropriate functions g and h as in the proof of Lemma 8.1.1, or that lemma can be used directly on the function $(f - m)^2$. We omit the details.

Exercise 8.1.2 (Cauchy-Schwarz inequality) Let f and g be two real-valued measurable functions on $(\Omega, \mathcal{A}, \mu)$. Then

$$\Big(\int |fg| d\mu\Big)^2 \leq \Big(\int f^2 d\mu\Big)\Big(\int g^2 d\mu\Big). \tag{8.3}$$

In particular, if f^2 and g^2 are integrable, then so is fg.

8.2 Jensen's inequality

Let $\varphi(x) = x^2$. Then Eqn. (8.2) is the same as saying

$$\int \varphi(f) dP \geq \varphi\Big(\int f dP\Big). \tag{8.4}$$

It turns out that (8.4) holds whenever $\varphi(\cdot)$ is a convex function and both integrals are assumed to be finite. This result has been extended in many directions, and there are lot of applications of these results in statistics (see Ferguson [1967]), probability, and functional analysis.

Definition 8.2.1. (**Convex set and function**) A set $G \subseteq \mathbb{R}^d$ is said to be *convex* if for all $x, y \in G$ and $0 \leq \alpha \leq 1$, $\alpha x + (1-\alpha)y$ is also in G. A function $f : G \to \mathbb{R}$, where G is convex, is said to be a *convex function* if, for all $x, y \in G$ and $0 \leq \alpha \leq 1$,

$$f(\alpha x + (1-\alpha)y) \leq \alpha f(x) + (1-\alpha)f(y)\,. \qquad \diamond$$

Exercise 8.2.1 Show that G is a convex set if and only if, for all $n > 1$, and for all choices of non-negative $\alpha_1, \ldots, \alpha_n$, $\sum_{i=1}^n \alpha_i = 1$, and $x_1, \ldots x_n \in G$, we have $\sum_{i=1}^n \alpha_i x_1 \in G$.

Exercise 8.2.2 Suppose G is a convex set in $\mathbb{R}^d$ for some $d > 1$.

(a) Exhibit such a set which is not a Borel set.
Hint: The boundary of the unit disc has Lebesgue measure 0.

(b) Show that G is a Lebesgue set.
[For more information on the measurablity of convex sets when we have "product measures" on $\mathbb{R}^d$, see Lang [1986].]

Exercise 8.2.3 Show that any convex function on G is continuous on the interior G^o of G.

Exercise 8.2.4 Let $f : I \to \mathbb{R}$ where I is an open interval of $\mathbb{R}$. Suppose that f'' is non-negative for all $x \in I$. Show that then f is convex. In particular, $-\log(\cdot)$ is a convex function on $(0, \infty)$.

We shall focus on dimension $d = 1$ for which we give all the details.

8.2.1 Jensen's inequality in $\mathbb{R}$

Exercise 8.2.5 Let $f : (\Omega, \mathcal{A}, P) \to (\mathbb{R}, \mathcal{B}(\mathbb{R}))$ be integrable. Let G be an interval such that $P\{f \in G\} = 1$. If f is not degenerate (that is, $P\{f = x\} < 1$ for all $x \in G$), then show that $\int f \in G^o$.

To state and prove Jensen's inequality, we need some properties of convex functions.

Lemma 8.2.1. Let $G \subset \mathbb{R}$ be convex and $\phi : G \to \mathbb{R}$ be a convex function. Then,

(a) the left and right derivatives, respectively $\phi'_-(x_0)$ and $\phi'_+(x_0)$, exist at all $x_0 \in G^o$. They are non-decreasing in x_0, $\phi'_-(x_0) \le \phi'_+(x_0)$, and

$$\phi(x_0) + (x - x_0)\phi'_+(x_0) \le \phi(x) \quad \text{for all} \quad x, x_0 \in G\,. \tag{8.5}$$

(b) for all $z \in G$, we have

$$\varphi(x) - \varphi(z) \le d(x - z) \quad \text{for all} \quad d \in [\varphi'_-(x),\ \varphi'_+(x)]\,. \tag{8.6}$$ ♦

Proof. For all $x, y, z \in G$, $x < y < z$, we can write

$$y = \frac{z-y}{z-x}x + \frac{y-x}{z-x}z = \frac{z-y}{z-x}x + \left(1 - \frac{z-y}{z-x}\right)z\,.$$

Note that $\frac{z-y}{z-x} \in (0,1)$. Hence by convexity of φ,

$$\varphi(y) \le \frac{z-y}{z-x}\varphi(x) + \frac{y-x}{z-x}\varphi(z)\,.$$

This is equivalent to saying that

$$\frac{\varphi(y) - \varphi(x)}{y - x} \le \frac{\varphi(z) - \varphi(x)}{z - x} \le \frac{\varphi(y) - \varphi(z)}{y - z}, \quad \text{for all } x < y < z\,. \tag{8.7}$$

In particular, $D(x) := \frac{\varphi(y)-\varphi(x)}{y-x}$ is an increasing function of x. Let $y \in G^o$ and $\{x_n\} \subseteq G$ such that $x_n \uparrow y$. Then $D(x_n)$ is an increasing sequence and from (8.7) is bounded above by $\inf_{z \in G, z > y}\left(\frac{\varphi(z)-\varphi(y)}{z-y}\right)$.

Hence, for every $y \in G^o$ the *left derivative* $\varphi'_-(y)$ of φ at y is finite and satisfies

$$\frac{\varphi(y) - \varphi(x)}{y - x} \uparrow \varphi'_-(y) \text{ as } x \uparrow y\,. \tag{8.8}$$

Similarly, the *right derivative* $\varphi'_+(y)$ of φ at any $y \in G^o$ is finite and satisfies

$$\frac{\varphi(y) - \varphi(z)}{y - z} \downarrow \varphi'_+(y) \text{ as } z \downarrow y\,. \tag{8.9}$$

It also follows from (8.7), (8.8) and (8.9) that, $\varphi'_+(y)$ and $\varphi'_-(y)$ are non-decreasing and

$$\varphi'_+(y) \geq \varphi'_-(y), \text{ for all } y \in G^o. \tag{8.10}$$

Now (8.5) follows for all $x > x_0$ using (8.9) (with z and y replaced respectively by x and x_0). For all $x < x_0$, it follows from (8.8) and 8.10 (with y replaced by x_0).

(b) This can be proved by considering the cases $z > x$ and $z < x$ separately and using the above relations. We omit the details. ■

Theorem 8.2.1. *(Jensen's inequality in* $\mathbb{R}$*) Let G be a convex set. Let $f : (\Omega, \mathcal{A}, P) \to (\mathbb{R}, \mathcal{B}(\mathbb{R}))$ be integrable and $P\{f \in G\} = 1$. Let $\varphi : G \to \mathbb{R}$ be a convex function such that $\int |\varphi(f)|dP < \infty$. Then $\int \varphi(f)dP \geq \varphi(\int f dP)$.* ◆

Proof. We exclude the trivial case $P\{f = x_0\} = 1$ for some $x_0 \in G$. Since $P\{f \in G\} = 1$, we have $\int f dP \in G^o$ by Exercise 8.2.5. In view of (8.6), for all $d \in [\varphi'_-(\int f dP), \varphi'_+(\int f dP)] \subsetneq (-\infty, \infty)$ we have

$$\varphi(f) - \varphi(\int f dP) \geq d(f - \int f dP).$$

Integrating both sides of the above inequality with respect to the probability measure P leads to the desired result. ■

Exercise 8.2.6 Using Jensen's inequality, show that arithmetic mean is at least as large as the geometric mean for non-negative numbers.

Exercise 8.2.7 Let $f : (\Omega, \mathcal{A}, P) \to (\mathbb{R}^+, \mathcal{B}(\mathbb{R}^+))$. Show that

(a) $\int \sqrt{f} dP) \leq \sqrt{\int f dP}$.

(b) $\int (f \log f) dP \geq (\int f dP) \log (\int f dP)$.

(c) $\int e^f dP \geq e^{\int f dP}$.

8.2.2 Jensen's inequality in $\mathbb{R}^d$

How does Jensen's inequality extend to functions f which are $\mathbb{R}^d$ valued? We shall state the result without proof.

Definition 8.2.2. Let $f : (\Omega, \mathcal{A}, P) \to (\mathbb{R}^d, \mathcal{B}(\mathbb{R}^d))$. Then $\int f dP$ is the vector $(\int (p_i \circ f) dP, 1 \leq i \leq d)$ whenever all the integrals exist. ◇

Theorem 8.2.2. ***(Jensen's inequality in*** $\mathbb{R}^d$*)* *Let* $G \subset \mathbb{R}^d$ *be a convex Borel set. Let* $f : (\Omega, \mathcal{A}, P) \to (G, \mathcal{B}(G))$.
(a) Suppose that

$$\int |p_i \circ f| dP < \infty \quad \text{for all} \quad 1 \le i \le d\,. \tag{8.11}$$

Then $\int f dP \in G$.
(b) Let $\varphi : G \to \mathbb{R}$ *be a convex function. Suppose that (8.11) holds and* $\int |\varphi(f)| dP < \infty$. *Then*

$$\int \varphi(f) dP \ge \varphi\Big(\int f dP\Big)\,.$$

◆

We remark that Jensen's inequality holds in more abstract spaces.

8.3 L^p space: definition

The L^p spaces are of prime importance, especially in real and functional analysis. We shall state and prove two inequalities, namely Hölder's and Minkowski's inequalities on these spaces. *In the next chapter we shall discuss the metric properties of the* L^p *spaces.*

Definition 8.3.1. (L^p **space**) Let $(\Omega, \mathcal{A}, \mu)$ be a measure space. For any $p > 0$, define

$$||f||_p := \Big(\int |f|^p d\mu\Big)^{1/p}, \quad f : (\Omega, \mathcal{A}, \mu) \to (\mathbb{C}, \mathcal{B}(\mathbb{C})), \tag{8.12}$$

$$L^p(\mu) := L^p(\Omega, \mathcal{A}, \mu), \tag{8.13}$$

$$:= \{f : (\Omega, \mathcal{A}, \mu) \to (\mathbb{C}, \mathcal{B}(\mathbb{C})), \int |f|^p d\mu < \infty\}. \tag{8.14}$$

We write L^p for $L^p(\mu)$ if the underlying measure μ is evident. ◇

Recall from Exercise 6.2.5(b) the equivalence classes of functions that are equal a.e. $[\mu]$. Note that if $f = g$ a.e. μ, then $||f - g||_p = 0$. Hence for every p, the space L^p is really a space of equivalence classes of functions.
While the notation $||f||_p$ may suggest it is a norm, but that is true only when $p \ge 1$. We shall show this soon.

8.4 Hölder's inequality

Lemma 8.4.1. (a) Suppose $a, b > 0$, $0 < \alpha < 1$. Then

$$a^{\alpha} b^{1-\alpha} \leq \alpha a + (1-\alpha) b \,. \tag{8.15}$$

(b) If $c, d > 0$ $p, q > 1$ such that $\dfrac{1}{p} + \dfrac{1}{q} = 1$. Then $cd \leq \dfrac{c^p}{p} + \dfrac{d^q}{q}$. ♦

Proof. (a) Take log on both sides of (8.15) and apply convexity of the function $-\log(\cdot)$.

(b) We apply (a) after choosing $\alpha = 1/p$, $1-\alpha = 1/q$, $a = c^p, b = d^q$. ∎

Definition 8.4.1. (**Conjugate pair**) (p, q) is a *conjugate pair* if $p, q \geq 1$ and $p^{-1} + q^{-1} = 1$. This includes the pairs $(1, \infty)$ and $(\infty, 1)$. ◇

Theorem 8.4.1 (Hölder's inequality). *Let $p, q > 1$ be a conjugate pair. If $f \in L^p$ and $g \in L^q$ then*

$$||fg||_1 \leq ||f||_p| \cdot |g||_q \,.$$ ♦

Proof. If $||f||_p \cdot ||g||_q = 0$ then one of f and g is 0 a.e. μ. Then the inequality is trivial. So assume that $||f||_p \, ||g||_q \neq 0$. Let

$$c = \frac{|f|}{||f||_p}, \quad d = \frac{|g|}{||g||_q} \,.$$

Then by Lemma 8.4.1(b),

$$\begin{aligned} \int \frac{|fg|}{||f||_p ||g||_q} d\mu &\leq \int \Big[\frac{|f|^p}{p||f||_p^p} + \frac{|g|^q}{q||g||_q^q} \Big] d\mu \\ &= \frac{1}{p} + \frac{1}{q} = 1. \end{aligned}$$ ∎

Exercise 8.4.1 Show that Cauchy-Schwarz inequality (8.3) follows from Hölder's inequality (8.15).

Exercise 8.4.2 Show that equality holds in Hölder's inequality if and only if $A|f|^p = B|g|^q$ a.e. μ for some constants A and B, not both zero.

Exercise 8.4.3 Let μ be a *finite* measure and $0 < r < s < \infty$.

(a) Show that for all f measurable, $||f||_r \leq k||f||_s$, for some constant k. Further, $k = 1$ is sufficient if μ is a probability measure. Give a value of k when μ is not necessarily a probability measure.

(b) Show that it follows from (a) that, $L^s(\mu) \subset L^r(\mu)$ whenever μ is a finite measure. Show that this is not true if μ is not finite.

8.5 Minkowski's inequality

Lemma 8.5.1. If $a, b \geq 0$ and $p \geq 1$, then $(a+b)^p \leq 2^{p-1}(a^p + b^p)$. ♦

Proof. Consider the function

$$f(x) = (a+x)^p - 2^{p-1}(a^p + x^p), \ x > 0\,.$$

Then it is easy to see that the derivative of f satisfies

$$f'(x) = p(a+x)^{p-1} - 2^{p-1}px^{p-1} \begin{cases} > 0, & \text{if } \ x < a, \\ = 0 & \text{if } \ x = a, \\ < 0 & \text{if } \ x > a. \end{cases}$$

Hence the maximum of f occurs at $x = a$. But $f(a) = 0$. ■

Exercise 8.5.1 Show that Lemma 8.5.1 is not true if $p < 1$.

Theorem 8.5.1 (Minkowski's inequality). *Let $p \geq 1$. If $f, g \in L^p$, then $f + g \in L^p$ and*

$$||f+g||_p \leq ||f||_p + ||g||_p. \tag{8.16}$$ ♦

Proof. We shall use Hölder's inequality and Lemma 8.5.1. Clearly (8.16) is true for $p = 1$, and so we assume that $p > 1$. By Lemma 8.5.1,

$$|f+g|^p \leq (|f| + |g|)^p \leq 2^{p-1}(|f|^p + |g|^p)\,.$$

Hence $f + g \in L^p$. Then $(p, q = \dfrac{p}{p-1})$ is a conjugate pair and

$$|f+g|^p = |f+g|\ (|f+g|^{p-1}) \leq |f|\ (|f+g|^{p-1}) + |g|\ (|f+g|^{p-1}). \tag{8.17}$$

Clearly

$$\int [|f+g|^{p-1}]^q d\mu = \int |f+g|^p d\mu < \infty\,.$$

Since $f \in L^p$ and $|f+g|^{p-1} \in L^q$, by Hölder's inequality,

$$\int |f|\,(|f+g|^{p-1})d\mu \le ||f||_p\Big[\int [|f+g|^{p-1}]^q d\mu\Big]^{1/q}$$

$$= ||f||_p|\,||f+g||_p^{p/q} \text{ as } q = \frac{p}{p-1}. \tag{8.18}$$

Similarly

$$\int |g|(|f+g|^{p-1})d\mu \le ||g||_p\,||f+g||_p^{p/q}. \tag{8.19}$$

Using (8.17), and adding over the two inequalities (8.18) and (8.19),

$$||f+g||_p^p \le (||f||_p + ||g||_p)(||f+g||_p^{p/q}).$$

Taking the second factor from the right to the left, we get the result. ∎

Exercise 8.5.2 Show that Minkowski's inequality is not true if $p < 1$.

8.6 Exercises

Exercise 8.6.1 Let $f : (\Omega, \mathcal{A}, P) \to (\mathbb{R}, \mathcal{B}(\mathbb{R}))$ be a positive function. Show that, $(\int (1/f^p)dP)(\int f dP)^p \ge 1$ for all $p > 0$.

Exercise 8.6.2 Suppose $p > 1$. Show that equality holds in Minkowski's inequality (8.16) if and only if $Af = Bg$ a.e. μ for some constants A and B, not both zero. Find condition for equality when $p = 1$.

Exercise 8.6.3 Let $f, g : (\Omega, \mathcal{A}, P) \to (\mathbb{R}, \mathcal{B}(\mathbb{R}))$ be positive functions, such that $fg \ge 1$ a.e. P. Show that $(\int f dP)(\int g dP) \ge 1$.

Exercise 8.6.4 Let $(\Omega, \mathcal{A}, \mu)$ be a measure space and $0 < p < q < r$. Show that $L^q \subset L^p + L^r$, in the sense that any $f \in L^q$ can be written as $f = f_1 + f_2$ where $f_1 \in L^p$ and $f_2 \in L^r$.

Exercise 8.6.5 Let $(\Omega, \mathcal{A}, \mu)$ be a measure space and $0 < p < q < r$. Show that $L^p \cap L^r \subset L^q$, and

$$||f||_q \le (||f||p)^{\lambda}\,(||f||_r)^{1-\lambda}, \text{ where } \lambda = \frac{q^{-1} - r^{-1}}{p^{-1} - r^{-1}}.$$

Hint: Use Hölder's inequality for $|f|^q$ with an appropriate choice of a conjugate pair.

Chapter 9

L^p spaces: topological properties

In this chapter we make a detailed study of L^p spaces. In particular, for $1 \leq p < \infty$, L^p is a complete normed vector space, and for $0 < p < 1$, it is a complete metric space. We shall also give meaning to L^p for $p = \infty$, and learn other interesting properties of these spaces.

9.1 Topology of L^p, $0 < p < \infty$

Let $(\Omega, \mathcal{A}, \mu)$ be a measure space. Recall the definitions of $||\cdot||_p$ and L^p from Definition 8.3.1:

$$||f||_p = \Big(\int |f|^p d\mu\Big)^{1/p}, \quad L^p(\mu) = \{f : \int |f|^p d\mu < \infty\}.$$

The nature of L^p spaces differ depending on the value of p as well as whether μ is finite or not. We distinguish between three different cases: $0 < p < 1$; $1 < p < \infty$; and $p = \infty$.

(i) First, suppose that $0 < p < 1$.

Exercise 9.1.1 Show that, for $a, b \geq 0$ and $0 < p < 1$, $(a+b)^p \leq a^p + b^p$

Theorem 9.1.1. *Fix* $0 < p < 1$. *Then:*

(a) L^p *is a vector space;*

(b) $d(f, g) := \int_\Omega |f - g|^p d\mu$ *defines a metric on* L^p*;*

(c) $||f||_p$ *is not a norm on* L^p. ◆

A. Bose, *Elements of Measure and Probability*, Texts and Readings in Mathematics 88,
https://doi.org/10.1007/978-981-95-2758-8_9

Proof. Parts (a) and (b) follow from Exercise 9.1.1.

(c) For $a, b > 0$, by Exercise 9.1.1, $(a+b)^{1/p} > a^{1/p} + b^{1/p}$ whenever $0 < p < 1$. Hence, the condition for norm in Definition 1.4.2 is violated. ■

(ii) For $1 \le p < \infty$ we have the following result. Its proof is an immediate consequence of Minkowski's inequality (8.16).

Theorem 9.1.2. *Suppose that* $1 \le p < \infty$. *Then*

(a) L^p *is a vector space;*

(b) $||f||_p = [\int |f|^p d\mu]^{1/p}$ *is a norm on* L^p. ◆

So, for all $1 \le p < \infty$, L^p is a metric space and we arrive at the following definition of convergence in these spaces.

Definition 9.1.1. Fix $0 < p < \infty$. Let $\{f_n\}, f \in L^p(\Omega, \mathcal{A}, \mu)$. We say that f_n converges to f in L^p, if $\int_\Omega |f_n - f|^p d\mu \to 0$. We write $f_n \stackrel{L^p}{\to} f$. ◇

Exercise 9.1.2 Let $\{f_n\}$ be Borel measurable functions dominated by $g \in L^p$ for some $p > 0$, and $f_n \stackrel{a.s}{\to} f$. Show that $f \in L^p$ and $f_n \stackrel{L^p}{\to} f$. Hint: Use DCT.

Exercise 9.1.3 (a) Consider the Lebesgue measure λ on $[0, \infty)$. Let

$$f_n(x) = \begin{cases} 1/n & \text{if } 0 \le x \le e^n, \\ 0 & \text{otherwise.} \end{cases}$$

Check that $f_n \to 0$, uniformly (hence a.e. λ). However, f_n does not converge to 0 in L^p for any $0 < p < \infty$.

(b) Consider the Lebesgue measure on $[0, \infty)$. Let

$$f_n(x) = \begin{cases} 1 & \text{if } n \le x \le n + \dfrac{1}{n}, \\ 0 & \text{otherwise.} \end{cases}$$

Check that $f_n \to 0$ a.s., and in L^p for any $0 < p < \infty$.

Theorem 9.1.3 (Denseness of simple functions in L^p). *Suppose* $f \in L^p(\Omega, \mathcal{A}, \mu)$ *for some* $p > 0$. *Fix any* $\epsilon > 0$. *Then there exists a simple function* g *such that* $|g| \le |f|$, $g \in L^p$ *and* $||f - g||_p < \epsilon$. ◆

Proof. Without loss of generality we can assume that f is a real-valued function.

Recall construction of simple functions $\{s_n\}$ from Lemma 6.1.3(b). These s_n satisfy, $s_n \to f$ a.e.$[\mu]$, and $|s_n| \le |f|$ for all n. Then by Exercise 9.1.2, $||s_n - f||_p \to 0$. ■

We shall now prove that L^p is complete for $0 < p < \infty$. We need the following lemma.

Lemma 9.1.1. Let $\{f_n\} \in L^p$ for some $p > 0$ and $||f_k - f_{k+1}||_p < 4^{-k}$ for all $k \ge 1$. Then $\{f_n\}$ converges a.e. μ to some measurable f. ♦

Proof. (This lemma should remind the reader of Exercise 1.3.1(e)).

Define

$$A_k := \{\omega : |f_k(\omega) - f_{k+1}(\omega)| \ge 2^{-k}\}.$$

Then by Markov's inequality,

$$\mu(A_k) \le 2^{pk}||f_k - f_{k+1}||_p^p < 2^{-pk}.$$

Hence $\sum_{k=1}^{\infty} \mu(A_k) < \infty$.

By the First Borel-Cantelli Lemma 3.3.1, $\mu(\limsup A_n) = 0$. Now if $\omega \notin \limsup A_n$, then there is an $N(\omega)$ such that for all $k \ge N(\omega)$, $|f_k(\omega) - f_{k-1}(\omega)| \le 2^{-k}$. Hence, by Exercise 1.3.1(e), $\{f_k(\omega)\}$ converges. Since this is true for all $\omega \notin \limsup A_n$, the proof is complete. [The reader should complete the definition of f on Ω so that it becomes measurable]. ■

Theorem 9.1.4 (**L^p is complete**). *Let $\{f_n\}$ be Cauchy in L^p for a $0 < p < \infty$. Then there is an f such that $||f_n - f||_p \to 0$ as $n \to \infty$.* ◆

Proof. (a) First assume that $p \ge 1$. Since $\{f_n\}$ is Cauchy in L^p, we can choose increasing integers $\{n_k\}$ such that

$$||f_n - f_m||_p < \frac{1}{4^k} \quad \text{for all} \quad m, n \ge n_k.$$

Let $g_k = f_{n_k}$, $k \ge 1$. By Lemma 9.1.1, $g_k \overset{a.s.}{\to} f$ (measurable).

We now show that $f \in L^p$ and $||f_n - f||_p \to 0$. Fix $\epsilon > 0$. Since $\{f_n\}$ is Cauchy in L^p, choose N such that $||f_n - f_m||_p^p < \epsilon$ for all $n, m \ge N$. Fix $n \ge N$, and let $m \to \infty$ through the sub-sequence $\{n_k\}$.

Then

$$\begin{aligned}\epsilon &\geq \liminf_{k\to\infty} \int_\Omega |f_n - f_{n_k}|^p d\mu \\ &\geq \int_\Omega \liminf_{k\to\infty} |f_n - f_{n_k}|^p d\mu \quad \text{by Fatou's Lemma 7.2.1} \\ &= ||f_n - f||_p^p \quad \text{since} \quad f_{n_k} \to f \quad \text{a.e.}\end{aligned}$$

Now let $n \to \infty$. Then $\limsup_{n\to\infty} ||f_n - f||_p \leq \epsilon$. Since ϵ was arbitrary, this proves that $f_n - f \stackrel{L^p}{\to} 0$.

We have not yet shown that $f \in L^p$. But $f = (f - f_n) + f_n$, and hence, $f \in L^p$ by Minkowski's Inequality.

(b) Now let $0 < p < 1$. From the above proof (we used $p \geq 1$ only in the final step) $f_n - f \stackrel{L^p}{\to} 0$.

To show that $f \in L^p$, we write $f = (f - f_n) + f_n$, and instead of Minkowski's inequality, now use Exercise 9.1.1. ■

Theorem 9.1.5 (Denseness of continuous functions in L^p). *Consider $(\mathbb{R}^d, \mathcal{B}(\mathbb{R}^d), \mu)$ where μ is a Lebesgue-Stieltjes measure. Let $f \in L^p$ for some $0 < p < \infty$. Then given any $\epsilon > 0$, there is a continuous $g \in L^p$ such that $||f - g||_p < \epsilon$.* ◆

Proof. By Theorem 9.1.3, simple functions are dense in L^p. Hence, it suffices to approximate any indicator function $\mathbf{1}_A \in L^p$ by a continuous function.

Now, $\mathbf{1}_A \in L^p$ implies $\mu(A) < \infty$. Hence by Theorem 5.2.4, there exists C closed and V open such that $C \subset A \subset V$ and $\mu(V - C) < \epsilon^p 2^{-p}$.

By Urysohn's Lemma 1.5.1, get a continuous function $g : \Omega \to [0,\ 1]$ such that $g = 1$ on C and $g = 0$ on V^c. Then

$$\begin{aligned}\int_{\mathbb{R}^d} |\mathbf{1}_A - g|^p d\mu &= \int_{\{\omega:\mathbf{1}_A(\omega)\neq g(\omega)\}} |\mathbf{1}_A - g|^p d\mu \\ &\leq \int_{\{\omega:\mathbf{1}_A(\omega)\neq g(\omega)\}} 2^p d\mu \\ &\leq 2^p \mu(V - C) < \epsilon^p.\end{aligned}$$

Hence $||\mathbf{1}_A - g||_p < \epsilon$, and the theorem is proved. ■

Exercise 9.1.4 By following the steps in the above proof, show that the approximating function g can be chosen to have compact support.

9.2 The space L^∞

Definition 9.2.1. (**Essential supremum**) Let $(\Omega, \mathcal{A}, \mu)$ be a measure space. For any real-valued measurable function f, define

$$\begin{aligned} \text{ess sup } f &:= \inf\{c \in \bar{\mathbb{R}} : \mu\{\omega : f(\omega) > c\} = 0\}, \\ ||f||_\infty &:= \text{ess sup } |f|, \\ L^\infty = L^\infty(\mu) &:= \{f : \ ||f||_\infty < \infty\}. \end{aligned}$$

$||f||_\infty$ is called the L^∞ norm of f. For complex-valued Borel measurable function f, $||f||_\infty$ is defined as the L^∞-norm of $|f|$. ◇

$||f||_\infty$ is the smallest c such that $|f| \leq c$ a.e. μ. So, $f \in L^\infty$ if and only if $|f|$ is *essentially bounded*, that is, bounded outside a measure 0 set.

Example 9.2.1. (a) If $f \equiv -1$, then ess sup $f = -1$.

(b) If $f = \sum_{i=1}^n c_i \mathbf{1}_{A_i}$ is a simple function, then

$$\text{ess sup } f = \sup\{c_i : \mu(A_i) > 0\}\,.$$

(c) If $f \leq g$ a.e. then ess sup $f \leq$ ess sup g. ▲

Theorem 9.2.1 (Properties of L^∞). *(a) L^∞ is a vector space and $||\cdot||_\infty$ is a norm on it.*

(b) Let $\{f_n\}$ be in L^∞. Then $||f_n - f||_\infty \to 0$ if and only if there exists $A \in \mathcal{A}$ such that $\mu(A) = 0$ and $f_n \to f$ uniformly on A^c.

(c) L^∞ is complete with respect to $||\cdot||_\infty$. ◆

Proof. (a) This is easy and we skip the proof.

(b) Suppose $||f_n - f||_\infty \to 0$. For positive integer m, $||f_n - f||_\infty \leq 1/m$ for all large n. Hence there exists A_m such that for $\mu(A_m) = 0$ and for all $\omega \notin A_m$, $|f_n - f| \leq 1/m$. Let $A = \cup_{m=1}^\infty A_m$. Then $\mu(A) = 0$ and $f_n \to f$ uniformly on A^c.

Conversely, let $A \in \mathcal{A}$, $\mu(A) = 0$, and $f_n \to f$ uniformly on A^c. Then given $\epsilon > 0$, for all large n, $|f_n - f| \leq \epsilon$ on A^c. This implies $|f_n - f| \leq \epsilon$ a.e. μ. That is, $||f_n - f||_\infty < \epsilon$.

(c) This is left as an exercise. ■

Exercise 9.2.1 Consider the Lebesgue measure on $\mathbb{R}$. Let

$$f_n(x) = \begin{cases} 1/n & \text{if } 0 \le x \le e^n, \\ 0 & \text{otherwise.} \end{cases}$$

Show that f_n does not converge to 0 in L^∞.

Exercise 9.2.2 (a) Show that Hölder's inequality holds for $p = 1, q = \infty$ and Minkowski's inequality holds for $p = \infty$.

(b) Show that (finite valued) simple functions are dense in L^∞.

(c) Consider $(\mathbb{R}^d, \mathcal{B}(\mathbb{R}^d), \mu)$ where μ is a Lebesgue-Stieltjes measure. Show that continuous functions $g \in L^\infty$ are not dense in L^∞.

9.3 l^p spaces

Definition 9.3.1. (**The little l^p**) Let Ω be a countable set. Without loss of generality, $\Omega = \{1, 2, \ldots, \}$. Equip Ω with the σ-field $\mathcal{P}(\Omega)$, and the counting measure μ. Then it is conventional to write $L^p(\Omega, \mathcal{P}(\Omega), \mu)$ as $l^p(\Omega)$, or simply as l^p and it is referred to as "*little l^p*". These are spaces of sequences (of complex numbers), and

$$l^p = \{\{x_n\} : \sum_{n=1}^{\infty} |x_n|^p < \infty\},\ 1 \le p < \infty,$$
$$l^\infty = \{\{x_n\} : \{x_n\} \text{ is a bounded sequence}\}.$$

The metrics and norms on l^p, $0 < p < 1$ (respectively $1 \le p \le \infty$ are defined as before with the measure now being the counting measure. ◇

Of course all results derived so far for L^p spaces with possibly infinite measures, are applicable to these spaces. As a sample we rewrite Minkowski's and Hölder's inequalities in l^p:

Theorem 9.3.1 (**Hölder's and Minkowski's inequalities for sequences**). *Let (p, q) be a conjugate pair. For complex sequences $\{x_n\}$ and $\{y_n\}$,*

(a) $\sum_{n=1}^{\infty} |x_n y_n| \le \big(\sum_{n=1}^{\infty} |x_n|^p\big)^{1/p} \big(\sum_{n=1}^{\infty} |y_n|^q\big)^{1/q}$.

(b) $\big(\sum_{n=1}^{\infty} |x_n + y_n|^p\big)^{1/p} \le \big(\sum_{n=1}^{\infty} |x_n|^p\big)^{1/p} + \big(\sum_{n=1}^{\infty} |y_n|^p\big)^{1/p}$. ◆

9.4 Exercises

Exercise 9.4.1 Let $(\Omega, \mathcal{A}, \mu)$ be a finite measure space. Show that if $r < s$, then $L^s(\mu) \subset L^r(\mu)$.

Exercise 9.4.2 Let $f : (\Omega, \mathcal{A}, \mu) \to (\mathbb{R}, \mathcal{B}(\mathbb{R}))$.

(a) Show that,

$$\liminf_{p\to\infty} ||f||_p \geq ||f||_\infty .$$

(b) Show that if μ is a finite measure, then

$$\lim_{p\to\infty} ||f||_p = ||f||_\infty .$$

(c) Give an example of an infinite measure and a function f such that the conclusion in (b) fails.

Exercise 9.4.3 Show that the space of (equivalence classes) of Riemann integrable functions on $[0,\ 1]$ with the L^1 metric is not complete.

Exercise 9.4.4 Let $\{f_n\} : (\Omega, \mathcal{A}, \mu) \to (\mathbb{C}, \mathcal{B}(\mathbb{C}))$ be in $L^1(\mu)$ where μ is a finite measure. Suppose $\{f_n\}$ converges to f uniformly. Show that $f \in L^1(\mu)$ and $\int |f_n - f| d\mu \to 0$.

Exercise 9.4.5 Let $(\Omega, \mathcal{A}, \mu)$ be a finite measure space. Show that if $f_n \overset{L^\infty}{\to} f$ then $f_n \overset{L^p}{\to} f$ for all $0 < p < \infty$.

Chapter 10

Product spaces and transition measures

After Definition 5.2.7, we alluded to the fact that the Lebesgue measure on $\mathbb{R}^d$ is 'd-fold product" of the (marginal) Lebesgue measure on $\mathbb{R}$. Indeed, we may construct product measures from marginal measures which are not necessarily identical. We can go further, and allow the marginal measure of a co-ordinate to depend on the measures for the previous co-ordinates. These are called transition measures. Once we have these product measures, a natural question is what are the relations between the iterated integrals and the integral on the product space, and we provide conditions for them to be equal. We shall work mainly with finite products. Results on infinite products will be discussed, but only briefly and without proofs.

10.1 Product σ-field

Definition 10.1.1. (**Product σ-field**) Suppose $(\Omega_i, \mathcal{A}_i)$, $1 \leq i \leq d$ are measurable spaces. Then $A \in \mathcal{A}_1 \times \cdots \times \mathcal{A}_d$ is called a **measurable rectangle**. The smallest σ-field of subsets of $\Omega = \Omega_1 \times \cdots \times \Omega_d$ that contains all measurable rectangles is called the *product σ-field*, and we write it as $\mathcal{A}_1 \otimes \cdots \otimes \mathcal{A}_d$ or $\otimes_{i=1}^{d} \mathcal{A}_i$. ◇

Exercise 10.1.1 Verify that $\mathcal{B}(\mathbb{R}^d)$ is the d-fold product σ-field of $\mathcal{B}(\mathbb{R})$.

Exercise 10.1.2 (a) Show that the collection of all measurable rectangles may not always be a field but is always a semi-field.

A. Bose, *Elements of Measure and Probability*, Texts and Readings in Mathematics 88,
https://doi.org/10.1007/978-981-95-2758-8_10

(b) Show that the collection of all finite disjoint unions of measurable rectangles is always a field but may not be a σ-field.

(c) Show that $(\mathcal{A}_1 \otimes \cdots \otimes \mathcal{A}_{d-1}) \otimes \mathcal{A}_d = \mathcal{A}_1 \otimes \cdots \otimes \mathcal{A}_d$.

Definition 10.1.2. (**Sections of a set**) Let $A \in \mathcal{A}_1 \otimes \mathcal{A}_2$. Let

$$A_1(\omega_2) = \{\omega_1 \in \Omega_1 : (\omega_1, \omega_2) \in A\} \subseteq \Omega_1, \text{ for } \omega_2 \in \Omega_2,$$
$$A_2(\omega_1) = \{\omega_2 \in \Omega_2 : (\omega_1, \omega_2) \in A\} \subseteq \Omega_2, \text{ for } \omega_1 \in \Omega_1.$$

These sets are called *sections* of A. Sections of d-fold products are defined in a similar way. ◇

Note that sections can be empty.

Exercise 10.1.3 (a) If A is a measurable rectangle, identify its sections.

(b) Show that $A_1(\omega_2) \in \mathcal{A}_1$, $A_2(\omega_1) \in \mathcal{A}_2$ for all $\omega_1 \in \Omega_1$ and $\omega_2 \in \Omega_2$.

(c) If $A, B \in \mathcal{A}_1 \otimes \mathcal{A}_2$ are disjoint, show that their sections are disjoint.

(d) Let $A_n \in \mathcal{A}_1 \otimes \mathcal{A}_2$ increase (resp. decrease) to A. Show that $A_{n1}(\omega_2)$ increases (resp. decreases) to $A_1(\omega_2)$. Likewise for $\{A_{n2}(\omega_1)\}$.

Exercise 10.1.4 Let $(\Omega_i, \mathcal{A}_i)$, $i = 1, 2$ be measurable spaces, and let $(\Omega_1 \times \Omega_2, \mathcal{A}_1 \otimes \mathcal{A}_2)$ be the product measurable space. Let $f : \Omega_1 \times \Omega_2 \to \bar{\mathbb{R}}$ be a Borel measurable function.

(a) Show that for every fixed $\omega_1 \in \Omega_1$, $f_1(\omega_2) = f(\omega_1, \omega_2)$ is a Borel measurable function on $(\Omega_2, \mathcal{A}_2)$. This function is called the first section of f. The second section is defined similarly.

(b) If f is simple, then show that its sections are also simple.

(c) Extend the above to d-fold products.

10.2 Transition and product measures

Transition measures are generalisation of transition probabilities which are fundamental in Markov chains. A marginal measure and a transition measure can be paired to yield a combined measure on the product space. This will lead to the construction of product measures.

Definition 10.2.1. (**Transition measure**) Let $(\Omega_i, \mathcal{A}_i)$, $i = 1, 2$ be measurable spaces. Let $\mu_2 : \Omega_1 \times \mathcal{A}_2 \to [0, \infty]$ be a function such that

(i) For every $\omega_1 \in \Omega_1$, $\mu_2(\omega_1, \cdot)$ is a measure on $\mathcal{A}_2$.

(ii) For every $A_2 \in \mathcal{A}_2$, $\omega_1 \to \mu_2(\omega_1, A_2)$ is a (Borel) measurable function (with respect to $\mathcal{A}_1$).

Then μ_2 is called a *transition measure.* If the measures in (i) are probability measures, then μ_2 is a **transition probability measure**. ◇

Exercise 10.2.1 Here is a toy example. Let $\Omega_1 = \{\text{Box 1, Box 2}\}$. Think of them as two values of ω_1. Suppose Box 1 has 3 red balls and 3 white balls. Box 2 has 3 reds, 2 white and 5 black balls. I pick a box by some mechanism (this mechanism could also be a probability measure on the two values, say measure μ_1). Once I pick a box, I pick one ball "at random" from that box (that is, all balls in that box have equal chance to come into my sample). That is, there are *two* probability measures, on $\Omega_2 := \{R, B, W\}$ depending on which box (ω_1) is chosen. These are the two distinct transition probability measures given below:

$$P_1(R) = 1/2,\ P_1(W) = 1/2,\ P_1(B) = 0 \ \text{ if } \ \omega_1 = \text{ Box } 1\,;$$
$$P_2(R) = 3/10,\ P_2(W) = 1/5,\ P_1(B) = 1/2 \ \text{ if } \ \omega_1 = \text{ Box } 2.$$

We investigate how to combine μ_1 and $\{\mu_2(\omega_1, \cdot) : \omega_1 \in \Omega_1\}$ into one measure on $\Omega_1 \times \Omega_2$. We need the following notion.

Definition 10.2.2. (Uniformly σ-finite) A transition measure is said to be *uniformly σ-finite* on $\mathcal{A}_2$, if there exists $\{B_n\}$ from $\mathcal{A}_2$ such that

$$\cup_{n=1}^{\infty} B_n = \Omega_2, \text{ and } \sup_{\omega_1 \in \Omega_1} \mu_2(\omega_1, B_n) < \infty \ \text{ for all } \ n \geq 1.$$

It is called **uniformly finite** if $\sup_{\omega_1 \in \Omega_1} \mu_2(\omega_1, \Omega_2) < \infty$. ◇

Theorem 10.2.1 (Transition and marginal measures combined). *Let μ_2 be a uniformly σ-finite transition measure and μ_1 be a σ-finite measure. Then there is a unique measure μ on $\mathcal{A} = \mathcal{A}_1 \otimes \mathcal{A}_2$ such that,*

$$\mu(A_1 \times A_2) = \int_{A_1} \mu_2(\omega_1, A_2)\mu_1(d\omega_1), \quad A_i \in \mathcal{A}_i,\ i = 1, 2, \tag{10.1}$$

$$\mu(A) = \int_{\Omega_1} \mu_2(\omega_1, A_2(\omega_1))\mu_1(d\omega_1), \ A \in \mathcal{A}. \tag{10.2}$$

Further, μ is σ-finite. If μ_1 and the transition measures are probability measures, then μ is also a probability measure. ◆

We note that the measurability Condition (i) in Definition 10.2.1 assures that the intergal in (10.1) is defined.

Proof. (a) First assume that $\mu_2(\omega_1, \cdot)$ are uniformly finite.

(i) If $A \in \mathcal{A}$, then by Exercise 10.1.3(b), we know that $A_2(\omega_1) \in \mathcal{A}_2$.

(ii) We show that $f(\omega_1) := \mu_2(\omega_1, A_2(\omega_1))$ is a measurable function for all $A \in \mathcal{A}$. To show this, let $\mathcal{C}$ be the class of sets in $\mathcal{A}$ for which this is true. Suppose $A = A_1 \times A_2 \in \mathcal{A}$. Then

$$\mu_2(\omega_1, A_2(\omega_1)) = \begin{cases} \mu_2(\omega_1, A_2) & \text{if } \omega_1 \in A_1, \\ \mu_2(\omega_1, \emptyset) = 0 & \text{if } \omega_1 \notin A_1. \end{cases}$$

Thus $\mu_2(\omega_1, A_2(\omega_1)) = \mu_2(\omega_1, A_2) I_{A_1}(\omega_1)$ which is measurable by assumption. Thus $\mathcal{C}$ contains all measurable rectangles. Since finite sums of measurable functions are measurable, $\mathcal{C}$ contains the field of all finite disjoint union of rectangles. Using finiteness of the transition measures and Exercise 10.1.3(d), it follows that $\mathcal{C}$ is a monotone class. Thus by monotone class Theorem 2.6.1, $\mathcal{C} = \mathcal{A}$.

(iii) Define

$$\mu(A) = \int_{\Omega_1} \mu_2(\omega_1, A_2(\omega_1)) \mu_1(d\omega_1), A \in \mathcal{A}.$$

For all $A \in \mathcal{A}$, the integrand is non-negative, and is measurable by (ii). Hence the integral is defined and satisfies (10.2). Letting $A = A_1 \times A_2$, it is easily seen that (10.1) holds. We now prove that μ is a measure. Suppose $\{A_n\}$ is a sequence of disjoint measurable sets from $\mathcal{A}$. Then

$$\begin{aligned} \mu(\cup_{n=1}^{\infty} A_n) &= \int_{\Omega_1} \mu_2(\omega_1, (\cup_{n=1}^{\infty} A_n)_2(\omega_1)) \mu_1(d\omega_1) \\ &= \int_{\Omega_1} \sum_{n=1}^{\infty} \mu_2(\omega_1, A_{n2}(\omega_1)) \mu_1(d\omega_1), \ \ (\mu_2(\omega_1, \cdot) \text{ a measure}) \\ &= \sum_{n=1}^{\infty} \int_{\Omega_1} \mu_2(\omega_1, A_{n2}(\omega_1)) \mu_1(d\omega_1) \\ &= \sum_{n=1}^{\infty} \mu(A_n). \end{aligned}$$

This proves the theorem in the finite case.

(b) Now assume that $\mu_2(\omega_1, \cdot)$ is uniformly σ-finite. Accordingly, split up Ω_2 into $\{B_n\}$, and proceed by considering the restricted transition measures. We omit the details.

To prove uniqueness, suppose ν is a measure on $\mathcal{A}$ for which (10.1) holds. Then $\nu = \mu$ on the field, $\mathcal{F}$, of finite disjoint unions of measurable rectangles. As $\mu_2(\omega_1, \cdot)$ is uniformly σ-finite, get a measurable partition $\{B_n\}$ of Ω_2 such that $\sup_{\omega_1} \mu_2(\omega_1, B_n) < \infty$ for all n. As μ_1 is σ-finite, get a measurable partition $\{A_m\}$ of Ω_1 such that $\mu_1(A_m) < \infty$ for all m. Then $\nu(A_m \times B_n) = \mu(A_m \times B_n) < \infty$ for all m, n. Thus μ is σ-finite on $\mathcal{F}$ and hence so is ν. So by Caratheodory Theorem 4.4.1, they must be equal on $\mathcal{A}$.

Finally, it is easy to see that if μ_1 and the transition measures $\mu_2(\omega_1, \cdot)$ are probability measures, then μ is a probability measure. ■

Definition 10.2.3. (**Product measure**) Let μ_i be σ-finite measures on $\mathcal{A}_i$, $i = 1, 2$. The unique measure μ on $\mathcal{A}_1 \otimes \mathcal{A}_2$ which satisfies $\mu(A_1 \times A_2) = \mu_1(A_1)\mu_2(A_2)$ for all $A_i \in \mathcal{A}_i$, $i = 1, 2$ is known as the *product measure*. We write μ as $\mu_1 \otimes \mu_2$ or $\mu_1 \times \mu_2$. The definition extends to finitely many measures in the obvious way. ◇

Exercise 10.2.2 Show that μ in Definition 10.2.3 does exist uniquely, and is σ-finite. Use it to construct the Lebesgue measure on $\mathbb{R}^2$.

Exercise 10.2.3 Let $(\Omega_i, \mathcal{A}_i, \mu_i)$, $i = 1, 2$ be σ-finite measure spaces. Let $\mu = \mu_1 \otimes \mu_2$ and $A \in \mathcal{A}_1 \otimes \mathcal{A}_2$. Show that the following are equivalent.

(i) $\mu(A) = 0$.

(ii) $\mu_1(A_1(\omega_2)) = 0$ a.e.$[\mu_2]$.

(iii) $\mu_2(A_2(\omega_1)) = 0$ a.e.$[\mu_1]$.

10.3 Fubini's theorem

Having constructed the product measures, a natural question is what are the relations between the integral on the product space, and the two iterated integrals. The next result provides the answer.

Theorem 10.3.1 (Fubini's theorem). *Let μ_1 on $(\Omega_1, \mathcal{A}_1)$ be a σ-finite measure, and μ_2 on $\Omega_1 \times \mathcal{A}_2$ be a uniformly σ-finite transition measure. Let $\Omega = \Omega_1 \times \Omega_2$, $\mathcal{A} = \mathcal{A}_1 \otimes \mathcal{A}_2$, and $f : \Omega \to \bar{\mathbb{R}}$ be Borel measurable.*

(a) (This is due to Tonelli [1909]) If f is non-negative, then

$$g(\omega_1) = \int_{\Omega_2} f(\omega_1, \omega_2)\mu_2(\omega_1, d\omega_2), \ \omega_1 \in \Omega_1, \tag{10.3}$$

exists, and defines a Borel measurable function on Ω_1. Further,

$$\begin{aligned} \int_{\Omega} f d\mu &= \int_{\Omega_1} g(\omega_1)\mu_1(d\omega_1) \\ &= \int_{\Omega_1} \Big[\int_{\Omega_2} f(\omega_1, \omega_2)\mu_2(\omega_1, d\omega_2)\Big]\mu_1(d\omega_1). \end{aligned} \tag{10.4}$$

(b) If $\int_\Omega f d\mu$ exists (resp. finite), then $g(\omega_1)$ as in (10.3) exists (resp. finite) a.s.$[\mu_1]$, and defines a Borel measurable function, if it is defined to be say 0 on the null set. Further, (10.4) holds. ◆

Proof. (a) Since f is jointly measurable, by Exercise 10.1.4, fixing ω_1, $f_1(\omega_2) = f(\omega_1, \omega_2)$ is a non-negative measurable function of ω_2. Hence $g(\omega_1)$ exists (yet to prove it is measurable). Now choose $f = I_A$. Then

$$\begin{aligned} \int_{\Omega_2} I_A(\omega_1, \omega_2)\mu_2(\omega_1, d\omega_2) &= \int_{\Omega_2} I_{A(\omega_1)}(\omega_2)\mu_2(\omega_1, d\omega_2) \\ &= \mu_2(\omega_1, A(\omega_1)), \end{aligned}$$

and by Part (ii) of the proof of Theorem 10.2.1(a), we know that this is a Borel measurable function. Moreover,

$$\begin{aligned} \int_\Omega I_A d\mu &= \mu(A) \\ &= \int_{\Omega_1} \mu_2(\omega_1, A(\omega_1))\mu_1(d\omega_1), \quad \text{by (10.2)} \\ &= \int_{\Omega_1} \Big[\int_{\Omega_2} I_A(\omega_1, \omega_2)\mu_2(\omega_1, d\omega_2)\Big]\mu_1(d\omega_1). \end{aligned}$$

So (10.4) and Part (a) is true for indicator functions. Hence Part (a) is also true for non-negative simple functions.

For arbitrary non-negative measurable f, get simple functions $\{f_n\}$ such that $f_n \uparrow f$. Then by MCT,

$$\int_{\Omega_2} f_n(\omega_1, \omega_2)\mu_2(\omega_1, d\omega_2) \uparrow \int_{\Omega_2} f(\omega_1, \omega_2)\mu_2(\omega_1, d\omega_2) = g(\omega_1).$$

By arguments given so far, left side is measurable in ω_1, hence so is g.

Now,

$$\begin{aligned}\int_\Omega f d\mu &= \lim_{n\to\infty}\int_\Omega f_n d\mu \quad \text{(by MCT)}\\ &= \lim_{n\to\infty}\int_{\Omega_1}\Big[\int_{\Omega_2} f_n(\omega_1,\omega_2)\mu_2(\omega_1,d(\omega_2))\Big]\mu_1(d\omega_1)\\ &\qquad\qquad \text{(already shown for simple functions)}\\ &= \int_{\Omega_1}\Big[\int_{\Omega_2} f((\omega_1,\omega_2)\mu_2(\omega_1,d\omega_2)\Big]\mu_1(d\omega_1) \quad \text{(use MCT twice).}\end{aligned}$$

This proves (a) completely.

(b) Since $\int_\Omega f d\mu$ exists, one of $\int_\Omega f^+ d\mu$ or $\int_\Omega f^- d\mu$ is finite. Suppose, without loss, $\int_\Omega f^- d\mu < \infty$. Then by Part (a),

$$\int_{\Omega_1}\int_{\Omega_2} f^-(\omega_1, d\omega_2)\mu_2(\omega_1, d\omega_2)\mu_1(d\omega_1) = \int_\Omega f^- d\mu < \infty\,.$$

Hence $\int_\Omega f^-(\omega_1, d\omega_2)\mu_2(\omega_1, d\omega_2)$ is μ_1 integrable. Thus it is finite a.e. μ_1. Consequently,

$$\begin{aligned}&\int_{\Omega_2} f(\omega_1,\omega_2)\mu_2(\omega_1,d\omega_2) =\\ &\qquad \int_{\Omega_2} f^+(\omega_1,\omega_2)\mu_2(\omega_1,d\omega_2) - \int_{\Omega_2} f^-(\omega_1,\omega_2)\mu_2(\omega_1,d\omega_2), \ \text{a.e. } \omega_1.\end{aligned}$$

If $\int_\Omega f d\mu$ is finite then both integrals on the right side above are finite a.e. μ_1. We can define all three integrals above to be 0 on the exceptional set. Then the above equation is valid now for *all* ω_1, and all functions are Borel measurable. Integrating both sides with respect to μ_1, and using Part (a) and additivity of integrals,

$$\begin{aligned}\int_{\Omega_1}\Big[\int_{\Omega_2} f(\omega_1,\omega_2)\mu_2(\omega_1,d\omega_2)\Big]\mu_1(d\omega_1) &= \int_\Omega f^+ d\mu - \int_\Omega f^- d\mu\\ &= \int_\Omega f d\mu.\end{aligned}$$

■

Exercise 10.3.1 Suppose in the above set up, the iterated intergal

$$\int_{\Omega_1}\Big[\int_{\Omega_2} |f(\omega_1,\omega_2)|\mu_2(\omega_1,d\omega_2)\Big]\mu_1(d\omega_1) < \infty\,.$$

Then show that $\int_\Omega f d\mu$ is finite and Fubini's theorem applies.

Exercise 10.3.2 (Fubini's theorem for product of two measures) Let $(\Omega_i, \mathcal{A}_i, \mu_i)$, $i = 1, 2$ be σ-finite measure spaces. Let $\Omega = \Omega_1 \times \Omega_2$, $\mathcal{A} = \mathcal{A}_1 \otimes \mathcal{A}_2$, and $\mu = \mu_1 \otimes \mu_2$. Let $f : \Omega \to \bar{\mathbb{R}}$ be a Borel measurable function. If $\int_\Omega f d\mu$ exists, then show that

$$\int_\Omega f d\mu = \int_{\Omega_1} \Big[\int_{\Omega_2} f d\mu_2\Big] d\mu_1 \tag{10.5}$$

$$= \int_{\Omega_2} \Big[\int_{\Omega_1} f d\mu_1\Big] d\mu_2. \tag{10.6}$$

Theorem 10.2.1 can be extended to product of d spaces, $d \geq 3$, when we have a sequence of transition measures. We state this result, without proof, only for the case where all measures are probability measures.

Theorem 10.3.2 (Measure on a finite product). *Suppose* $(\Omega_j, \mathcal{A}_j)$, $1 \leq j \leq d$ *are measurable spaces. Let* Q_1 *be a probability measure on* $\mathcal{A}_1$ *and let* $Q_{j+1} : \Omega_1 \times \cdots \times \Omega_j \times \mathcal{A}_{j+1} \to [0, 1]$ *be a transition probability measure for* $1 \leq j \leq d - 1$. *Then there is a unique probability measure* P_d *on* $\mathcal{A}_1 \otimes \cdots \otimes \mathcal{A}_d$ *such that for all* $A_1 \times \cdots \times A_d \in \mathcal{A}_1 \otimes \cdots \otimes \mathcal{A}_d$,

$$P_d(\prod_{i=1}^d A_i) = \int_{A_1} Q_1(d\omega_1) \int_{A_2} Q_2(\omega_1, d\omega_2) \cdots \int_{A_d} Q_d(\omega_1, \ldots \omega_{d-1}, d\omega_d). \tag{10.7}$$

As a consequence, for all $B^d \in \mathcal{A}_1 \otimes \cdots \otimes \mathcal{A}_d$,

$$P_d(B^d) = \int_{\Omega_1} Q_1(d\omega_1) \int_{\Omega_2} Q_2(\omega_1, d\omega_2) \cdots \int_{\Omega_d} I_{B^d} Q_d(\omega_1, \ldots \omega_{d-1}, d\omega_d). \tag{10.8}$$

◆

Remark 10.3.1. (a) Right side of (10.7) is read as follows: Integrate the constant function 1 with respect to $Q_d(\omega_1, \ldots \omega_{d-1}, \cdot)$ on the set A_d to get $Q_d(\omega_1, \ldots \omega_{d-1}, A_d)$. By assumption, this is a measurable function of $(\omega_1, \ldots, \omega_{d-1})$. Then integrate this function on A_{d-1} with respect to the transition measure $Q_d(\omega_1, \ldots \omega_{d-2}, \cdot)$ to get a measurable function of $(\omega_1, \ldots \omega_{d-2})$. Continue to finally get a measurable function of ω_1 which is then integrated with respect to the measure Q_1 on the set A_1. We read (10.8) similarly, using the appropriate sections (which are measurable).

(b) Fubini's theorem remains valid and hence integrals with respect to P_d can be computed in an iterative way.

(c) A special case of Theorem 10.3.2 is when the transition probability measures Q_{j+1}, $1 \leq j \leq d - 1$ do not depend on $(\omega_1, \ldots \omega_j)$. In that case P_d is the **product probability measure** $Q_1 \otimes \cdots \otimes Q_d$. ●

Exercise 10.3.3 (a) Formulate a version of Theorem 10.3.2 when Q_1 is a σ-finite measure and Q_{j+1}, $1 \le j \le d-1$ are uniformly σ-finite transition measures. What are the anticipated hurdles in extending your proof to infinitely many transition measures?

(b) Show that Fubini's theorem also remains valid for any $d \ge 3$.

(c) Show that the Lebesgue measure λ_d on $\mathbb{R}^d$ is the d-fold product of the Lebesgue measure λ on $\mathbb{R}$.

10.4 Infinite product

We shall now see how to define measures on infinite product spaces. We shall mostly deal with product measures, and in addition, we shall restrict our discussion to only probability measures and a countable product. For sets Ω_j, $j \ge 1$, let

$$\Omega := \prod_{j=1}^{\infty} \Omega_j = \{(\omega_1, \omega_2, \ldots) : \omega_j \in \Omega_j \ \text{ for all } \ j \ge 1\}.$$

Definition 10.4.1. (**Measurable rectangles and cylinders**) Let $(\Omega_j, \mathcal{A}_j)$ be measurable spaces. For any $B^n \subseteq \Omega_1 \times \cdots \times \Omega_n$, the set

$$\begin{aligned} B_n &= B^n \times \Omega_{n+1} \times \Omega_{n+2} \times \cdots \\ &= \{\omega \in \Omega : (\omega_1, \omega_2, \ldots, \omega_n) \in B^n\} \subset \Omega, \end{aligned}$$

is called a *cylinder* with n-**dimensional base** B^n. The set B_n is said to be a *measurable cylinder* if $B^n \in \mathcal{A}_1 \otimes \cdots \otimes \mathcal{A}_n$. If $B^n = A_1 \times \cdots \times A_n$, then B_n is called a (finite dimensional) *rectangle*. It is called a *measurable rectangle* if $A_i \in \mathcal{A}_i$, $1 \le i \le n$. ◇

Exercise 10.4.1 Show that:

(a) an n-dimensional cylinder is also an $(n+1)$ dimensional cylinder;

(b) the finite disjoint unions of all measurable rectangles forms a field;

(c) the set of measurable cylinders is a field.

Definition 10.4.2. (**Infinite product σ-field**) The smallest σ-field generated by all measurable cylinders is called the **product σ-field**. It is written as $\otimes_{j=1}^{\infty} \mathcal{A}_j$ or as $\prod_{j=1}^{\infty} \mathcal{A}_j$. If all $\mathcal{A}_j$ are equal to $\mathcal{A}$, then we write the product σ-field as $\mathcal{A}^{\infty}$. ◇

Exercise 10.4.2 Show that $\otimes_{j=1}^{\infty}\mathcal{A}_j$ is also the smallest σ-field generated by all finite dimensional measurable rectangles.

We now state the product measure theorem for infinite products, restricting to only probability and transition probability measures.

Theorem 10.4.1 (Measure on infinite product). *Let $\{(\Omega_j, \mathcal{A}_j)\}$ be measurable spaces. Let Q_1 be a probability measure on $\mathcal{A}_1$ and let $Q_{j+1} : \Omega_1 \times \cdots \times \Omega_j \times \mathcal{A}_{j+1} \to [0,1]$ be transition probability measures for $j \geq 1$. Then there is a unique probability measure P on $\otimes_{j=1}^{\infty}\mathcal{A}_j$ that agrees with the probability measures $\{P_n\}$ given in (10.8). That is,*

$$(10.9)\quad P(B_n) = P_n(B^n),, \quad B_n = B^n \times \Omega_{n+1} \times \cdots\,, \quad B^n \in \mathcal{A}_1 \otimes \cdots \otimes \mathcal{A}_n. \;\blacklozenge$$

Proof. From Exercise 10.4.1(c), the set of measurable cylinders, say $\mathcal{F}$, is a field. Define $P(\cdot)$ on $\mathcal{F}$ by Eqn. (10.9). Since cylinders may have different representations, we first show that this definition is unambiguous. Let,

$$B_n = \{\omega \in \Omega : (\omega_1, \ldots, \omega_m) \in C^m\}, \text{ where } C^m \in \mathcal{A}_1 \otimes \cdots \otimes \mathcal{A}_m\,.$$

Without any loss, assume that $m < n$. Then clearly,

$$(\omega_1, \ldots, \omega_m) \in C^m \text{ if and only if } (\omega_1, \ldots, \omega_n) \in B^n\,,$$

and hence, $B^n = C^m \times \Omega_{m+1} \times \cdots \times \Omega_n$. Since each transition measure is a probability measure, $P_n(B^n) = P_m(C^m)$.

Since each P_n is a probability measure on $\mathcal{A}_1 \otimes \cdots \otimes \mathcal{A}_n$, it easily follows that $P(\cdot)$ on $\mathcal{F}$ is *finitely additive.*

We now show that $P(\cdot)$ is countably additive on $\mathcal{F}$ by using Exercise 3.5.1(b). Consider a sequence of measurable cylinders that decreases to $\emptyset$. Without loss we can take it to be $\{B_n\}$ where B_n is an n-dimensional measurable cylinder, $n \geq 1$. Then $P(B_n)$ is a decreasing sequence. Suppose if possible $\lim P(B_n) > 0$. From (10.8),

$$P(B_n) = \int_{\Omega_1} f_n^{(1)} Q_1(d\omega_1) \text{ where,}$$
$$f_n^{(1)}(\omega_1) = \int_{\Omega_2} Q_2(\omega_1, d\omega_2) \cdots \int_{\Omega_n} I_{B^n} Q_n(\omega_1, \ldots \omega_{n-1}, d\omega_n)\,.$$

Note that $\{f_n^{(1)}(\cdot)\}$ is a sequence of measurable functions, uniformly bounded by 1.

As $B_{n+1} \subseteq B_n$, we have $B^{n+1} \subseteq B^n \times \Omega_{n+1}$, and hence,

$$\mathbf{1}_{B^{n+1}}(\omega_1, \ldots, \omega_{n+1}) \le \mathbf{1}_{B^n}(\omega_1, \ldots, \omega_n)\,.$$

This implies that $\{f_n^{(1)}(\cdot)\}$ decreases, to $f_1(\cdot)$ say. By DCT Theorem 7.2.7, $\int_{\Omega_1} f_1(\omega_1) Q_1(d\omega_1) = \lim P(B_n) > 0\,.$

As a consequence $f_1(\tilde{\omega}_1) > 0$ for some $\tilde{\omega}_1 \in \Omega_1$. But this $\tilde{\omega}_1 \in B^1$ (otherwise $\mathbf{1}_{B^n}(\tilde{\omega}_1, \ldots, \omega_n) = 0$ for all n and $f_1(\tilde{\omega}_1) = 0$ which would be a contradiction).

Now note that

$$\begin{aligned} f_n^{(1)}(\tilde{\omega}_1) &= \int_{\Omega_2} f_n^{(2)}(\omega_2) Q_2(\tilde{\omega}_1, d\omega_2) \ \text{ where,} \\ f_n^{(2)}(\omega_2) &= \int_{\Omega_3} Q_3(\tilde{\omega}_1, \omega_2, d\omega_3) \cdots \int_{\Omega_n} I_{B^n} Q_n(\tilde{\omega}_1, \ldots \omega_{n-1}, d\omega_n)\,. \end{aligned}$$

Repeating the arguments given for $f_n^{(1)}$, $f_n^{(2)}(\cdot)$ decreases to f_2 say, and there is a $\tilde{\omega}_2 \in \Omega_2$ such that, $f_2(\tilde{\omega}_2) > 0$ and $(\tilde{\omega}_1, \tilde{\omega}_2) \in B^2$.

In general, we get a sequence $\{\tilde{\omega}_n\}$ such that $(\tilde{\omega}_1, \ldots \tilde{\omega}_n) \in B^n$ for all n.

But then $(\tilde{\omega}_1, \tilde{\omega}_2, \ldots) \in \cap B_n = \emptyset$, which is a contradiction. This proves that P is countably additive on $\mathcal{F}$.

Now, since P is a probability measure on $\mathcal{F}$ by Carathédory's extension Theorem 4.4.1, P has unique extension to $\sigma(\mathcal{F}) = \otimes_{n=1}^{\infty} \mathcal{A}_n$ and is a probability measure.

This completes the proof. ■

Then the following corollary is easy to establish.

Corollary 10.4.1. *Let* $(\Omega_j, \mathcal{A}_j, P_j)$, $j \ge 1$ *be probability spaces. Then there is a unique probability measure* P *on* $\otimes_{j=1}^{\infty} \mathcal{A}_j$ *such that for all* $n \ge 1$ *and* $A_j \in \mathcal{A}_j$, $1 \le j \le n$,

$$P\{\omega \in \Omega : \omega_1 \in A_1, \ldots, \omega_n \in A_n\} = \prod_{j=1}^{n} P_j(A_j), \tag{10.10}$$

◆

We denote the probability in (10.10) as $P = \prod_{j=1}^{\infty} P_j$ or $\otimes_{j=1}^{\infty} P_j$.

If all P_j are equal to P say, then their product probability is written as P^{∞}. If there is no scope for confusion, we may write it also as P.

Remark 10.4.1. (**Probability measure on arbitrary products**)

(a) Corollary 10.4.1 can be extended to arbitrary products of probability spaces $(\Omega_t, \mathcal{A}_t, P_t), t \in I$. See for example Section 38 of Halmos [1950] or Chapter V of Neveu [1965].

(b) **Kolmogorov's extension theorem** starts with given probability measures $\{_n\}$ on all finite product spaces, which are not necessarily product measures or have arisen from transition probability measures, but must satisfy an obvious consistency condition (for example, $P_n(B^n) = P_{n+1}(B^n \times \Omega_{n+1})$), and the spaces must have suitably "nice" toplological properties. Then, there there is a unique extension to the infinite product. See Theorem 11.3.1 for this result when I is countable and $\Omega_t = \mathbb{R}$ for all t. See Billingsley [1995], page 508 for a proof when I is arbitrary and $\Omega_t = \mathbb{R}$ for all t. ●

Exercise 10.4.3 (**The coin tossing space**) Let for all $j \geq 1$, $\Omega_j = \{H, T\}, \mathcal{A}_j = \mathcal{P}(\Omega)$, with the probability $P_j\{H\} = p, P_j\{T\} = 1 - p$, $0 < p < 1$. [Note that $p = 1$ and $p = 0$ are uninteresting cases]. Consider $\Omega = \prod_{j=1}^{\infty} \Omega_j$, $\mathcal{A} = \otimes_{j=1}^{\infty} \mathcal{A}_j$. Note that $\mathcal{A}$ is uncountable. The product probability P^∞ is called the **coin tossing probability**, and $(\Omega, \mathcal{A}, P^\infty)$ is called the **coin tossing space**. It is called **symmetric** or **fair** if $p = 1/2$. Show that

(a) $\mathcal{A} \neq \mathcal{P}(\Omega)$.

(b) every sequence of H and T is an element of $\mathcal{A}$.

(c) σ-field generated by the sequences in (b) is strictly smaller than $\mathcal{A}$.

(d)

$$\begin{aligned} &P^\infty(\omega) = 0 \quad \text{for all} \quad \omega \in \Omega, \\ &P^\infty(A) = 0 \quad \text{if, for every} \quad \omega \in A, \text{ only finitely many } \omega_i = H. \end{aligned}$$

10.5 Exercises

Exercise 10.5.1 Show that if $\mathcal{A}_i = \mathcal{A}$, $i \geq 1$ are countably generated σ-fields, then so is $\bigotimes_{i=1}^{\infty} \mathcal{A}_i$.

Exercise 10.5.2 Formulate and then prove a d-factor version of Fubini's Theorem 10.3.1.

Exercise 10.5.3 Give a proof of Theorem 10.3.2.

Exercise 10.5.4 Let Ω_i, $i = 1, 2$ be the set of positive integers, with the power σ-fields and counting measures. Define the function f as

$$f(i,j) = \begin{cases} n \ \text{ if } \ i = j = n, \\ -n \ \text{ if } \ i = n, j = n+1, \\ 0 \ \text{ otherwise.} \end{cases}$$

Compute the two iterated integrals as defined in (10.5) and (10.6) and verify that they are not equal. Why does Fubini's theorem fail?

Exercise 10.5.5 Let $\Omega_i := \mathbb{R}$, $i = 1, 2$, with the Borel σ-field. Let μ_1 and μ_2 be the Lebesgue and the counting measures. Compute the two iterated integrals for $\mathbf{1}_A$ where $A = \{(\omega_1, \omega_2) : \omega_1 = \omega_2\}$. Show that Theorem 10.3.2 and Fubini's theorem fail.

Exercise 10.5.6 Using Fubini's theorem and change of variable, show that (λ is the Lebesgue measure),

$$\int_{\mathbb{R}} e^{-x^2} \lambda(dx) = \sqrt{\pi}\,.$$

Exercise 10.5.7 Show that (λ is the Lebesgue measure),

$$\lim_{t\to\infty} \int_{(0,\ t)} \frac{\sin x}{x} \lambda(dx) = \frac{\pi}{2}\,.$$

Hint: Verify that

$$\int_{(0,\ t)} e^{-ux} \sin x \lambda(dx) = \frac{1}{1+u^2}\left[1 - e^{-ut}(u \sin t + \cos t)\right], \ x > 0\,.$$

Then use Fubini's theorem after noting that $\dfrac{1}{x} = \int_{(0,\infty)} e^{-tx} \lambda(dt)$, $x > 0$.

Exercise 10.5.8 Let F, G be bounded distribution functions on $[a,\ b]$. Show that

$$\int_{(a,\ b]} G(x)F(dx) + \int_{(a,\ b]} F(x)G(dx) = \sum_x (F(x) - F(x-))(G(x) - G(x-)) + F(b)G(b) - F(a)G(a)\,.$$

In particular, if F and G have no common discontinuities, then the first term of the right side in the above relation vanishes.

Exercise 10.5.9 Let F be a probability distribution function. Show that (λ is the Lebesgue measure),

$$\int_{\mathbb{R}} [F(x+c) - F(x)]\lambda(dx) = c, \text{ for any } c \in \mathbb{R}.$$

Exercise 10.5.10 Let $(\Omega, \mathcal{A}, P^\infty)$ be the fair coin tossing space as described in Exercise 10.4.3. Define $X : \Omega \to \mathbb{R}$ as $X(\omega) = \sum_{i=1}^{\infty} \frac{\mathbf{1}_{\{\omega_i = H\}}}{2^i}$.
(a) Show that X is a measurable function. Calculate $\int X(\omega)P^\infty(d\omega)$ and $\int X^2(\omega)P^\infty(d\omega)$.
(c) Find the induced (push forward) probability measure P_X of X. Hint. First find $P_X\{X \leq d\}$ for diadic rationals d between 0 and 1.
(d) Compute $\int X^n(\omega)P^\infty(d\omega)$ for $n \geq 3$. Hint: Use change of variable.

Exercise 10.5.11 (An inequality of Hardy) Let f be a non-negative continuous function on $[0, \infty)$. Let $F(t) := \frac{1}{t}\int_{(0,\, t)} f(x)\lambda(dx), \quad t > 0.$ Show that for every $1 < p < \infty$,
(a) $||F||_p \leq \frac{p}{p-1}||f||_p$,
(b) if $||f||_p < \infty$, then equality holds in (a) if and only if $f \equiv 0$. Hint: Use integration by parts and then Hölder's inequality.

Exercise 10.5.12 Let $f, g : (\mathbb{R}^d, \mathcal{B}(\mathbb{R}^d), \lambda_d) \to (\mathbb{C}, \mathcal{B}(\mathbb{C}))$. Define

$$(f * g)(x) = \int_{\mathbb{R}} f(x-y)g(y)\lambda_d(dy),$$

if the integral is finite. $f * g$ is called the L^1-**convolution** of f and g.
(a) Check that the convolution $*$ is defined in the following cases:
(i) f and g are both in $L^2(\mathbb{R}^d, \mathcal{B}(\mathbb{R}^d), \lambda)$
(ii) one of the functions among f and g is bounded, and the other belongs to $L^1(\mathbb{R}^d, \mathcal{B}(\mathbb{R}^d), \lambda)$
(iii) f and g are both in $L^1(\mathbb{R}^d, \mathcal{B}(\mathbb{R}^d), \lambda)$. Hint: Use Fubini's theorem 10.3.1 and the translation invariance of the Lebesgue measure λ.
(b) Show that in any of the above cases, $f * g = g * f$.
(c) Show that for f and g in $L^1(\mathbb{R}^d, \lambda)$, $||f * g||_1 \leq ||f||_1 \cdot ||g||_1$.

Chapter 11

Random variables and vectors

In the next few chapters, our focus will be on probability measures. We begin by familiarising ourselves with the language of probability, such as random experiments, events, random variables and vectors (measurable functions on a probability space), and probability distributions.

11.1 Random variables and induced measures

Definition 11.1.1. (**Random experiment, events, outcomes**) Any probability space $(\Omega, \mathcal{A}, P)$ is considered as a *random experiment*. The elements of Ω are called *outcomes* and the elements of $\mathcal{A}$ are *events*. ◇

Example 11.1.1. Consider the two-fold product in the coin-tossing Example 10.4.3. Then, $\Omega = \{HH, HT, TH, TT\}$. It is equipped with the power σ-field, and

$$P(HH) = p^2,\ P(TT) = (1-p)^2,\ P(TH) = (1-p)p,\ P(HT) = p(1-p).$$

The random experiment here is tossing of a coin two times. The four **possible results** (outcomes) are listed as the elements of Ω. The experiment is "random" as its result is not certain but is governed by the above probabilities. This probability assignment of course could be different from the one given above that used the product measure. Once the experiment is performed (the two tosses are complete), we are able to determine if a specified event has occurred or not. ▲

A. Bose, *Elements of Measure and Probability*, Texts and Readings in Mathematics 88,
https://doi.org/10.1007/978-981-95-2758-8_11

Definition 11.1.2. (**Random variable**) A *random variable* (in short rv) is a real-valued measurable function on a probability space. ◇

For instance, in Example 11.1.1, the number of heads, X, is a rv:

$$X(HH) = 2, X(TT) = 0, X(HT) = X(TH) = 1\,.$$

Definition 11.1.3. (**Distribution of a rv**) If X is a rv defined on some probability space $(\Omega, \mathcal{A}, P)$ then the probability measure P_X on the Borel σ-field is the push forward measure defined by

$$P_X(B) := P\{\omega : X(\omega) \in B\},\ \ B \in \mathcal{B}(\mathbb{R}),$$

is called the **probability distribution** or in short, the *distribution* of X. Often we suppress ω and write $P\{X \in B\}$. The **cumulative distribution function** (or **cdf**) of X is defined by

$$\begin{aligned} F_X(x) &:= P_X(-\infty,\ x],\ x \in \mathbb{R} \\ &= P\{\omega : X(\omega) \leq x\} = P\{X \leq x\}. \end{aligned}$$

We often say **distribution function** or **df**, F. ◇

Exercise 11.1.1 If X is a rv, then $F_X(-\infty) = 0$ and $F_X(\infty) = 1$. Conversely, let F on $\mathbb{R}$ be a distribution function (right continuous, non-decreasing) such that $F(-\infty) = 0, F(\infty) = 1$. Show that we can construct a probability space and a rv X such that its cdf is F.

Remark 11.1.1. (a) A rv X is **discrete** if there is $\{x_n\}$ such that $\sum_{n=1}^{\infty} P\{X = x_n\} = 1$. The number $p_n = P\{X = x_n\}$ is called the **mass** of X at x_n and $\{p_n\}$ is collectively called the **probability mass function** or **pmf** of X. The pmf which puts the entire mass 1 at a fixed x is denoted by δ_x, and is referred to as the **point mass at** x.

(b) A rv X or its cdf F is **absolutely continuous** if there is a non-negative Borel measurable f such that (λ is the Lebesgue measure)

$$P\{X \leq x\} = F(x) = \int_{(-\infty,\ x]} f(t)\lambda(dt),\ \text{for all } x \in \mathbb{R}\,.$$

The function f is called a **probability density function** (**pdf**) or, simply a density function of X or F_X. A pdf is not unique. However, if f and g are two density functions of X, then $f = g$ a.e. λ.

(c) A rv X is **continuous** if its distribution function F is continuous. Equivalently, $P\{X = x\} = 0$ for all $x \in \mathbb{R}$. If X is absolutely continuous, then it is continuous. The converse is false: recall the Cantor distribution function F from Exercise 5.3.17. Using Exercise 11.1.1, create an rv X with cdf F. Then X is continuous, but is not absolutely continuous. ●

Example 11.1.2. (a) X is said to be a **Bernoulli** rv with parameter p if $P\{X = 1\} = p,\ P\{X = 0\} = q,\ p + q = 1$. We write $X \sim Ber(p)$.

(b) X is said to be a **gamma** rv if it has the pdf:

$$f(x) = \frac{\lambda^p}{\Gamma p} e^{-\lambda x} x^{p-1}, \ x > 0. \tag{11.1}$$

Here $p > 0, \lambda > 0$ are parameters. We write $X \sim Gamma(p, \lambda)$. Its cdf is as in Example 5.1.2 If $p = 1$, X is called an **exponential** rv and the distribution is known as the exponential distribution. We write $X \sim Exp(\lambda)$. Its cdf when $\lambda = 1$ is the same as given in Example 5.1.3

(c) X is said be a **normal** or a **Gaussian** rv with parameters (m, σ^2), $\sigma^2 > 0$ if its density function is given by

$$f(x) = \frac{1}{\sqrt{2\pi}\sigma} e^{-\dfrac{(x-m)^2}{2\sigma^2}}, \ x \in \mathbb{R}.$$

From Exercise 10.5.6, f is a pdf. This distribution is known as the normal or the Gaussian distribution. We write $X \sim N(m, \sigma^2)$. If $m = 0$ and $\sigma^2 = 1$, then the distribution is called **standard Gaussian** or **standard normal**. Its cdf is as given in Example 5.1.1. In case $\sigma^2 = 0$, X is still called a Gaussian variable, and has a point mass at m.

(d) A rv X is said to have a **Poisson distribution** with parameter λ, $0 < \lambda < \infty$ (we write $X \sim Poi(\lambda)$) if its pmf is

$$P\{X = k\} = e^{-\lambda} \frac{\lambda^k}{k!}, \ k = 0, 1, 2, \ldots .$$

(e) X is said be a **uniform** rv on $(a,\ b)$ (we write $X \sim U(a,\ b)$) if its density function is given by

$$f(x) = \frac{1}{b-a}, \ a < x < b.$$

(f) X is said to be a (standard) Cauchy rv if it has the pdf

$$f(x) = \frac{1}{\pi}\frac{1}{1+x^2}, \quad x \in \mathbb{R}. \tag{11.2}$$ ▲

Exercise 11.1.2 If $X \sim N(m, \sigma^2)$, show that $aX+b \sim N(am+b, a^2\sigma^2)$ for any $a, b \in \mathbb{R}$.

11.2 Random vectors

The concepts of real-valued random variables and of cdf on $\mathbb{R}$ can be extended to vectors and cdf on $\mathbb{R}^d$ in the natural way.

Definition 11.2.1. (**Random vector, cdf**) Any measurable function $X : (\Omega, \mathcal{A}, P) \to (\mathbb{R}^d, \mathcal{B}(\mathbb{R}^d))$ is called a *random vector* of dimension d. Its probability distribution and cdf are defined in the natural way:

$$\begin{aligned} P_X(B) &= P\{\omega : X(\omega) \in B\}, \ B \in \mathcal{B}(\mathbb{R}^d), \\ F_X(x) &= P_X(-\infty, \ x], \ x \in \mathbb{R}^d \\ &= P\{\omega : X(\omega) \le x\} \ \ \text{(co-ordinate wise)} \\ &= P\{X \le x\} \quad \text{(common way of writing)}. \end{aligned}$$

Random variables and vectors are denoted by $X, Y, \ldots$. For any d-dimensional random vector X, write $X = (X_1, \ldots, X_d)'$, where X_i are real-valued and are called the *co-ordinate variables.* Random variables or vectors, X, Y, possibly defined on different probability spaces, are said to be equal in distribution if $F_X \equiv F_Y$, and we write $X \stackrel{D}{=} Y$. ◇

Exercise 11.2.1 Let $X = (X_1, \ldots, X_d)' : (\Omega, \mathcal{A}, P) \to \mathbb{R}^d$ be a function and p_i be the projection mappings (see Definition 6.1.3). Show that X is a random vector if and only if $X_i = p_i \circ X$ is an rv for each i.

Definition 11.2.2. (**Marginal and joint distribution**) Suppose that $X = (X_1, \ldots, X_d)'$ is a random vector. Measure P_X is called the *joint probability distribution* of $(X_1, \ldots, X_d)'$. Measures $\{P_{X_i}\}$, $1 \le i \le d$ are called the *marginal distributions.* ◇

A random vector $X = (X_1, \ldots, X_d)'$ is **continuous** if its cdf $F_X(\cdot)$ is continuous. It is **absolutely continuous** if there is a non-negative

Borel measurable function $f : \mathbb{R}^d \to [0, \infty)$ (called the pdf of X, and which is unique a.e. λ_d) such that for all $x_1, \ldots x_d \in \mathbb{R}$,

$$P\{X_i \leq x_i,\ 1 \leq i \leq d\} = \int_{\prod_{i=1}^{d}(-\infty,\ x_i]} f(y_1, \ldots, y_d)\lambda_d(dy_1, \ldots, dy_d)\,.$$

X is **discrete** if, there exists $\{x_{ij}\},\ 1 \leq i \leq d,\ j \geq 1$ such that

$$\sum_{j_1, \ldots j_d = 1}^{\infty} P\{X_1 = x_{1j_1}, \ldots, X_d = x_{dj_d}\} = 1\,.$$

Exercise 11.2.2 Show that, $X = (X_1, \ldots, X_d)$ is absolutely continuous implies that each X_i is also so. Show that the converse is false.

Exercise 11.2.3 Show that $X = (X_1, \ldots, X_d)'$ is discrete if and only if each X_i is discrete.

Exercise 11.2.4 Show by an example that if a random vector is absolutely continuous, its density is not determined by the marginal densities.

Exercise 11.2.5 Consider $(\Omega := \{(x, y) : x^2 + y^2 \leq 1\}, \mathcal{B}(\Omega), P)$, where $P \equiv \pi^{-1}\lambda_2$ and λ_2 is the restriction of the Lebesgue measure to Ω. Consider the random vector $X(x, y) = (x, y)'$ on Ω. Denote the coordinate rvs by X_1 and X_2. Find the pdf of $R = X_1^2 + X_2^2$.

Definition 11.2.3. (Multivariate Gaussian) Let $X = (X_1, \ldots, X_d)'$ be a random vector. It is said to be *multivariate normal/Gaussian* if for any real vector $t = (t_1, \ldots, t_d)'$, $t'X$ is a Gaussian rv. ◊

Exercise 11.2.6 (a) Let X be a Gaussian rv, and $a_1, b_1, a_2, b_2 \in \mathbb{R}$. Show that $(a_1 X + b_1, a_2 X + b_2)$ is multivariate Gaussian.

(b) Let $X = (X_1, \ldots, X_d)'$ be multivariate Gaussian. Let $\{a_i\}$ be d-dimensional real vectors and $\{b_i\}$ be real numbers. Let $Y = (Y_1, \ldots, Y_n)'$ where each $Y_i = a_i' X + b_i$. Show that Y is multivariate Gaussian.

11.3 Kolmogorov's extension theorem

The *Kolmogorov's extension theorem* guarantees the existence of collections of rvs with specified joint distributions. We shall state this result only for a countable collection. A proof can be based on arguments given in the proof of Theorem 10.4.1 and the details are left to the reader.

Theorem 11.3.1. ***(Kolmogorov's extension theorem)*** *Let F_n be a cdf on $\mathbb{R}^n$, $n \geq 1$. If $\{F_n\}$ satisfy the consistency condition so that,*

$$F_n(x_1, \ldots, x_n) = F_{n+m}(x_1, \ldots, x_n, \underbrace{\infty, \ldots, \infty}_{m \text{ times}}), \text{ for every } n, m\,.$$

Then we can construct a probability space $(\mathbb{R}^{\mathrm{N}}, \mathcal{B}(\mathbb{R}^{\mathrm{N}}), P)$ such that if

$$X_i(\omega_1, \omega_2, \ldots) = \omega_i, \ (\omega_1, \omega_2, \ldots) \in \mathbb{R}^{\mathrm{N}}, \ i \geq 1\,,$$

then for every n, $(X_1, \ldots, X_n)$ has the distribution F_n. ◆

Remark 11.3.1. (a) Theorem 11.3.1 guarantees existence of rvs with given consistent joint distributions. In Theorem 10.4.1, transition probabilities yield measures $\{P_n\}$ that are automatically consistent.

(b) See Billingsley [1995], page 508 for statement and proof of this theorem when we have cdfs $\{F_\alpha\}, \alpha \in I$ for an arbitrary index set I. ●

Exercise 11.3.1 Let $\{F_n\}$ be probability distribution functions on $\mathbb{R}$. Show that there is a probability space and a random vector X whose co-ordinates $\{X_n\}$ have distributions $\{F_n\}$.

11.4 Exercises

Exercise 11.4.1 If F is a continuous cdf, evaluate $\int_{\mathbb{R}} F(x)F(dx) = 1/2$.

Exercise 11.4.2 Let $X : (\Omega, \mathcal{A}, P) \to (\mathbb{R}, \mathcal{B}(\mathbb{R}))$ with cdf F. Recall F^{-1} and its properties from Exercise 5.3.19. Show that

(a) $P(F(X) \leq t) \leq t$ for all $0 \leq t \leq 1$.

(b) $P(F(X) \leq t) < t$ if and only if t is not in closure of range of F.

(c) If F is continuous, then $Y := F(X) \sim U(0,\ 1)$.

(d) $P(F^{-1}(F(X)) \neq X) = 0$ if X is distributed as F.

(e) If F is continuous and Y is such that $F(Y) \sim U(0,\ 1)$, then $Y \sim F$.

Exercise 11.4.3 (**Mills ratio**) Let $X \sim N(0,\ 1)$ with density $\varphi(\cdot)$. Show that

(a) $\dfrac{x}{1+x^2}\varphi(x) < P(X > x) < \dfrac{1}{x}\varphi(x)$ for all $x > 0$.

(b) $P(|X| > x) \sim \dfrac{2}{x}\varphi(x)$ as $x \to \infty$.

Chapter 12

Moments and cumulants

This chapter is on moments, cumulants, and their generating functions. We shall explore the relation between the moments and cumulants using the Möbius function on the set of all partitions. One consequence of this will be Isserlis' formula, discussed in Section 18.7, that gives a method to compute mixed moments of Gaussian rvs. Another consequence will be a proof of the central limit theorem, given in Section 19.1.1, that avoids the use characteristic functions.

12.1 Expectation and moments

Definition 12.1.1. (Restatement and extension of Definition 8.1.1) Let $X : (\Omega, \mathcal{A}, P) \to (\mathbb{R}, \mathcal{B}(\mathbb{R}))$. Its **expectation** or **mean** is defined as

$$m := \mathrm{E}(X) := \int_\Omega X(\omega)P(d\omega), \quad \text{whenever the integral exists.}$$

If $\mathrm{E}(X)$ is finite then the **variance** of X is defined as

$$\sigma_X^2 := \mathrm{V}(X) := \int_\Omega [X(\omega) - \mathrm{E}(X)]^2 P(d\omega).$$

For any non-negative integer n, the n**-th moment** of X is defined as

$$\mathrm{E}(X^n) = \int_\Omega X^n(\omega)P(d\omega), \text{ if the integral exists}.$$

A. Bose, *Elements of Measure and Probability*, Texts and Readings in Mathematics 88,
https://doi.org/10.1007/978-981-95-2758-8_12

For random vector $X = (X_1, \ldots, X_d)'$ and non-negative integers $\{k_i\}$ $\mathrm{E}(X_1^{k_1} X_2^{k_2} \cdots X_d^{k_d})$, provided they exist, are the moments of X. ◊

Note that $\mathrm{V}(X)$ may equal ∞ and $\mathrm{E}(X^n)$ may equal $\pm\infty$.

Exercise 12.1.1 Suppose X is a rv with mean $\mathrm{E}(X)$ and variance σ_X^2. If $\sigma_X^2 = 0$, then show that $P\{X = \mathrm{E}(X)\} = 1$.

Using the change of variable formula of Exercise 7.3.3, the mean, moments and variance of X can be calculated using P_X as

$$\begin{aligned}
\mathrm{E}(X) &= \int_{\mathbb{R}} x P_X(dx), \text{ whenever } \mathrm{E}(X) \text{ exists,} \\
\mathrm{E}(X^n) &= \int_{\mathbb{R}} x^n P_X(dx) \text{ whenever } \mathrm{E}(X^n) \text{ exists,} \\
\mathrm{V}(X) &= \int_{\mathbb{R}} (x - m)^2 P_X(dx) \text{ whenever } m \text{ is finite.}
\end{aligned}$$

Exercise 12.1.2 For rvs X, Y, if $\mathrm{E}(X)$ and $\mathrm{E}(Y)$ are finite, then show that $\mathrm{E}(X + Y) = \mathrm{E}(X) + \mathrm{E}(Y)$. Compare with Theorem 7.2.4.

Exercise 12.1.3 Suppose X is a rv such that $\mathrm{E}(X)$ is finite. Show that

(a) $\mathrm{V}(aX + b) = a^2\,\mathrm{V}(X)$ for all real numbers a and b.

(b) $\mathrm{V}(X) = \mathrm{E}(X^2) - [\mathrm{E}(X)]^2$. Caution: Both sides may equal ∞.

Exercise 12.1.4 (a) Find $\mathrm{E}(X^n)$ when $X \sim Ber(p)$ or $X \sim Exp(\lambda)$.

(b) Show that if $X \sim N(m, \sigma^2)$, then $\mathrm{E}(X) = m$ and $\mathrm{V}(X) = \sigma^2$.

(c) If $X \sim N(0,\ 1)$, show that its moments are given by

$$\begin{aligned}
m_{2k}(X) &:= \mathrm{E}(X^{2k}) = \frac{(2k)!}{2^k k!}, \\
m_{2k+1}(X) &:= \mathrm{E}(X^{2k+1}) = 0, \text{ for } k = 0, 1, \ldots.
\end{aligned} \tag{12.1}$$

Definition 12.1.2. (Covariance, correlation and dispersion) Let $(X_1, \ldots, X_d)'$ be a random vector where each co-ordinate has finite positive variance. Then the *covariance* Cov, and the *correlation* ρ, between X_i and X_j, and the **dispersion matrix** of X are defined as

$$\begin{aligned}
\mathrm{Cov}(X_i, X_j) &= \mathrm{E}\left[(X_i - \mathrm{E}(X_i))(X_j - \mathrm{E}(X_j))\right], \ 1 \le i, j \le d, \\
\rho(X_i, X_j) &= \frac{\mathrm{Cov}(X_i, Y_j)}{\sqrt{\mathrm{V}(X_i)\,\mathrm{V}(X_j)}}, \ 1 \le i, j \le d, \\
\Sigma(X) &:= ((\mathrm{Cov}(X_i, X_j))_{1 \le i,j \le d} \,.
\end{aligned}$$ ◇

Exercise 12.1.5 Let $\{X_i\}$ be rvs with finite variances.

(a) Show that the covariances are finite.

(b) Show that $-1 \le \rho(X_i, X_j) \le 1$, When do any of the equalities hold?

(c) Show that $\mathrm{Cov}(X_i, X_i) = \mathrm{V}(X_i)$ and $\Sigma(X)$ is non-negative definite.

(d) Show that

$$\mathrm{V}(X_1 + \cdots + X_d) = \sum_{i=1}^{d} \mathrm{V}(X_i) + 2 \sum_{1 \le i < j \le d} \mathrm{Cov}(X_i, X_j)\,.$$

12.2 Moment generating function

Definition 12.2.1. (**Moment generating function**) The *moment generating function* (mgf) of a rv is defined as

$$M_X(t) = \mathrm{E}\left(e^{tX}\right), \ t \in \mathbb{R}\,.$$

The *mgf* of $X = (X_1, \ldots, X_d)'$ is defined as

$$M_X(t) := M_{X_1,\ldots,X_d}(t_1, \ldots, t_d) := \mathrm{E}\left(e^{t'X}\right), \ t = (t_1, \ldots, t_d)' \in \mathbb{R}^d.$$ ◇

Exercise 12.2.1 If X is a real-valued rv then check that $M_X(0) = 1$ and the values of t for which $M_X(t)$ is finite, is always an interval that includes 0. Give an example where this interval contains only 0.

Exercise 12.2.2 For a real-valued rv X, show that for $a, b \in \mathbb{R}$,

$$M_{aX+b}(t) = M_X(at)e^{bt}, \ t \in \mathbb{R} \text{ (both sides may equal } \infty)\,.$$

Exercise 12.2.3 Let X be a rv such that $\{t : M_X(t) < \infty\}$ includes a non-trivial interval around zero. Show that all moments of X are finite.

Exercise 12.2.4 Suppose $X \sim N(0,\ 1)$. Show that its mgf is given by

$$M_X(t) = e^{t^2/2},\ t \in \mathbb{R}. \tag{12.2}$$

Exercise 12.2.5 Let $X = (X_1, \ldots, X_d)'$ be a random vector. Suppose that $M_X(t)$ is finite in a non-trivial interval around the origin in $\mathbb{R}^d$.

(a) Show that $\mathrm{E}(|X_1^{k_1} X_2^{k_2} \cdots X_d^{k_d}|) < \infty$ for all non-negative $k_1, \ldots, k_d$.

(b) Find a formula for computing the above moments using the derivatives of $M_X(t)$ at $t = 0$.

Exercise 12.2.6 (Moments do not determine distributions) Let

$$f_0(x) = (2\pi)^{-1/2} x^{-1} e^{-(\log x)^2/2},\ x > 0,$$
$$f_a(x) = f_0(x)[1 + a \sin(2\pi \log x)],\ 0 \le a \le 1\,.$$

$f_0(\cdot)$ is called the *log-normal density.* Show that

(i) For every $0 \le a \le 1$, $f_a(\cdot)$ is a pdf.

(ii) All the densities $f_a(\cdot)$, $0 \le a \le 1$ have identical moments given by

$$\int_{\mathbb{R}} x^n f_a(x) \lambda(dx) = e^{n^2/2} \ \text{for all}\ \ 0 \le a \le 1\,.$$

If mgfs of X and Y are finite and equal in $(-r, r)$, then $P_X \stackrel{D}{=} P_Y$. In Chapter 18 we shall discuss such uniqueness issues, and sufficient conditions for moments to identify a probability measure uniquely.

12.3 Cumulants

Definition 12.3.1. (Cumulant generating function) If $M_{X_1,\ldots,X_d}$ is finite in an open neighborhood N of the origin, then the joint *cumulant generating function* (cgf), is defined as

$$C_{X_1,\ldots,Xd}(t_1, \ldots, t_d) := \log M_{X_1,\ldots,X_d}(t_1, \ldots, t_d),\ t \in N\,. \tag{12.3}$$

It also has a power series expansion

$$C_{X_1,\ldots,X_d}(t_1, \ldots, t_d) = \sum_{k_1,\ldots,k_d=0}^{\infty} \frac{t_1^{k_1} \cdots t_d^{k_d}}{k_1! \cdots k_d!} c_{k_1,\ldots,k_d}(X_1, \ldots, X_d),\ t \in N.$$

Numbers $c_{k_1,\ldots,k_d}(X_1,\ldots,X_d)$ are called cumulants of $\{X_i : 1 \le i \le d\}$. If $k_j \neq 0$ for at least two indices j, then $c_{k_1,\ldots,k_d}(X_1,\ldots,X_d)$ is called a *mixed cumulant* of $\{X_i : 1 \le i \le d\}$. ◇

It is clear from (12.3) that the moments and cumulants must be in $1-1$ correspondence. We shall soon discover the nature of this relation. For now, it is apparent that $c_{k_1,\ldots,k_d}$ involves only those moments $\mathrm{E}(X_1^{t_1}\cdots X_d^{t_d})$ where $t_j \le k_j$ for all j.

Example 12.3.1 Suppose $X \sim N(\mu, \sigma^2)$. Then

$$M_X(t) = \exp\left(t\mu + \frac{1}{2}t^2\sigma^2\right) \text{ and } C_X(t) = t\mu + \frac{1}{2}t^2\sigma^2,\ t \in \mathbb{R}.$$

Hence

$$c_1(X) = \mu,\quad c_2(X) = \sigma^2 \text{ and } c_n(X) = 0,\ n \ge 3.$$

A theorem of Marcinkiewicz [1938], says that there exists no other probability distribution which has only finitely many non-vanishing cumulants. For a proof, see page 213 of Lukacs [1970]. ▲

Example 12.3.2 If $X \sim Poi(\lambda)$ then, for all $t \in \mathbb{R}$,

$$M_X(t) = \exp\left(\lambda(e^t - 1)\right),$$

$$C_X(t) = \lambda(e^t - 1) = \lambda\sum_{j=1}^{\infty}\frac{t^j}{j!},$$

$$c_j(X) = \lambda \text{ for all } j \ge 1. \tag{12.4}$$

Conversely, a rv that satisfies (12.4) is distributed as $Poi(\lambda)$. ▲

Exercise 12.3.1 Show that for any rv X, and any real constants a and b, $c_j(aX+b) = a^j c_j(X)$ for all $j \ge 2$.

We shall unearth a linear relation between the *multiplicative extension* of moments and cumulants via the *Möbius function* on the set of all partitions. This has interesting consequences. One of them is Isserlis' formula, popularly known as Wick's formula, which gives a method to compute mixed moments of Gaussian rvs. Another will be a moment/cumulant based proof of the central limit theorem in Chapter 19.

12.4 Moments from cumulants

Going from cumulants to moments is relatively easy. First consider a single rv X, with mgf $M_X(\cdot)$ and cgf $C_X(\cdot)$ which are finite in a neighborhood N of the origin. Let us consider the relation between the first few moments and cumulants directly. By direct calculations, the first three moments and cumulants satisfy:

$$\begin{aligned} &m_1 = c_1,\ m_2 = c_2 + c_1^2, \\ &m_3 = c_3 + 3c_1c_2 + c_1^3. \end{aligned} \tag{12.5}$$

Observe that $\{1,2,3\}$ has five partitions, $\{\{1\},\{2\},\{3\}\}$, $\{\{1,2\},\{3\}\}$, $\{\{1,3\},\{2\}\}$, $\{\{1\},\{2,3\}\}$ and $\{\{1,2,3\}\}$. Each of these five partitions contributes a term, in a *multiplicative* way, in the formula (12.5).

We now show how to extend this idea to moments and cumulants of all orders. Define

$$\mathcal{P}_n := \text{Set of all partitions of } \{1,\ldots,n\},\ \mathcal{P} := \cup_{n=1}^{\infty}\mathcal{P}_n.$$

Any $\pi \in \mathcal{P}_n$ is written as $\{V_1,\ldots,V_k\}$ where $V_1,\ldots,V_k$ are the *blocks* of π. Let $|\pi|$ denote the number of blocks of π. For any block V let $\#V$ or $|V|$ be the number of elements in V. For instance, $\pi = \{\{1,3\},\{2\}\}$ has two blocks, $V_1 = \{1,2\}$ and $V_2 = \{3\}$, $|\pi| = 2$, $|V_1| = 2$ and $\#V_2 = 1$. π is a **pair-partition** if $|V_i| = 2$ for all i. For $\pi, \sigma \in \mathcal{P}_n$, we write $\pi \leq \sigma$ if each block of π is contained in some block of σ. This is known as the *reverse refinement partial order.* Partitions $\{\{1\},\ldots,\{n\}\}$ and $\{1,\ldots,n\}$ shall be denoted by $\mathbf{0}_n$ and $\mathbf{1}_n$ respectively. They are the smallest and the largest elements of $\mathcal{P}_n$.

Definition 12.4.1. (**Multiplicative extension**) Let $\{a_n : n \geq 1\}$ be a sequence of real numbers. For any $\pi \in \mathcal{P}$, let

$$a_\pi := \prod_{i=1}^{k} a_{|V_i|}, \quad \text{whenever} \quad \pi = \{V_1,\ldots,V_k\}.$$

Then $\{a_\pi : \pi \in \mathcal{P}\}$ is the *multiplicative extension* of $\{a_n : n \geq 1\}$. ◇

We are now ready to state and prove the formula that gives $\{m_\pi\}$ for every $\pi \in \mathcal{P}$ as a linear function of $\{c_\sigma,\ \sigma \leq \pi\}$.

Lemma 12.4.1. Let X be a rv with a finite mgf in a neighbourhood N of 0, and with moment and cumulant sequences $\{m_n\}$ and $\{c_n\}$. Then,

$$m_\pi = \sum_{\sigma \in \mathcal{P}_n:\ \sigma \le \pi} c_\sigma \quad \text{for all} \quad \pi \in \mathcal{P}_n,\ n \ge 1, \tag{12.6}$$

$$m_n = \sum_{\sigma \in \mathcal{P}_n} c_\sigma,\ n \ge 1. \tag{12.7}$$

♦

Proof. Let M and C respectively be the mgf and cgf of X. Note that

$$M(t) = \exp\left(C(t)\right),\ t \in N.$$

Using an induction argument, it is easy to see that the derivatives of M and C for $t \in N$ are related as:

$$\begin{aligned}
M^{(1)}(t) &= C^{(1)}(t) \exp\left(C(t)\right), \\
M^{(2)}(t) &= C^{(2)}(t) \exp\left(C(t)\right) + \left(C^{(1)}(t)\right)^2 \exp\left(C(t)\right), \\
&= \exp\left(C(t)\right) \left[C^{(2)}(t) + \left(C^{(1)}(t)\right)^2\right],
\end{aligned}$$

$$M^{(n)}(t) = \exp\left(C(t)\right) \sum_{\pi = \{V_1, \ldots, V_k\} \in \mathcal{P}_n} C^{(|V_1|)}(t) \cdots C^{(|V_k|)}(t). \tag{12.8}$$

(12.8) with $t = 0$ gives (12.7). The proof of (12.6) is an exercise. ■

Example 12.4.1 Let $X \sim N(0, 1)$. Recall from Exercise 12.3.1 that

$$c_2(X) = 1, \text{ and } c_j(X) = 0 \quad \text{for all} \quad j \neq 2.$$

Hence

$$c_\sigma = \begin{cases} 1 & \text{if } \sigma \text{ is a pair-partition,} \\ 0 & \text{otherwise.} \end{cases}$$

As a consequence,

$$\begin{aligned}
m_n &= \text{the number of pair-partitions in } \mathcal{P}_n \\
&= \begin{cases} \frac{(2k)!}{2^k k!} & \text{if } n = 2k \text{ is even,} \\ 0 & \text{if } n \text{ is odd.} \end{cases}
\end{aligned} \tag{12.9}$$

Example 12.4.2 Let $X \sim Poi(\lambda)$. From Exercise 12.3.2, $c_j = \lambda$ for all $j = 1, 2, \ldots$. Hence,

$$c_\sigma = \lambda^{|\sigma|} \quad \text{for any partition} \quad \sigma.$$

As a consequence,

$$m_n = \sum_{j=1}^{n} \lambda^j S(n,j)\,,$$

where,

$$S(n,j) := \#\{\pi \in \mathcal{P}_n : |\pi| = j\}\,.$$

In particular, if $\lambda = 1$, then

$$m_n = \#\mathcal{P}_n = \sum_{j=1}^{n} S(n,j).$$

The numbers $S(n,j)$ are the *Stirling numbers of the second kind* and $\#\mathcal{P}_n = \sum_{j=1}^{n} S(n,j)$ are the *Bell numbers.* ▲

12.5 Cumulants from moments

Relation (12.6) turns out to be invertible, via the Möbius function.

12.5.1 Möbius function

Let P be a finite partially ordered set (poset) with partial order $\leq$. For any $\pi, \tau \in P$, consider the sets

$$V_1 := \{\sigma : \pi \geq \sigma \text{ and } \tau \geq \sigma\},\ V_2 := \{\sigma : \pi \leq \sigma \text{ and } \tau \leq \sigma\}\,.$$

If there are unique largest and smallest elements respectively of V_1 and V_2, then we write them as $\pi \wedge \tau$ and $\pi \vee \tau$ respectively.

Definition 12.5.1. (**Lattice**) A poset $(P,\ \leq)$ is called a *lattice* if $\pi \wedge \tau$ and $\pi \vee \tau$ exist for all π, τ in P. ◇

Clearly for every n, $\mathcal{P}_n$ with the reverse refinement partial order, is a lattice.

For $\pi \leq \sigma$, $(\pi,\ \sigma)$ denotes the (ordered) pair π, σ. The *interval* $[\pi,\ \sigma]$ is defined as:

$$[\pi,\ \sigma] := \{\tau : \pi \leq \tau \leq \sigma\}.$$

The set of all intervals of P will be written as

$$P^{(2)} := \{[\pi,\ \sigma] : \pi, \sigma \in P, \pi \leq \sigma\}\,.$$

For any two complex valued functions $F, G : P^{(2)} \to \mathbb{C}$, their *convolution* $F * G : P^{(2)} \to \mathbb{C}$ is defined by:

$$F * G\,[\pi,\ \sigma] := \sum_{\substack{\rho \in P \\ \pi \le \rho \le \sigma}} F[\pi,\ \rho] G[\rho,\ \sigma].$$

F is called *invertible* if there is a (unique) G (the inverse of F), for which

$$F * G\,[\pi,\ \sigma] = G * F\,[\pi,\ \sigma] = \mathbf{1}_{\{\pi = \sigma\}},\ \forall \pi \le \sigma \in P\,.$$

Lemma 12.5.1. Suppose $(P,\ \le)$ is a finite lattice. Then the function F on $P^{(2)}$ is invertible if and only if $F[\pi,\ \pi] \ne 0$ for every $\pi \in P$. ♦

Proof. Let $n = \#P$. Then P can be enumerated (in many possible ways) as $E = \{\pi_1, \pi_2, \ldots, \pi_n\}$ where, for any $i < j$, either π_i, π_j are not ordered or $\pi_i \le \pi_j$. The proof of this is left as an exercise. We fix one such enumeration. Now, let $F_E = ((F_{ij}))$ where

$$F_{ij} = \begin{cases} F[\pi_i,\ \pi_j] & \text{if} \quad i \le j \text{ and } \pi_i \le \pi_j, \\ \\ 0 & \text{otherwise.} \end{cases} \tag{12.10}$$

Then F_E is an upper triangular matrix with $\{F[\pi_i, \pi_i]\}$ on the diagonal.

Now suppose $F[\pi,\ \pi] \ne 0$ for all π. Then the inverse of F_E, say G_E is upper triangular. Construct a function G on $P^{(2)}$ using the matrix G_E by reversing the process described above. Then the matrix relation $F_E G_E = G_E F_E = I$ is same as saying that G is the inverse of F.

For the converse, if F has an inverse G, then by definition

$$F * G\,[\pi,\ \pi] = F[\pi,\ \pi] G[\pi,\ \pi] = \mathbf{1}_{\{\pi = \pi\}} = 1, \text{ for all } \pi \in P\,.$$

Hence $F[\pi,\ \pi] \ne 0$ for all π. ■

Definition 12.5.2. (**Zeta and Möbius functions**) Suppose $(P,\ \le)$ is a finite lattice. The *Zeta function* ξ of P is defined by

$$\xi[\pi,\ \sigma] := 1 \text{ for all } [\pi,\ \sigma] \in P^{(2)}.$$

The *Möbius function* μ of P is the inverse of ξ. ◇

The Möbius function exists by Lemma 12.5.1. Moreover, it does not depend on the enumeration that we used in the proof of Lemma 12.5.1. As a consequence,

$$\mu * \xi\ [\pi,\ \sigma] = \xi * \mu\ [\pi,\ \sigma] = \mathbf{1}_{\{\pi=\sigma\}} \text{ for all } \pi \leq \sigma \in P.$$

12.5.2 Cumulants from moments

For the time being, let us continue to assume that the mgf is finite in an open neighbourhood of 0. Recall $\mathcal{P}_n$ with the reverse refinement partial order. It is a finite lattice and μ_n be its Möbius function. The moment-cumulant relation can be written in terms of the Möbius function as follows. Recall the linear relation (12.6),

$$m_\pi = \sum_{\sigma \in \mathcal{P}_n:\ \sigma \leq \pi} c_\sigma \quad \text{for all} \quad \pi \in \mathcal{P}_n. \tag{12.11}$$

This is a convolution relation: for each fixed n, identify $m_\pi, \pi \in \mathcal{P}_n$ with the function m on $\mathcal{P}_n^{(2)}$ which is given by $m[\mathbf{0}_n,\ \pi] = m_\pi$ and zero otherwise. We can define a function c via $c_\pi, \pi \in \mathcal{P}_n$. Then the above relation can be written as

$$c * \zeta = m\,. \tag{12.12}$$

Convolving both sides of (12.12) with μ_n on the right, we get

$$c_\pi = \sum_{\sigma \leq \pi} m_\sigma \mu_n[\sigma,\ \pi],\ \pi \in \mathcal{P}_n. \tag{12.13}$$

This, when specialized to $\pi = \mathbf{1}_n$, gives c_n in terms of m_k, $k \leq n$, and the Möbius function μ_n. Incidentally, μ_n is defined on $\mathcal{P}_n^{(2)}$ for each fixed n. We can define μ on $\cup_n \mathcal{P}_n^{(2)}$ in the obvious way.

So far, we have worked under the assumption that, the mgf is finite in a neighbourhood of 0. However, it is possible that the mgf does not satisfy this condition but all moments are finite, or even only up to the nth moment of X is finite. Then how do we define cumulants?

Its answer lies in the relation (12.13). If the nth moment is finite, then for $k \leq n$, this relation defines $c_k(X)$ in terms of $m_j(X),\ j \leq k$ in a unique way. Then by using the inverse, for any k, $m_k(X) = \sum_{\pi \leq \mathbf{1}_k} c_\pi$, and the right side involves $c_j(X), j \leq k$. Thus, the moment-cumulant relations are available for $k \leq n$ if the nth moment is finite.

12.6 Moment-cumulant relation for vectors

Now suppose, we have n real-valued r.vs. $X_1, \ldots, X_n$. Then we need multi-indexes to write the moments and cumulants, We may then define the multiplicative extension of the moment and cumulant sequences as before. Then Lemma 12.4.1 can be easily extended to many rvs. Moreover, this relation can be inverted. We shall not write the general statement. Instead we state the following particular consequence which we shall put to use in Section 18.7 to Gaussian rvs.

$$\mathrm{E}(X_1 X_2 \cdots X_n) = \sum_{\sigma \in \mathcal{P}_n} c_\sigma, \tag{12.14}$$

where c_σ are defined in a multiplicative way, using the cumulants of the random vector $(X_1, \ldots, X_n)'$, *taking care to preserve the appropriate indices.* Note that if we take $X_i \equiv X$ for all i, then we get back (12.7).

12.7 Exercises

Exercise 12.7.1 Let X, Y be non-negative rvs with densities f and g. If there exists y such that $f(x) \leq g(x)$ for $x \leq y$, and $f(x) \geq g(x)$ for $x \geq y$, then show that $\mathrm{E}(X) \geq \mathrm{E}(Y)$.

Exercise 12.7.2 Suppose X is a non-negative rv.

(a) Show that

$$\mathrm{E}(X) = \int_{(0,\ \infty)} P\{X \geq x\} \lambda(dx)\,.$$

(b) (Generalisation of (a)) Let $G : [0,\ \infty) \to [0,\ \infty)$ be non-decreasing and right continuous with $G(0) = 0$. Show that

$$\begin{aligned}\mathrm{E}(G(X)) &= \int_{(0,\ \infty)} [1 - F_X(x)] G(dx) \\ &= \int_{(0,\ \infty)} P(X > x) G(dx) = \int_{(0,\ \infty)} P(X \geq x) G(dx).\end{aligned}$$

(c) What happens in (b) when $G(x) = x$ and $G(x) = x^2$?

(d) Show that

$$\sum_{n=1}^{\infty} P\{X \geq n\} \leq \mathrm{E}(X) \leq 1 + \sum_{n=1}^{\infty} P\{X \geq n\}\,.$$

(e) Suppose $\sum_{n=0}^{\infty} P\{X = n\} = 1$. Show that

$$\mathrm{E}(X) = \sum_{n=1}^{\infty} P\{X \geq n\}\,.$$

(f) Suppose X is an integrable rv. Show that

$$\mathrm{E}(X) = \int_{(0,\ \infty)} P(X > t)\lambda(dt) - \int_{(-\infty,\ 0)} P(X < t)\lambda(dt)\,.$$

Exercise 12.7.3 If $\sum_{k=0}^{\infty} P\{X = k\} = 1$, then show that,

$$\mathrm{E}(X^2) = \sum_{k=0}^{\infty} (2k+1) P(X > k)\,.$$

Exercise 12.7.4 Suppose $\mathrm{E}(X)$ is finite. Show that for every $\alpha \geq 0$,

$$\int_{\{X>\alpha\}} X dP = \alpha P\{X > \alpha\} + \int_{(\alpha,\ \infty)} P\{X > t\}\lambda(dt)\,.$$

Exercise 12.7.5 Let $h(x) \geq 0$ be Borel measurable and let

$$H(x) := \int_{(-\infty,\ x]} h(y)\lambda(dy),\ x > 0\,.$$

Show that for any rv X,

$$\mathrm{E}\, H(X) = \int_{\mathbb{R}} h(y) P(X \geq y)\lambda(dy)\,.$$

Exercise 12.7.6 For rvs X, Y on $(\Omega, \mathcal{A}, P)$, $\mathrm{E}(|X| + |Y|) < \infty$, show

$$\mathrm{E}(Y) - \mathrm{E}(X) = \int_{\mathbb{R}} [P\{X < t \leq Y\} - P\{Y < t \leq X\}]\lambda(dt)\,.$$

Exercise 12.7.7 Let X be a rv with distribution function F. Show that

$$\mathrm{E}[F(X)] = \frac{1}{2} + \frac{1}{2}\sum_x [P\{X = x\}]^2 .$$

[In particular, if F is continuous, then $\mathrm{E}[F(X)] = 1/2$.]

Exercise 12.7.8 Let X be a rv with distribution function F. Show that if $a > 0$ and $F(a) > 0$, then
(a)

$$\int_{\{X>\alpha\}} X dP \le \alpha(-\log F(\alpha)) + \int_{(\alpha,\ \infty)} (-\log F(t))\lambda(dt) \le \frac{1}{F(\alpha)} \int_{\{X>\alpha\}} X dP.$$

(b) $\mathrm{E}\,X^+ < \infty$ if and only if $\int_{(\alpha,\ \infty)}(-\log F(t))\lambda(dt) < \infty$ for some α.

Exercise 12.7.9 Recall F^{-1} from Exercise 5.3.19. Show that

$$\mathrm{E}\,|X| < \infty \quad \text{if and only if} \quad \int_0^1 F_X^{-1}(t)\lambda(dt) < \infty .$$

Exercise 12.7.10 On a probability space $(\Omega, \mathcal{A}, P)$, for $A, B \in \mathcal{A}$, let $d(A, B) = P(A\Delta B)$.Call sets A and B to be equivalent if $d(A, B) = 0$

(i) Show that this identification defines an equivalence relation.

(ii) Show that d is a metric on the equivalence classes.

(iii) Let X be a real-valued rv on $(\Omega, \mathcal{A}, P)$ and $\mathrm{E}\,|X| < \infty$. Show that the map $A \to \int_A X dP$ is continuous in the metric d.

Exercise 12.7.11 Show that for any rv X with a finite mgf in a non-trivial neighbourhood of 0, and any real constants a and b,

$$c_j(aX + b) = a^j c_j(X) \quad \text{for all} \quad j \ge 2 .$$

Exercise 12.7.12 Identify the cumulants of a $U(0,\ 1)$ and a $Exp(\lambda)$ rv.

Exercise 12.7.13 Show that if X has the log-normal density f_0 given in Exercise 12.2.6, then $Y = \log X$ is a standard Gaussian rv.

Exercise 12.7.14 If X has a finite mgf in a non-trivial neighbourhood of 0, show that all odd cumulants are 0 iff all odd moments are 0.

Exercise 12.7.15 Show that any mixed cumulant $c_{k_1,\ldots,k_n}(X_1, \ldots, X_n)$ involves only those moments $\mathrm{E}(X_1^{t_1} \cdots X_n^{t_n})$ where $t_j \le k_j$ for all j.

Exercise 12.7.16 In Lemma 12.4.1, establish (12.6) using (12.7).

Exercise 12.7.17 Show that, number of pair-partitions of $\{1, \ldots, 2n\}$ equals $\frac{(2n)!}{2^n n!}$.

Exercise 12.7.18 Prove that the enumeration claimed in the beginning of the proof of Lemma 12.5.1 can always be achieved. Show by example that this enumeration need not be unique. Prove that nevertheless, the Möbius function is independent of the specific choice of the enumeration.

Exercise 12.7.19 Find the Möbius function for $\mathcal{P}_3$.

Exercise 12.7.20 Möbius function was used to move between moments and cumulants in (12.11) and (12.13). Show that the following general version can be proved by the same arguments. Let P be a finite lattice and let μ be the Möbius function on $P^{(2)}$. Suppose $f, g : P \to \mathbb{C}$ are two functions. Then the following two relations are equivalent.

$$f(\pi) = \sum_{\sigma \in P, \sigma \leq \pi} g(\sigma), \quad g(\pi) = \sum_{\sigma \in P, \sigma \leq \pi} f(\sigma)\mu[\sigma, \ \pi].$$

Exercise 12.7.21 State and prove an extension of Lemma 12.4.1 to several rvs so that mixed moments of any order are expressed as a sum of the product of suitable cumulants of equal or lower order. Use this relation and the Möbius function on $\mathcal{P}_n$ to express any cumulant in terms of the moments of equal and lower orders.

Exercise 12.7.22 Sequentially arrange the numbers $1, \ldots, n$ on a circle. For any block of $\pi \in \mathcal{P}_n$, all its elements are joined by arcs. Then π is called *non-crossing* if the arcs of different blocks do not cross. The set of non-crossing partitions is denoted by $NC(n)$.
(a) Identify the elements of $NC(n)$ for $n = 2, 3, 4$.

(b) Show that $NC(n)$ is a lattice (with same partial ordering as $\mathcal{P}_n$).

(c) Consider the multiplicative extension of $\{m_n\}$, now restricted to $NC(n)$. Show that there exists a sequence $\{k_n\}$ such that for every n,

$$m_\pi = \sum_{\tau \in NC(n), \tau \leq \pi} k_\tau, \ \pi \in NC(n),$$

where k_τ is the multiplicative extension of $\{k_n\}$, restricted to $NC(n)$. The numbers k_n's are known as *free cumulants*, and are crucial in the theory of *non-commutative probability spaces*. See Bose [2021]

Chapter 13

Further modes of convergence of functions

We are already familiar with two notions of convergence for measurable functions, namely convergence almost surely and convergence in L^p. We shall now see several other notions of convergence of measurable functions and their interrelations.

13.1 Convergence almost surely

We recall the definition of almost sure convergence.

Definition 13.1.1. (**Convergence almost surely**) Let $\{f_n\}$ and f be complex-valued measurable functions on a measure space $(\Omega, \mathcal{A}, \mu)$.

(a) We say f_n converges to f almost surely, if there is a null set A such that $f_n(\omega) \to f(\omega)$ for all $\omega \notin A$. We write $f_n \stackrel{a.s.}{\to} f$ or $f_n \stackrel{a.e.}{\to} f$.

(b) We say f_n is Cauchy almost surely, if there is a null set A such that $\{f_n(\omega)\}$ is a Cauchy sequence for all $\omega \notin A$. ◇

Lemma 13.1.1. If μ is a finite measure, then $f_n \stackrel{a.e.}{\to} f$ if and only if,

$$\lim_{n\to\infty} \mu\big(\cup_{k=n}^{\infty} \{\omega : |f_k(\omega) - f(\omega)| \geq \delta\}\big) = 0 \text{ for every } \delta > 0. \qquad \blacklozenge$$

Proof. Let

$$B_{n,\delta} = \{\omega : |f_n(\omega) - f(\omega)| \geq \delta\}, \quad B_\delta = \limsup B_{n,\delta}\,.$$

A. Bose, *Elements of Measure and Probability*, Texts and Readings in Mathematics 88,
https://doi.org/10.1007/978-981-95-2758-8_13

Then

$$\mu(\cup_{k=n}^{\infty} B_{k,\delta}) \downarrow \mu(B_\delta) \text{ as } n \to \infty.$$

Note that

$$\{\omega : f_n(\omega) \nrightarrow f(\omega)\} = \cup_{\delta>0}^{\infty} B_\delta = \cup_{m=1}^{\infty} B_{1/m}.$$

Hence $f_n \xrightarrow{a.e.} f$, if and only if $\mu(B_\delta) = 0$ for all $\delta > 0$, which in turn holds if and only if $\mu(\cup_{k=n}^{\infty} B_{k,\delta}) \to 0$ for all $\delta > 0$. ■

Exercise 13.1.1 If μ is a finite measure, show that $\{f_n\}$ is Cauchy a.e. if and only if, for all $\delta > 0$,

$$\lim_{m,n\to\infty} \mu\left(\cup_{j=m,k=n}^{\infty} \{\omega : |f_k(\omega) - f_j(\omega)| \geq \delta\}\right) = 0.$$

13.2 Convergence in measure and probability

Definition 13.2.1. (Convergence in measure/probability) The sequence $\{f_n\} : (\Omega, \mathcal{A}, \mu) \to (\mathbb{C}, \mathcal{B}(\mathbb{C}))$ is said to converge in μ-measure to f (write $f_n \xrightarrow{\mu} f$) if, for every $\epsilon > 0$,

$$\mu\{\omega : |f_n(\omega) - f(\omega)| > \epsilon\} \to 0 \text{ as } n \to \infty.$$

If μ is a probability measure, say P, then we say f_n converges to f in probability and write $f_n \xrightarrow{P} f$. ◇

Exercise 13.2.1 Suppose $f_n \xrightarrow{\mu} f$ and $f_n \xrightarrow{\mu} g$. Show that $f = g$ a.e.

Exercise 13.2.2 Consider the Lebesgue measure λ on $[0, \infty)$. Let

$$f_n(x) = \begin{cases} 1 & \text{if } n \leq x \leq n+1, \\ 0 & \text{otherwise.} \end{cases}$$

Check that $f_n \to f$ a.e. λ, but f_n does not converge in measure.

Exercise 13.2.3 Suppose $(\Omega, \mathcal{A}, \mu)$ is a finite measure space. Show that if $f_n \xrightarrow{a.e} f$ then $f_n \xrightarrow{\mu} f$.

Just like we have the notion of a sequence of elements in a metric space to be a Cauchy sequence, we have the notion of a sequence of measurable functions to be Cauchy in measure.

Definition 13.2.2. (**Cauchy in measure/probability**) A sequence $\{f_n\} : (\Omega, \mathcal{A}, \mu) \to (\mathbb{C}, \mathcal{B}(\mathbb{C}))$ is *Cauchy in measure* if, given any $\epsilon > 0$,

$$\mu\{\omega : |f_n(\omega) - f_m(\omega)| > \epsilon\} \to 0, \text{ as } m, n \to \infty. \qquad \diamond$$

Exercise 13.2.4 Let $\{f_n\}$ and f be measurable functions on $(\Omega, \mathcal{A}, \mu)$. Show that if $f_n \overset{\mu}{\to} f$, then $\{f_n\}$ is Cauchy in measure.

Exercise 13.2.5 Suppose $\{f_n\}$ is a sequence of functions in $L^p(\mu)$.

(a) Show that if $f_n \overset{L^p}{\to} f$ for some $0 < p < \infty$, then $f_n \overset{\mu}{\to} f$.

(b) Show that if $\{f_n\}$ is Cauchy in L^p for some $0 < p < \infty$, then $\{f_n\}$ is Cauchy in measure.

Theorem 13.2.1. *Let $f_n \overset{\mu}{\to} f$. Then there is a sub-sequence $\{n_k\}$ such that $f_{n_k} \overset{a.s.}{\to} f$.* ◆

Proof. First note that from Exercise 13.2.4, $\{f_n\}$ is Cauchy in measure. Now for $k = 1, 2, \ldots$, choose $\{n_k\}$ strictly increasing such that

$$\mu\{\omega : |f_m(\omega) - f_n(\omega)| \geq 2^{-k}\} \leq 2^{-k}, \text{ for all } n, m \geq n_k.$$

Let $g_k = f_{n_k}$, and

$$A_k = \{\omega : |g_k(\omega) - g_{k+1}\omega)| \geq 2^{-k}\}, \ A = \limsup A_k. \tag{13.1}$$

Then $\mu(A_k) \leq 2^{-k}$. o, by First Borel-Cantelli Lemma 3.3.1, $\mu(A) = 0$. But for all $\omega \notin A$, $\omega \in A_k$ for only finitely many k, and hence $\{g_k(\omega)\}$ is Cauchy. So for all $\omega \notin A$, $g_k(\omega) \to g(\omega)$ for some g. Since $\mu(A) = 0$, $g_k \overset{a.e.}{\to} g$. But this g must equal f a.e. since $g_k \overset{\mu}{\to} f$. Thus $g_k \overset{a.e.}{\to} f$. ■

Exercise 13.2.6 Show that $\{f_n\}$ is Cauchy in measure if and only if f_n converges in measure. Hint. Borrow from the above proof.

13.3 Almost uniform convergence

This notion of convergence is stronger than a.e. convergence. However, for finite measure spaces, the two notions are equivalent.

Definition 13.3.1. (**Almost uniform convergence**) The sequence $\{f_n\} : (\Omega, \mathcal{A}, \mu) \to (\mathbb{C}, \mathcal{B}(\mathbb{C}))$ is said to converge *almost uniformly* to f if, given any $\epsilon > 0$, there is a set $A \in \mathcal{A}$ so that $\mu(A) < \epsilon$ and $f_n \to f$ uniformly on A^c. We write $f_n \stackrel{a.u}{\to} f$. ◇

Exercise 13.3.1 Consider the Lebesgue measure on $[0, \infty)$. Let

$$f_n(x) = \begin{cases} 1 & \text{if } n \le x \le n + 2n, \\ 0 & \text{otherwise.} \end{cases}$$

Check that $f_n \to 0$ a.s., but does not converge almost uniformly.

Theorem 13.3.1. *(a) If $f_n \stackrel{a.u.}{\to} f$ then $f_n \stackrel{a.s}{\to} f$ and $f_n \stackrel{\mu}{\to} f$.*
(b) Let $f_n \stackrel{\mu}{\to} f$. Then there is a sub-sequence $\{n_k\}$ such that $f_{n_k} \stackrel{a.u.}{\to} f$, and hence $f_{n_k} \stackrel{a.s.}{\to} f$. ◆

Proof. (a) Fix $\epsilon > 0$. Let A be such that $\mu(A) < \epsilon$ and $f_n \to f$ uniformly on A^c. Fix $\delta > 0$. Then for all large n, $|f_n(\omega) - f(\omega)| < \delta$ for all $\omega \in A^c$. Therefore for such n, $\{\omega : |f_n(\omega) - f(\omega)| > \delta\} \subseteq A$. This implies that

$$\mu\{\omega : |f_n(\omega) - f(\omega)| > \delta\} \le \mu(A) \le \epsilon \text{ for all large } n.$$

So $f_n \stackrel{\mu}{\to} f$.

To prove that $f_n \stackrel{a.e.}{\to} f$, for every integer $k > 1$, choose a set A_k such that $\mu(A_k) < 1/k$ and $f_n \to f$ uniformly on A_k^c. Let $B = \cup_{k=1}^{\infty} A_k^c$. Then $f_n \to f$ on B. Moreover,

$$\mu(B^c) = \mu(\cap_{k=1}^{\infty} A_k) \le \mu(A_k) \to 0 \text{ as } k \to \infty.$$

(b) By Exercise 13.2.6, $\{f_n\}$ is Cauchy in measure. Construct $\{g_k\}$ and $\{A_k\}$ as in the proof of Theorem 13.2.1. We already know that $g_k \stackrel{a.e.}{\to} f$.

To prove that $\{g_k\}$ converges almost uniformly, fix $\epsilon > 0$.and let $B_r := \cup_{k=r}^{\infty} A_k$. Then $\mu(B_r) < \epsilon$ for all large r. If $\omega \notin B_r$, by (13.1),

$$|g_k(\omega) - g_{k+1}(\omega)| < 2^{-k}, k = r, r+1, \ldots.$$

By using an extension of Exercise 1.3.1(e), $\{g_k\}$ converges uniformly on B_r^c. Again, the limit must be equal to f a.e. ■

Theorem 13.3.2 (**Egoroff's Theorem**). *If μ is a finite measure, and $f_n \stackrel{a.e.}{\to} f$, then $f_n \stackrel{a.u.}{\to} f$.* ◆

Proof. Fix $\epsilon > 0$ and integer $j > 1$. From Lemma 13.1.1, for sufficiently large $n = n(j)$,

$$\mu(A_j) = \mu(\cup_{k=n(j)}^{\infty}\{|f_k - f| \geq 1/j\} \leq \epsilon/2^j.$$

Let $A = \cup_{j=1}^{\infty} A_j$. Then $\mu(A) < \epsilon$. For $\delta > 0$, choose $j > 1/\delta$. Then for any $k \geq n(j)$ and $\omega \in A^c$,

$$|f_k(\omega) - f(\omega)| < 1/j < \delta.$$

Thus $f_n \to f$ uniformly on A^c. ■

13.4 Uniform integrability

Integrability of a rv X on $(\Omega, \mathcal{A}, P)$ is equivalent to any one of the following conditions:

(i) Given $\varepsilon > 0$, there is a $\delta > 0$ so that $P(A) < \delta$ implies $\mathrm{E}(|X|\mathbf{1}_A) < \varepsilon$.

(ii) $\lim_{T\to\infty} \int_{\{|X|>T\}} |X|\, dP = 0$.

Definition 13.4.1. (**Uniform integrability**). Let $\{X_\alpha\}$, $(\alpha \in I)$ be a family of rvs on $(\Omega, \mathcal{A}, P)$. It is said to be *uniformly integrable* (ui) if

$$\lim_{T\to\infty} \sup_{\alpha\in I} \int_{\{|X_\alpha|>T\}} |X_\alpha|\, dP = 0\,.$$ ◇

We will not mention the index set I unless it is necessary.

Exercise 13.4.1 Show that $\{X_\alpha\}$ is ui if and only if the following two conditions hold:

(a) $\sup_{\alpha\in I} \mathrm{E}(|X_\alpha|) < \infty$ and,

(b) given $\varepsilon > 0$, $\exists\, \delta > 0$ so that $P(A) < \delta \Rightarrow \sup_{\alpha\in I} \mathrm{E}(|X_\alpha|\mathbf{1}_A) < \varepsilon$.

Exercise 13.4.2 (a) Show that a finitely family of integrable rvs is ui.

(b) If $\{X_\alpha\}$ and $\{Y_\alpha\}$ are ui, then show that $\{X_\alpha + Y_\alpha\}$ is ui.

(c) Give an example where $\sup_{\alpha\in I} \mathrm{E}(|X_\alpha|) < \infty$ but $\{X_\alpha\}$ is not ui.

(d) If Y is integrable and $|X_\alpha| \leq |Y|$ for all α, show that $\{X_\alpha\}$ is ui.

Recall that if $\{X_n\}$ converges to X a.s., or in probability, then, $\mathrm{E}(X_n) \to \mathrm{E}(X)$ need not hold. A sufficient condition for this to hold is given by the domination condition in DCT. We shall show below a stronger result that uses ui.

Theorem 13.4.1. *Let $\{X_n\} : (\Omega, \mathcal{A}, P) \to (\mathbb{R}, \mathcal{B}(\mathbb{R}))$. The following are equivalent.*

(i) $X_n \to X$ in L^1.

(ii) $X_n \xrightarrow{P} X$ and $\{X_n\}$ is ui. ◆

Remark 13.4.1. Theorem 13.4.1 implies that, if $X_n \xrightarrow{P} X$ and $\{X_n\}$ is ui, then $\mathrm{E}(X_n) \to \mathrm{E}(X)$. Exercise 13.4.2(d) shows that this is a stronger statement than DCT. ●

Proof of Theorem 13.4.1. First suppose (ii) holds. We show that then $\mathrm{E}(|X|) < \infty$. Since $X_n \xrightarrow{P} X$, by Theorem 13.3.1(b), pick a sub-sequence $\{X_{n_k}\}$ such that $X_{n_k} \to X$ a.s. Then

$$\begin{aligned}\mathrm{E}(|X|) &= \mathrm{E}\left(\liminf_{k\to\infty} |X_{n_k}|\right)\\ &\le \liminf_{k\to\infty} \mathrm{E}\left(|X_{n_k}|\right) \quad \text{(by Fatou's Lemma 7.2.1)}\\ &\le \sup_{n\ge 1} \mathrm{E}\,|X_n|\\ &< \infty \quad \text{since} \quad \{X_n\} \quad \text{is ui}\,.\end{aligned}$$

Define

$$Y_n := |X_n - X|\,.$$

Clearly, $Y_n \xrightarrow{P} 0$ and $\{Y_n\}$ is ui. It remains to show $\lim_{n\to\infty} \mathrm{E}(Y_n) = 0$.

Fix $\varepsilon > 0$. As $\{Y_n\}$ is ui, get $T > 0$ so that $\sup_n \mathrm{E}\left(Y_n \mathbf{1}_{\{Y_n \ge T\}}\right) \le \varepsilon$. Then for this T,

$$\begin{aligned}\limsup_{n\to\infty} \mathrm{E}(Y_n) &\le \limsup_{n\to\infty} \mathrm{E}\left(Y_n \mathbf{1}_{\{Y_n < T\}}\right) + \limsup_{n\to\infty} \mathrm{E}\left(Y_n \mathbf{1}_{\{Y_n \ge T\}}\right)\\ &\le \limsup_{n\to\infty} \mathrm{E}\left(Y_n \mathbf{1}_{\{Y_n < T\}}\right) + \varepsilon = \varepsilon\,.\end{aligned}$$

Now, $\{Y_n \mathbf{1}_{\{Y_n < T\}}\}$ is a sequence of uniformly bounded rvs which converges to 0 in probability. Hence arguing through subsequences and DCT Theorem 7.2.7, the first term converges to 0. Since ε is arbitrary, (i) is established.

Now suppose that (i) holds. Then, $X_n \xrightarrow{P} X$. Let $Y_n := |X_n - X|$. Then it suffices to show that $\{Y_n\}$ is ui. Fix $\varepsilon > 0$. By (i), for some N,

$$\sup_{n\ge N} \mathrm{E}(Y_n) \le \varepsilon\,.$$

For $i = 1, \ldots, N-1$, there exists $T_i > 0$ with $\mathrm{E}(Y_i \mathbf{1}_{\{Y_i > T_i\}}) \leq \varepsilon$. Let $T := \max\{T_1, \ldots, T_{N-1}\}$. Then it follows that

$$\mathrm{E}\left(Y_n \mathbf{1}_{\{Y_n > T\}}\right) \leq \varepsilon\,,\ n \geq 1\,.$$

That is, $\{Y_n\}$ is ui. This completes the proof. ■

Exercise 13.4.3 Supose $\{X_\alpha\}$ and $\{Y_\alpha\}$ are rvs where $|Y_\alpha| \leq |X_\alpha|$ for all α, and $\{X_\alpha\}$ is ui. Show that then $\{Y_\alpha\}$ is also ui.

The next result follows from Theorem 13.4.1.

Theorem 13.4.2. *If $\{X_n\}$ and X are in L^p for some $p \geq 1$, then the following are equivalent.*

(i) $X_n \to X$ in L^p.

(ii) The family $\{|X_n|^p : n \geq 1\}$ is ui and $X_n \xrightarrow{P} X$.

(iii) $X_n \xrightarrow{P} X$ and $|X_n|^p \to |X|^p$ in L^1. ◆

Proof. Equivalence of (i) and (ii). If (i) holds, then $|X_n - X|^p \to 0$ in L^1. Then Theorem 13.4.1 implies that $X_n \xrightarrow{P} X$ and $\{|X_n - X|^p : n \geq 1\}$ is ui. Note that since $p \geq 1$,

$$|X_n|^p \leq 2^p \left(|X_n - X|^p + |X|^p\right)\,.$$

Then Exercises 13.4.2 and 13.4.3 imply $\{|X_n - X|^p + |X|^p\}$, and hence $\{|X_n|^p\}$, is ui. Thus (ii) holds.

Conversely, if (ii) holds, then $|X_n - X|^p \xrightarrow{P} 0$ and since $p \geq 1$, $|X_n - X|^p \leq 2^p \left(|X_n|^p + |X|^p\right)$. Again, Exercises 13.4.2 and 13.4.3 imply that $\{|X_n - X|^p : n \geq 1\}$ is ui. This proves (i) after an appeal to Theorem 13.4.1. Equivalence of (ii) and (iii) follows from Theorem 13.4.1. ■

We end the section with a characterisation of ui rvs.

Definition 13.4.2. (**Test function of ui**) Let $G : [0, \infty) \to [0, \infty)$ be a function. It is called a *test function of uniform integrability* if it is non-decreasing, and

$$\lim_{x \to \infty} x^{-1} G(x) = \infty.$$ ◇

$G(x) = x^p$, $p > 1$, and $G(x) = x \log x$ are test functions of ui. The following criterion of uniform integrability is due to de La Vallée Pousssin (see Meyer [1966]). The simple proof is taken from Chafai [2014], who attributes a simplification in the argument to Nicolas Fournier.

Theorem 13.4.3. (Characterization of ui via test functions). *A nonempty family $\mathcal{X}$ of integrable functions is uniformly integrable if and only if there exists a test function of uniform integrability G such that*

$$\sup_{X\in\mathcal{X}} \mathrm{E}(G(|X|)) < \infty. \tag{13.2}$$

When G exists, it can be chosen to be non-decreasing and convex. ♦

Proof of Theorem 13.4.3. First suppose there is a test function G that satisfies (13.2). Fix $\varepsilon > 0$. Then there exists $T \geq 0$ such that $x \leq \varepsilon G(x)$ for every $x \geq T$. Therefore,

$$\sup_{X\in\mathcal{X}} \mathrm{E}(|X|\mathbf{1}_{\{|X|\geq T\}}) \leq \varepsilon \sup_{X\in\mathcal{X}} \mathrm{E}(G(|X|)\mathbf{1}_{\{|X|\geq T\}}) \leq \varepsilon \sup_{X\in\mathcal{X}} \mathrm{E}(G(|X|)).$$

This shows that $\mathcal{X}$ is ui.

Now suppose that $\mathcal{X}$ is ui. Then from the definition of ui, there exists $\{T_m\}$ increasing to ∞, such that

$$\sup_{X\in\mathcal{X}} \mathrm{E}\left(|X|\mathbf{1}_{\{|X|\geq T_m\}}\right) \leq 2^{-m}. \tag{13.3}$$

Define

$$G(x) = \sum_{m=1}^{\infty} (x - T_m)^+. \tag{13.4}$$

Then clearly G is convex and non-decreasing. Moreover,

$$x^{-1}G(x) = \sum_{m=1}^{\infty} (1 - \frac{T_m}{x})^+ \uparrow \sum_{m=1}^{\infty} 1 = \infty.$$

Hence G is a convex test function of ui. Now observe that for any $X \in \mathcal{X}$, using (13.3) and (13.4),

$$\mathrm{E}\left(G(|X|)\right) \leq \sum_{m=1}^{\infty} \mathrm{E}\left(|X|\mathbf{1}_{\{|X|\geq T_m\}}\right) \leq \sum_{m=1}^{\infty} 2^{-m} = 1.$$ ∎

Remark 13.4.2. For a useful extension see, Hu and Rosalsky [2011]. For ui in abstract spaces, see Alexopoulos [1994], Castaing and de Fitte [2000], Rivera [2000] and Martin and Milman [2001]. ●

13.5 Exercises

Exercise 13.5.1 Let $\{X_n\}$ be rvs on a probability space $(\Omega, \mathcal{A}, P)$ where Ω is countable. Show that $X_n \xrightarrow{a.e.} X$ if and only if $X_n \xrightarrow{P} X$.

Exercise 13.5.2 (**Convergence in measure DCT**) Suppose we have $\{f_n\} : (\Omega, \mathcal{A}, \mu) \to (\mathbb{R}, \mathcal{B}(\mathbb{R}))$. If $\sup_n |f_n| \le g$ where g is integrable, and $f_n \xrightarrow{\mu} f$, then show that $\int f_n d\mu \to \int f d\mu$. Hint: Use Theorem 13.3.1.

Exercise 13.5.3 (**Lévy metric for rvs**) [Compare this metric with the Lévy metric defined for probability distributions in Exercise 17.13.17.] Let $(\Omega, \mathcal{A}, P)$ be a probability space. For any rvs X and Y, let

$$d_L(X,Y) = \inf\{\epsilon : P(|X-Y| \ge \epsilon) \le \epsilon\,.$$

(a) Show that $d_L(X,Y) = 0$ if and only if $X = Y$ a.s.

(b) Identify rvs as equal if they are a.s. equal. Show that d_L is a metric.

(c) Show that $X_n \xrightarrow{P} X$ if and only if $d_L(X_n, X) \to 0$. Hence convergence in probability is metrizable.

(d) Show that in general almost sure convergence is not metrizable.

Exercise 13.5.4 (**Convergence in (finite) measure is metrizable**) Let $(\Omega, \mathcal{A}, \mu)$ be a measure space. Measurable functions that are equal a.s. are taken to be equal. Define d_1 as

$$d_1(f,g) = \int_\Omega \frac{|f-g|}{1+|f-g|} d\mu, \quad f, g \text{ measurable}\,.$$

(a) Show that d_1 is a metric.

(b) Show that $d_1(f_n, f) \to 0$ implies $f_n \xrightarrow{\mu} f$.

(c) Show that if μ is finite then $f_n \xrightarrow{\mu} f$ implies $d_1(f_n, f) \to 0$. Hence convergence in measure is metrizable when μ is finite.

(d) Give an example to show that (c) is not true when μ is not finite.

(e) If μ is a probability measure, then show that d_1 is equivalent to d_L defined in Exercise 13.5.3.

Exercise 13.5.5 Suppose $X_n \xrightarrow{P} X$, and there are continuous functions $g,\ h \ge 0$, with $g(x) > 0$ for large x, and $h(x)/g(x) \to 0$, as $|x| \to \infty$ and $\operatorname{E} g(X_n) \le K < \infty$ for all n. Then show that $\operatorname{E} h(X_n) \to \operatorname{E} h(X)$.

Exercise 13.5.6 Let $(\Omega, \mathcal{A}, P)$ be a probability space. Let $\{X_n\}$ be a non-decreasing sequence of rvs which converges to X in probability. Show that the convergence is a.s. P.

Exercise 13.5.7 Suppose $\{X_n\}$ and $\{Y_n\}$ are sequences of rvs such that $X_n \xrightarrow{P} X$ and $Y_n \xrightarrow{P} Y$. Show that $X_n + Y_n \xrightarrow{P} X + Y$.

Exercise 13.5.8 If $\{X_n\}$ is a sequence of rvs on $(\Omega, \mathcal{A}, P)$, then show that there is a non-random sequence c_n such that $X_n / c_n \to 0$ a.s. P.

Exercise 13.5.9 Suppose μ is a finite measure. Show that $f_n \xrightarrow{a.e.} f$ if any of the following hold:

(a) $\sum_{n=1}^{\infty} \mu\{|f_n - f| \geq \epsilon\} < \infty$ for every $\epsilon > 0$.

(b) $\sum_{n=1}^{\infty} ||f_n - f||_p^p < \infty$ some $0 < p < \infty$.

Exercise 13.5.10 Show that for any $p > 0$, $\{|X_\alpha|^p\}$ is ui, if for some $\varepsilon > 0$, $\sup_{\alpha \in I} \mathrm{E}(|X_\alpha|^{p+\varepsilon}) < \infty$.

Exercise 13.5.11 Show that $\{X_\alpha\}$ is ui if for a finite constant c, and an integrable Y, $\sup_{\alpha \in I} P(|X_\alpha| > x\} \leq cP(|Y| > x\}$, for all x.

Exercise 13.5.12 Give an example of a ui sequence of rvs, which is not dominated by any rv.

Exercise 13.5.13 Let $\{X_n\}$ be a sequence of rvs on $(\Omega, \mathcal{A}, P)$. Show that $\{X_n\}$ is ui if and only if the following two conditions hold:

(a) $\sup_n \mathrm{E}(|X_n|) < \infty$ and,

(b) for all $\varepsilon > 0$, there exists $\delta > 0$ such that for every sequence $\{A_n\}$ from $\mathcal{A}$ with $P(A_n) < \delta$, $n \geq 1$, we have $\sup_n \mathrm{E}(|X_n| \mathbf{1}_{A_n}) < \varepsilon$.

Exercise 13.5.14 If $\{X_n\}$, X are in L^p for $p \geq 1$, and $X_n \xrightarrow{P} X$ as $n \to \infty$, then show that the following are equivalent.

(a) $X_n \to X$ in L^p.

(b) $|X_n| \to |X|$ in L^p.

(c) $\limsup_{n\to\infty} \mathrm{E}(|X_n|^p) \leq \mathrm{E}(|X|^p)$.

(d) $|X_n|^p \to |X|^p$ in L^1.

(e) The family $\{|X_n|^p\}$ is ui.

Hint: Show (a)⇒(b)⇒(c)⇒(d)⇒(e)⇒(a). For showing (c)⇒(d), write

$$\big|\, |X_n|^p - |X|^p \,\big| = |X_n|^p + |X|^p - 2\left(|X_n|^p \wedge |X|^p\right).$$

Chapter 14

Independence and basic conditional probability

A key concept in probability theory is of *independence*. The idea is already germane in the product probability spaces that we constructed.

Let us revisit the elementary Example 11.1.1. Consider the events $A = \{HT, HH\}$ and $B = \{HH, TH\}$, verbally described as "the first toss is head", and "the second toss is head". The probabilities of A, B and $A \cap B$ are given by,

$$P(A) = P(B) = p, \quad P(A \cap B) = p^2 .$$

Note that the two tosses should not "influence" each other. As A and B depend, respectively, on the first and second tosses, how is this reflected in the probabilities that we have mentioned above?

Observe that $P(A \cap B) = P(A)P(B)$. This is the basic defining relation for "probabilistic independence" of events.

14.1 Independent events, Second Borel-Cantelli

Definition 14.1.1. (**Independence of two events**) Events A and B in $(\Omega, \mathcal{A}, P)$ are said to be *(probabilistically) independent* if

$$P(A \cap B) = P(A)P(B) . \qquad \diamond$$

Exercise 14.1.1 If A and B are independent, show that then each of the pair of events $\{A^c, B^c\}$, $\{A^c, B\}$ and $\{A, B^c\}$ are also independent.

A. Bose, *Elements of Measure and Probability*, Texts and Readings in Mathematics 88,
https://doi.org/10.1007/978-981-95-2758-8_14

Exercise 14.1.2 Let $(\Omega, \mathcal{A}, P)$ be a probability space.

(a) Show that any event with probability 1 or 0 is independent of any other event in $\mathcal{A}$. In particular, Ω and $\emptyset$ are always independent and each is independent of itself.

(b) Suppose A and B are disjoint. Show that they are independent if and only if at least one of them is a P-null set.

(c) If A is independent of itself, then show that $P(A) = 0$ or 1.

(d) Consider $\mathcal{C} = \{A \in \mathcal{A} : P(A) = 0 \text{ or } 1\}$. Show that $\mathcal{C}$ is a σ-field and any two events in this collection are independent.

In Example 11.1.1, "intuition" told us, events A and B cannot "influence" each other. Indeed, we constructed our product probability space in such a manner. Sometimes, the presence of independence is not that intuitive.

Exercise 14.1.3 In Example 11.1.1, in addition to the two events A and B, let $C = \{HH, TT\}$ be the event that "the outcome of both tosses are same". Show that A and C are independent, and so are B and C. Show that events $A \cap B$ and C are not independent.

If we have three events A, B and C that we wish to declare to be independent, then, for example, $A \cap B$ and C should also be independent. This leads us to the following definition.

Definition 14.1.2. (**Independence of collections of events**) Let $\{\mathcal{C}_i, i \in I\}$ be collections of events in a probability space where I is some index set. Then $\{\mathcal{C}_i\}$ are said to be independent, if for all distinct indices$\{i_j\}$ from I, and $A_{i_j} \in \mathcal{C}_{i_j}$, $1 \leq j \leq k$, $k \geq 1$,

$$P(A_{i_i} \cap A_{i_2} \cdots \cap A_{i_k}) = P(A_{i_1})P(A_{i_2}) \cdots P(A_{i_k}) \,. \tag{14.1}$$ ◇

Exercise 14.1.4 Suppose $\{A_i, i \in I\}$ are independent. Show that

(a) Any sub-collection is also independent.

(b) Eqn. (14.1) holds also for countable sub-collection of distinct indices.

(c) any collection $\{B_i, i \in I\}$ obtained by replacing some (or all) A_i by A_i^c are also independent.

Example 14.1.1. To claim independence, one needs to check a lot of conditions, none of which can be dropped. Consider the throw of two dice and the set of all 36 outcomes. Suppose the dice is **fair**. So, each of the 36 outcomes have the same probability 1/36.

Let

$$\begin{aligned} A &= \{\text{second throw shows 1, 2 or 5}\}, \\ B &= \{\text{second throw shows 4, 5 or 6}\}, \\ C &= \{\text{sum of the two throws equals 9}\}. \end{aligned}$$

Then $P(A) = 1/2, P(B) = 1/2,\ P(C) = 1/9$.

$$\begin{aligned} P(A) = 1/2, P(B) &= 1/2,\ P(C) = 1/9, \\ P(A \cap B) &= \frac{1}{6} \neq P(A)P(B) = \frac{1}{4}, \\ P(A \cap C) &= \frac{1}{36} \neq P(A)P(C) = \frac{1}{18}, \\ P(B \cap C) &= \frac{1}{12} \neq P(B)P(C) = \frac{1}{18}, \\ P(A \cap B \cap C) &= \frac{1}{36} = P(A)P(B)P(C). \end{aligned}$$

The last equality *does not* imply that A, B and C are independent. However, later we shall develop a notion of conditional independence where two events may be independent conditioned on a third. See Example 14.4.1. ▲

Lemma 3.3.1 said that $P(\limsup A_n) = 0$ if $\sum_{i=1}^{\infty} P(A_i) < \infty$. What happens if $\sum_{i=1}^{\infty} P(A_i) = \infty$? Second Borel-Cantelli Lemma provides an answer when $\{A_i\}$ are independent.

We will need the following inequality:

$$1 - x \leq e^{-x}, \text{ for every } x \in \mathbb{R}^+ . \tag{14.2}$$

Lemma 14.1.1. (**Second Borel-Cantelli Lemma**) Suppose $\{A_i\}$ are independent events. Then $\sum_{i=1}^{\infty} P(A_i) = \infty \Rightarrow P(\limsup_n A_n) = 1$. ♦

Proof. Recall that $\limsup_n A_n = P(\cap_{n=1}^{\infty} \cup_{k=n}^{\infty} A_k)$. Using this,

$$\begin{aligned} P(\cap_{n=1}^{\infty} \cup_{k=n}^{\infty} A_k) &= \lim_{n\to\infty} P(\cup_{k=n}^{\infty} A_k) \\ &= \lim_{n\to\infty} \lim_{m\to\infty} P(\cup_{k=n}^{m} A_k). \end{aligned}$$

Now

$$\begin{aligned} P(\cup_{k=n}^m A_k)^c &= \prod_{k=n}^m P(A_k^c), \text{ by independence} \\ &\le \prod_{k=n}^m \exp\{-P(A_k)\}, \text{ by Eqn 14.2} \\ &\to 0, \text{ since } \sum_{i=1}^\infty P(A_i) = \infty. \end{aligned}$$

■

14.2 Independent σ-fields and random vectors

Definition 14.2.1. Suppose $(\Omega, \mathcal{A}, P)$ is a probability space. Sub-σ-fields $\{\mathcal{A}_i, i \in I\}$ are said to be independent, if any collection of events $\{A_i \in \mathcal{A}_i\}$ are independent. ◇

Exercise 14.2.1 Consider two probability spaces $(\Omega_i, \mathcal{A}_i, P_i), i = 1, 2$. Suppose $A_i \in \mathcal{A}_i, i = 1, 2$. Check that the events $A_1 \times \Omega_2$ and $\Omega_1 \times A_2$ are independent in the product probability space. In what ways do you think can this be generalized to n-fold product and infinite product?

Definition 14.2.2. Suppose $(\Omega, \mathcal{A}, P)$ is a probability space. For any collection of random vectors $\{X_\alpha, \alpha \in I\}$, the **$\sigma$-field generated by** $\{X_\alpha, \alpha \in I\}$ is the smallest sub-σ-field of $\mathcal{A}$ that makes all these vectors measurable. It is written as $\sigma(X_\alpha, \alpha \in I)$. ◇

Definition 14.2.3. (**Independence of random vectors**) Random vectors $\{X_i, i \in I\}$ defined on $(\Omega, \mathcal{A}, P)$ are said to be *independent* if the σ-fields generated by them are independent. ◇

Independent random vectors may not have the same number of coordinates. Real-valued rvs X_1 and X_2 are independent if and only if

$$P\{X_1 \in B_1, X_2 \in B_2\} = P\{X_1 \in B_1\}P\{X_2 \in B_2\}, \text{ for all } B_1, B_2 \in \mathcal{B}(\mathbb{R}).$$

Independence of real-valued rvs $X_1, \ldots, X_d$, can be described in terms of the induced marginal and joint measures as

$$P_{(X_1, X_2, \ldots, X_d)} = P_{X_1} \otimes P_{X_2} \otimes \cdots \otimes P_{X_d}.$$

That is, the joint probability is product of the marginal probabilities.

We will write **iid** for "*independent and identically distributed*". For any cdf F on $\mathbb{R}$, by Kolmogorov's extension Theorem 11.3.1 there exists iid rvs $\{X_i\}$ each with cdf F.

Theorem 14.2.1 (Independence criteria). *Let $X = (X_1, \ldots, X_d)'$ be a random vector on $(\Omega, \mathcal{A}, P)$ with joint and marginal distribution functions F and $F_1, F_2, \ldots, F_d$.*

(a) Then $\{X_i, 1 \le i \le d\}$ are independent if and only if

$$F(x_1, \ldots, x_d) = \prod_{i=1}^{d} F_i(x_i), \quad \text{for all} \quad x_i \in \mathbb{R}. \tag{14.3}$$

(b) Suppose X is absolutely continuous with a density f, and let f_i be the density of X_i, $1 \le i \le d$. Then $\{X_i\}$ are independent if and only if

$$f(x_1, \ldots, x_d) = \prod_{i=1}^{d} f_i(x_i), \ \text{a.e}\ \lambda_d. \tag{14.4}$$

Conversely, if $\{X_i\}$ are independent rvs with densities $\{f_i\}$, then X is absolutely continuous, and f given by (14.4) is a density of X. ◆

Proof. (a) Suppose $\{X_i\}$ are independent. Then

$$\begin{aligned} F(x_1, \ldots, x_d) &= P\{X_1 \le x_1, \ldots X_d \le x_d\} \\ &= \prod_{i=1}^{d} P\{X_i \le x_i\} = \prod_{i=1}^{d} F_i(x_i)\,. \end{aligned}$$

Conversely, let (14.3) hold. Then for $a = (a_1, \ldots, a_d), b = (b_1, \ldots, b_d)$, $a_i \le b_i$ for all i, it can be checked that

$$P_X(a,\ b] = \prod_{i=1}^{n} [F_i(b_i) - F_i(a_i)] = \prod_{i=1}^{d} P_{X_i}(a_i,\ b_i].$$

That is, for all right semi-closed intervals $\{B_i\}$,

$$P\{X_i \in B_i, 1 \le i \le d\} = \prod_{i=1}^{d} P\{X_i \in B_i\}\,. \tag{14.5}$$

We wish to prove that (14.5) holds for all Borel sets $\{B_i\}$.

First fix Borel sets $B_2, \ldots, B_d$ and consider

$$\mathcal{C} = \{B_1 \in \mathcal{B}(\mathbb{R}) : \text{ Eqn. (14.5) holds}\}.$$

Then $\mathcal{C}$ is a monotone class containing the field of all finite disjoint union of right semi-closed intervals. Hence by Theorem 2.6.1, $\mathcal{C} = \mathcal{B}(\mathbb{R})$. So, (14.5) holds for all $B_1 \in \mathcal{B}(\mathbb{R})$ and all right semi-closed $B_2, \ldots, B_d$. Now consider the successive co-ordinates to complete the proof of (a).

(b)

$$\begin{aligned} F_1(x_1) &= P\{X_1 \le x_1\} \\ &= P\{X_1 \le x_1, X_2 \in \mathbb{R}, \ldots X_n \in \mathbb{R}\} \\ &= \int_{(-\infty,\ x_1]} \int_{\mathbb{R}} \cdots \int_{\mathbb{R}} f(t_1, \ldots, t_d)\lambda(dt_d) \cdots \lambda(dt_1). \end{aligned}$$

By definition, X_1 has pdf (by Fubini's theorem the integral is finite),

$$f_1(x_1) = \int_{\mathbb{R}} \cdots \int_{\mathbb{R}} f(x_1, t_2, \ldots, t_s)\lambda(dt_d) \cdots \lambda(dt_2).$$

Further, f_1 is measurable. Similarly each X_i has a pdf obtained by integrating out the other variables.

Now suppose (14.4) holds. Then

$$\begin{aligned} F(x_1, \ldots, x_d) &= \int_{(-\infty,\ x_1]} \cdots \int_{(-\infty,\ x_d]} f(t_i, \ldots, t_d)\lambda(dt_1) \cdots \lambda(dt_d) \\ &= \int_{(-\infty,\ x_1]} \cdots \int_{(-\infty,\ x_d]} (\prod_{i=1}^{d} f_i(t_i))\lambda(dt_1) \cdots \lambda(dt_d) \\ &= \prod_{i=1}^{d} F_i(x_i). \end{aligned}$$

Hence by (a) $\{X_i\}$ are independent.

Conversely, suppose $\{X_i\}$ are independent. Then

$$\begin{aligned} F(x_1, \ldots, x_d) &= \prod_{i=1}^{d} F_i(x_i) \\ &= \prod_{i=1}^{d} \int_{-\infty}^{x_1} \cdots \int_{-\infty}^{x_d} \prod_{i=1}^{d} f_i(t_i)\lambda(dt_1) \cdots \lambda(dt_d). \end{aligned} \tag{14.6}$$

Define $g(x_1, \ldots, x_d) = \prod_{i=1}^d f_i(x_i)$. Now it is easy to conclude that

$$P_X(B) = \int_B g(x)\lambda(dx) \text{ for all } B \in \mathcal{B}(\mathbb{R}).$$

Since $P_X(B)$ also equals $\int_B f(x)\lambda(dx)$ for all $B \in \mathcal{B}(\mathbb{R})$, it follows from Exercise 7.3.8 that $f = g$ a.e. λ. For the last part, start with Eqn. (14.6) and argue as above. ∎

Exercise 14.2.2 Suppose $X = (X_1, \ldots, X_d)'$ is a random vector where each X_i is discrete. Show that $\{X_i\}$ are independent if and only if

$$P\{X_1 = x_1, \ldots, X_n = x_d\} = \prod_{i=1}^d P\{X_i = x_i\} \text{ for all } x_i \in \mathbb{R}.$$

Exercise 14.2.3 Suppose X and Y are independent rvs.

(a) If X and Y are non-negative, then show that $\mathrm{E}(XY) = \mathrm{E}(X)\,\mathrm{E}(Y)$. The common value may equal ∞.

(b) If $\mathrm{E}(|X|+|Y|) < \infty$, then show that $\mathrm{E}(XY) = \mathrm{E}(X)\,\mathrm{E}(Y)$. Showing that the left side exists is a part of the problem.

The following lemma is often useful in proving independence. We state it without proof.

Lemma 14.2.1. Let the mgf $M_{X_1,\ldots,X_n}(t_1, \ldots, t_n)$ exist in a neighborhood N of the origin. Then, independence of $\{X_i : 1 \leq i \leq n\}$ is equivalent to (a) or (b) given below.

(a) $M_{X_1,\ldots,X_n}(t_1, \ldots, t_n) = \prod_{j=1}^n M_{X_j}(t_j)$ for all $(t_1, \ldots, t_n) \in N$.

(b) $C_{X_1,\ldots,X_n}(t_1, \ldots, t_n) = \sum_{j=1}^n C_{X_j}(t_j)$, for all $t_1, \ldots, t_n$. That is, all mixed cumulants are zero. ♦

Exercise 14.2.4 Show that for independent rvs X and Y (with finite mgfs), $c_j(X+Y) = c_j(X) + c_j(Y)$ for all $j \geq 1$.

Exercise 14.2.5 Show that sums of independent Gaussian rvs are again Gaussian, and likewise for Poisson.

Exercise 14.2.6 Let $\{X_i\}$ be independent random vectors (they may have different dimensions). Show that then $\{f_i \circ X_i\}$ are also independent for all Borel measurable functions in appropriate dimensions.

Exercise 14.2.7 (a) If X and Y are independent and $\mathrm{E}(X^2+Y^2)<\infty$, then show that $\mathrm{Cov}(X,Y)=0$, and $\mathrm{V}(X+Y)=\mathrm{V}(X)+\mathrm{V}(Y)$.
(b) If $\{X_i\}$ are independent with finite second moments, then show that

$$\mathrm{V}(X_1+\cdots+X_d)=\sum_{i=1}^{d}\mathrm{V}(X_i).$$

Exercise 14.2.8 (**Weak law of large numbers**) Let $\{X_i\}$ be iid with mean μ and finite variance σ^2. Using Chebyshev's inequality, show that

$$\frac{X_1+\cdots+X_n}{n}\xrightarrow{P}\mu\,.$$

Exercise 14.2.9 Let $\{X_i\}$ be independent rvs and $\{b_n\}$ be a sequence of real numbers such that $b_n\to\infty$ and $\sum b_n^{-2}Var(X_n)<\infty$. Show that

$$\frac{X_1+\cdots+X_n-\mathrm{E}(X_1+\cdots+X_n)}{b_n}\xrightarrow{P}0.$$

Exercise 14.2.10 Suppose $\{X_n\}$ is a sequence of independent rvs on $(\Omega,\mathcal{A},P)$ such that $\mathrm{E}(X_n)=0$ and $\mathrm{E}(X_n^4)<\infty$ for all n.
(a) Find a formula for $\mathrm{E}(X_1+\cdots+X_n)^4$ in terms of the individual moments of $\{X_i\}$.
(b) Now assume $\sup_{n\geq 1}\mathrm{E}(X_n^4)<\infty$. Show that $(X_1+\cdots+X_n)/n\to 0$ a.s. $[P]$. Hint: Use (a) and First Borel-Cantelli Lemma 3.3.1.

14.3 Basic conditional probability

Suppose C is an event in some probability space $(\Omega,\mathcal{A},P)$. If we are told that the event C has "occurred", how does that change the probabilities of events in $\mathcal{A}$? Intuitively, the original probabilities now need to be re-normalised in a way that there is no mass left outside C. This leads to the following definition.

Definition 14.3.1. (**Conditional probability measure**) If $(\Omega,\mathcal{A},P)$ is a probability space and $P(C)>0$, then the probability measure $P(\cdot|C)$ or $P_C(\cdot)$ is defined as

$$P(A|C)=P_C(A):=\frac{P(A\cap C)}{P(C)},\ A\in\mathcal{A},$$

is called the **conditional probability measure given** C. For every $A \in \mathcal{A}$, $P(A|C)$ is the **conditional probability of** A **given** C. ◇

The conditional probability measure depends on the "condition" C, and is defined only if $P(C) > 0$. Further, if $P(A) > 0$, the conditional probability of C given A equals $P(C|A) = P(A \cap C)/P(A)$. This is in general different from $P(A|C)$ unless $P(A) = P(C)$.

Based on the Radon-Nikodym derivative developed in Chapter 21, a more general notion of conditional probability is defined in Chapter 23.

Exercise 14.3.1 Check that P_C is a probability measure. Moreover, if $P(C) = 1$ then $P_C \equiv P$.

Exercise 14.3.2 Check that P_C can be considered as a probability measure on the restricted σ-field $\mathcal{A}_C$ (see Definition 2.4.3).

Exercise 14.3.3 (a) Let A_1 and A_2 be events with $P(A_2) > 0$. Then

$$P(A_1 \cap A_2) = P(A_2)P(A_1|A_2)\,.$$

Moreover, A_1 and A_2 are independent if and only if $P(A_1|A_2) = P(A_1)$.
(b) Let $\{A_i\}$ be events where $P(A_1 \cap A_2 \cap \cdots \cap A_{n-1}) > 0$. Then

$$P(A_1 \cap A_2 \cap \cdots \cap A_n) = P(A_1) \prod_{i=2}^{n} P(A_i|A_1 \cap A_2 \cap \cdots \cap A_{i-1})\,.$$

This is usually known as the **law of successive conditioning**.

Exercise 14.3.4 Let $\{B_i\}$ be a measurable partition. Show that,

$$P(A) = \sum_{i=1}^{\infty} P(A \cap B_i) = \sum_{\{i:P(B_i)>0\}}^{\infty} P(A|B_i)P(B_i) \text{ for any event } A\,.$$

This is usually known as the **law of total probability**.

14.4 Conditional independence given an event

Often independence may be present in a restricted (conditional) manner. For example, suppose X_1, X_2, X_3 are independent random variables. Let $Y_1 = X_1 + X_2, Y_2 = X_2 + X_3$. Then Y_1 and Y_2 are not independent, but "conditioned on X_2", Y_1 and Y_2 should be independent.

The concept of conditional independence is extremely useful, and has applications in many areas, in particular in the study of Markov chains and in graphical data analysis. Basic definition of conditional independence given an event is stated below. The more general case of conditional independence given a σ-field or a random variable or vector has to wait till Section 23.4 of Chapter 23 where we develop the idea of conditional probability and conditional independence given a σ-field.

Definition 14.4.1. **(Independence given an event)** Let $(\Omega, \mathcal{A}, P)$ be a probability space. Then we say that events A_1 and A_2 are conditionally independent given C if

$$P(A_1 \cap A_2 | C) = P(A_1|C)P(A_2|C) . \tag{14.7}$$

If $P(C) = 0$ then both sides of (14.7) are taken to be 0 and conditional independence holds. ◇

Definition 14.4.1 is same as saying that A_1 and A_2 are independent in the probability space $(\Omega, \mathcal{A}, P_C)$ where P_C is the conditional probability measure as in Definition 14.3.1. This has an obvious extension to conditional independence of several events $\{A_i\}$ given one event C.

Example 14.4.1. In Example 14.1.1, A and B are conditionally independent given C. ▲

Exercise 14.4.1 Let A, B, C be three events. Show that A and B are conditionally independent given C if

(a) $P(B \cap C) = 0$ or $P(A \cap C) = 0$.

(b) $P(B \cap C) > 0$ and $P(A|B \cap C) = P(A|C)$.

(c) $P(A \cap C) > 0$ and $P(B|A \cap C) = P(B|C)$.

Exercise 14.4.2 If A, B, C are three independent events show that any two of them are conditionally independent given the third.

Exercise 14.4.3 Construct an example where A and B are independent but they are not conditionally independent given C.

Exercise 14.4.4 You have two coins in your pocket. One of them is a fair coin and the other is a two-headed coin. You pick a coin with probabilities 1/2 each, and then toss the chosen coin twice. Let

$$\begin{aligned} A &:= \{\text{First toss is a head}\} \\ B &:= \{\text{Second toss is a head}\} \\ C &:= \{\text{You had chosen the fair coin}\}. \end{aligned}$$

Show that A and B are not independent but they are conditionally independent given C.

An equivalent definition of condition independence can be based on the idea of conditional probability given two events as follows. Let A, B, C be three events, $P(B \cap C) > 0$. Then we define the *conditional probability of A given B and C* as

$$P(A|B,C) := P(A|B \cap C) = P(A \cap B|C)/P(B|C).$$

[To anticipate the general definition, stated in Chapter 17, of conditional probability given a σ-field, this is indeed the conditional probability of A given $\sigma(B,C)$.] We can now state the following definition.

Definition 14.4.2. A and B are conditionally independent given C if

$$P(A|B,C) = P(A|C).$$

[In case $P(B \cap C) = 0$, we say that conditional independence holds.] ◇

Exercise 14.4.5 Show that Definitions 14.4.1 and 14.4.2 are equivalent.

Condition independence of sub-σ-fields and random variables given an event C is defined in the natural way. We restrict to two sub-σ-fields, extension to several sub-σ-fields being obvious.

Definition 14.4.3. (Independence of rvs and σ-fields given an event)
(i) Sub-σ-fields $\mathcal{G}_1$ and $\mathcal{G}_2$ are conditionally independent given C if

$$P(A \cap B|C) = P(A|C)P(B|C) \quad \text{for all} \quad A \in \mathcal{G}_1, B \in \mathcal{G}_2.$$

(ii) Random variables or vectors X and Y are conditionally independent given C if $\sigma(X)$ and $\sigma(Y)$ are conditionally independent given C. ◇

14.5 Exercises

Exercise 14.5.1 Suppose $\{A_1, \ldots, A_n\}$ are events. Show that they are independent if and only if for all 2^n possible choices of $B_i = A_i$ or A_i^c, $1 \le i \le n$, $P(B_1 \cap \cdots \cap B_n) = P(B_1) \cdots P(B_n)$.

Exercise 14.5.2 Let $\{A_n\}$ be independent events. Show that

$$P(\cap_{n=1}^{\infty} A_n) = \prod_{n=1}^{\infty} P(A_n),$$

$$P(\cup_{n=1}^{\infty} A_n) = 1 - \prod_{n=1}^{\infty} (1 - P(A_n)).$$

Exercise 14.5.3 Let $(\Omega, \mathcal{A}, P)$ be a probability space. Let $\mathcal{B}$ and $\mathcal{C}$ be independent sub semi-fields of $\mathcal{A}$. Show that $\sigma(\mathcal{B})$ and $\sigma(\mathcal{C})$ are independent.

Exercise 14.5.4 Let X_1 and X_2 be independent rvs with distribution functions F_1 and F_2 respectively. Show that

$$\mathrm{E}[F_2(X_1)] + \mathrm{E}[F_1(X_2)] = 1 + P\{X_1 = X_2\}.$$

Exercise 14.5.5 Suppose $\{X_i\}$, $1 \le i \le n$ are rvs on $(\Omega, \mathcal{A}, P)$.

(a) Show that $\{X_i\}$ are independent if and only if for all measurable functions $\{f_i\}$ for which $\mathrm{E}(|f_i(X_i)|) < \infty$, $1 \le i \le n$, we have

$$\mathrm{E}\Big[\prod_{i=1}^{n} f_i(X_i)\Big] = \prod_{i=1}^{n} \mathrm{E}[f_i(X_i)].$$

(b) Show that $\{X_i\}$ are independent if and only if for *all bounded* Borel measurable functions $\{g_i\}$ from $\mathbb{R}$ to $\mathbb{R}$,

$$\mathrm{E}\Big[\prod_{i=1}^{n} g_i(X_i)\Big] = \prod_{i=1}^{n} \mathrm{E}[g_i(X_i)].$$

Exercise 14.5.6 Consider the Lebesgue measure on $\mathcal{B}(0,\ 1]$. For every $\omega \in (0,\ 1]$ consider its non-terminating binary expansion. Let $X_i(\omega)$ be the i-th term in this expansion. Show that $\{X_i\}$ are iid Bernoulli rvs.

Exercise 14.5.7 Suppose $\{X_n\}$ are iid and $Y_k = X_k I_{|X_k| \le k}$, $k \ge 1$. Show that $\sum_{k=1}^{\infty} k^{-2}\, \mathrm{V}(Y_k) \le 4\, \mathrm{E}\,|X_1|$.

Exercise 14.5.8 Let X and Y be independent rvs with their induced probability measures μ_1 and μ_2, and distribution functions F_1 and F_2. Show that for any $t \in \mathbb{R}$, and any $B \in \mathcal{B}(\mathbb{R})$,

$$P\{X+Y \le t\} = \int_{\mathbb{R}} F_1(t-x)F_2(dx) = \int_{\mathbb{R}} F_2(t-x)F_1(dx),$$
$$P\{X+Y \in B\} = \int_{\mathbb{R}} \mu_2(B-x)\mu_1(dx)\,.$$

Exercise 14.5.9 Suppose X and Y are independent rvs where X is discrete and Y is absolutely continuous.

(a) Show that the distribution of $Z = X+Y$ is absolutely continuous, and find the density of Z in terms of the pmf of X and the pdf of Y.

(b) What can you say about the nature of the distribution of XY?

Exercise 14.5.10 Let X_1, X_2 be iid $U(0,\ 1)$ rvs. Show that the pdf of $Z_2 = X_1 + X_2$ is the *triangular density*

$$f_2(x) = \begin{cases} x \text{ if } 0 < x \le 1, \\ 2-x \text{ if } 1 \le x < 2, \\ 0 \text{ otherwise}\,. \end{cases}$$

Exercise 14.5.11 Let $\{X_i\}$ be iid $N(0,1)$. Show that

(a) $X_1 + X_2 \sim N(0,2)$.

(b) $(X_1 + X_2 + \cdots + X_n)/\sqrt{n} \sim N(0,1)$ for every $n \ge 1$.

Exercise 14.5.12 Let X_1 and X_2 be iid rvs with density f.

(a) Show that $P\{X_1 = X_2\} = 0$.

(b) Define $X_{(1)} = \min\{X_1, X_2\}$ and $X_{(2)} = \max\{X_1, X_2\}$. Find the densities of $X_{(1)}$, $X_{(2)}$ and $(X_{(1)}, X_{(2)})$ in terms of f. Hint: Start with sets for which probabilities are easier to calculate.

(c) Generalise (b) to the case where we have n iid absolutely continuous rvs. Note that you have to define second minimum....etc. These are called *ordered statistics.*

Exercise 14.5.13 Let $\{X_n\}$ be iid continuous rvs. Then X_n is said to be a **record** if $X_n > \max\{X_1, \ldots X_{n-1}\}$. Call this event A_n. Show that $A_2, A_3, A_4, \ldots$ are independent and $P(A_n) = 1/n$, for $n \ge 2$.

Exercise 14.5.14 Let $\{A_n\}$ be a sequence of independent events such that $P(A_n) < 1$ for all n. Show that $P(\cup A_n) = 1$ iff $P(\limsup A_n) = 1$.

Exercise 14.5.15 Let $\{X_n\}$ be iid $Ber(p)$ rvs, $0 < p < 1$. Let A be the event that in the sequence $\{X_n\}$, the triplet $(1, 0, 1)$ appears infinitely many times. Show that $P(A) = 1$.

Exercise 14.5.16 Let X be a random vector whose components have finite variances. If the components are independent, then the dispersion matrix, $\Sigma(X)$, is a diagonal matrix. Show by an example that the converse is not true.

Exercise 14.5.17 Let $\{X_n\}$ be iid. Show that

$$\mathrm{E}\,|X_1| < \infty \;\text{ iff }\; P\{|X_n| > n\epsilon \text{ infinitely often}\} = 0 \;\text{ for every }\; \epsilon > 0\,.$$

Exercise 14.5.18 Let $\{X_n\}$ be iid with distribution F. Find necessary and sufficient conditions in terms of properties of F for each of the following to hold, as $n \to \infty$:

(i) $X_n/n \to 0$ a.s.

(ii) $X_n/n \to 0$ in probability.

(iii) $\max_{1 \le m \le n} X_m/n \to 0$ a.s.

(iv) $\max_{1 \le m \le n} X_m/n \to 0$ in probability.

Exercise 14.5.19 Suppose $\{X_n\}$ is a sequence of independent rvs. Show that $P\{\sup_n X_n < \infty\} = 1$ iff $\sum_{n=1}^{\infty} P\{X_n > M\} < \infty$ for some $M > 0$.

Exercise 14.5.20 Let $S_i = X_1 + \cdots + X_i, i \ge 1$ where $\{X_n\}$ is iid, $P\{X_n = 1\} = p$, $P\{X_n = -1\} = 1 - p$. Show that if $p \ne 1/2$, then $P\{S_i = 0 \text{ infinitely often}\} = 0$.

Exercise 14.5.21 Let $\{X_n\}$ be iid rvs such that $\mathrm{E}\,X_1^2 < \infty$. Show that for any $\epsilon > 0$, $nP\{|X_n| > \epsilon\sqrt{n}\} \to 0$. Using this, show that $\max_{1 \le i \le n} |X_i|/\sqrt{n} \to 0$ a.s.

Exercise 14.5.22 Let $\{X_n\}$ be iid, $\mathrm{E}(X_1) = 0$, $\mathrm{E}(X_1^2) = 1$. Show that

(a) $\limsup S_n/\sqrt{n} = \infty$ almost surely.

(b) $S_n/\sqrt{n}$ does not converge in probability.

Exercise 14.5.23 Let $\{X_n\}$ be iid rvs. Show that $\mathrm{E}\,|X_1| < \infty$ if and only if $\mathrm{E}[\max_{1 \le k \le n} |X_k|/n] \to 0$.

Exercise 14.5.24 Let $\{X_n\}$ be iid rvs. Let $N = \inf\{n \geq 2 : X_n > X_1\}$. Show that $\mathrm{E}(N) = \infty$.

Exercise 14.5.25 Let $\{X_i\}$ be iid $Ber(p)$ rvs.

(a) Find the probability distribution of the rv N_1 defined below.

$$N_1 = \inf\{n \geq 1 : X_n = 1\}.$$

(b) Find the distribution of N_2 defined below.

$$N_2 = \inf\{n > N_1 : X_n = 1\}.$$

Hint: Use total and conditional probability, along with independence.

(c) Check whether N_1 and $N_2 - N_1$ are independent.

Exercise 14.5.26 Let $\{X_n\}$ be iid $U(0,\ 1)$ random variables. Define $N = \inf\{k : X_k \leq X_{k+1}\}$. Find $P(N > n)$ for every $n \geq 1$, and $\mathrm{E}(N)$.

Exercise 14.5.27 Let $X_n \sim Ber(p_n)$, $n \geq 1$ be independent. Show that

(a) $X_n \xrightarrow{P} 0$ if and only if $\lim p_n = 0$.

(b) $X_n \xrightarrow{a.e.} 0$ if and only if $\sum_{n=0}^{\infty} p_n < \infty$.

Exercise 14.5.28 Let $\{X_i\}$ be iid $Ber(p)$. Show that $X := X_1 + \cdots + X_n$ has the pmf

$$P\{X = k\} = \binom{n}{k} p^k (1-p)^{n-k},\ k = 0, \ldots, n\,. \tag{14.8}$$

This is known as the **binomial distribution** with parameters n and p and we write $X \sim Bin(n,p)$.

Exercise 14.5.29 Let X_1 and X_2 be iid $Exp(\lambda)$. Find the density of $X_1 + X_2$. Can you extend to $n \geq 3$ variables?

Exercise 14.5.30 (a) Let $X \sim Poi(\lambda)$. Show that $\mathrm{E}(X) = \mathrm{V}(X) = \lambda$.

(b) Suppose $X_i \sim Poi(\lambda_i)$, $i = 1, 2$ are independent. Show that then $X_1 + X_2 \sim Poi(\lambda_1 + \lambda_2)$.

Exercise 14.5.31 Let $\{X_n\}$ be iid $Poi(\lambda)$. Show that

$$\frac{\lambda^n}{n!} e^{-\lambda} \leq P\{X_n \geq n\} \leq \frac{\lambda^n}{n!}.$$

Using this, show that

$$P\{\limsup \frac{X_n}{\log\log n}\} = 1\,.$$

Exercise 14.5.32 Let $\{X_n\}$ be independent with distributions $\{F_n\}$.
(a) Show that

$$P\{\sup_n X_n < \infty\} = \begin{cases} 1 & \text{if for some } \ x,\ \sum_{n=1}^{\infty}(1 - F_n(x)) < \infty, \\ 0 & \text{otherwise}\,. \end{cases}$$

(b) Suppose $P\{\sup_n X_n < \infty\} = 1$. Let $X = \sup X_n$. Show that the distribution function of X is $\prod_{n=1}^{\infty} F_n$.
(c) In that case, prove $\mathrm{E}(X^+) < \infty$ iff $\sum \int_{X_n > x} X_n dP < \infty$ for some x.

Exercise 14.5.33 Suppose $\{X_n\}$ are non-negative. Show that
(a) if $\sum \mathrm{E}(X_n) < \infty$, then $\sum X_n < \infty$ a.s.
(b) if X_n are independent and uniformly bounded, and $\sum \mathrm{E}(X_n) = \infty$, then $\sum X_n = \infty$ a.s.

Exercise 14.5.34 Determine the behaviour of $\sum_{n=1}^{\infty} \frac{X_n}{n}$, where $\{X_n\}$ is iid with $P\{X_n = 1\} = \frac{1}{2} = 1 - P\{X_n = -1\}$.

Exercise 14.5.35 Let $\{X_n\}$ be independent zero mean rvs such that $\sum_{n=1}^{\infty} \mathrm{V}(X_n) < \infty$. Show that $\sum_{k=1}^{n} X_k$ converges in probability.
Hint: Show that this sequence is Cauchy in probability.

Exercise 14.5.36 $\{X_1, \ldots X_d\}$ is said to be **exchangeable** if its distribution is the same as that of $(X_{\pi(1)}, \ldots X_{\pi(d)})$ for any permutation $\pi(\cdot)$ of $\{1, \ldots, d\}$.
(a) Verify that in that case, $\mathrm{E}(X_i)$, $\mathrm{V}(X_i)$, $\mathrm{Cov}(X_i,\ X_j)$ $(i \neq j)$ are free of i and j (provided of course they exist).
(b) Check that a sequence of iid rvs is exchangeable.

Exercise 14.5.37 Let $(X_1, \ldots, X_d)'$ be a Gaussian random vector. Give a necessary and sufficient condition for $X_1, \ldots, X_d$ to be exchangeable.

Exercise 14.5.38 Let $X_1, \ldots, X_n$ be iid continuous rvs. Let R_i be the rank of X_i. So, R_1 is 1 if $X_1 = \max\{X_1, \ldots, X_n\}$ etc. Check that the probability of ties is 0. Is $R_1, \ldots, R_n$ exchangeable? Describe the probability distribution of the random vector $R := (R_1, \ldots, R_n)'$.

Exercise 14.5.39 An infinite sequence $\{X_i\}$ of random variables is called exchangeable if all finite sub-sequences are exchangeable. Suppose we have an infinite sequence of exchangeable rvs, each with variance one. What can you say about the covariances?

Exercise 14.5.40 Suppose Box 1 has 6 red and 4 blue marbles. Box 2 has 5 red, 4 blue and 10 white marbles. Suppose we pick Box 1 or 2 with probabilities 1/3 and 2/3. Once a box is picked, we draw a marble from that box giving equal chance to all the marbles in the box. Formulate this with appropriate probability spaces and use conditional and total probability to compute the probability that the marble drawn is red.

Exercise 14.5.41 (Key problem) Suppose you have n keys and exactly one of them can open a door. You try the keys one by one (at random!). Note that there are two ways of doing it: discard any key that has failed, before trying again or, always choose from the full set of keys. Let N_n be the number of tries needed to open the door. Derive probabilistic properties of N_n in each case and see what happens when $n \to \infty$.

Exercise 14.5.42 In Exercise 14.5.41, let X_i be the indicator variable of whether or not the door opens at the ith trial. Show that $X_1, \ldots, X_n$ is exchangeable, irrespective of with or without replacemnet of keys.

Exercise 14.5.43 (Polya's Urn scheme). Suppose an urn contains r red balls and b black balls. One ball is drawn at random. If it is red then c red balls are added to the urn. Likewise if it is black, then c black balls are added. The drawn ball is returned to the urn. This process is repeated. Let p_n be the probability that the ball drawn at the nth try is red.

(a) What can you say about the sequence p_n, $n \geq 1$?

(b) Would your answer in (a) change if you had not put the drawn balls back in the urn?

Exercise 14.5.44 Let $\{X_i\}$ be iid rvs with finite mean μ. Let the sequence of partial sums be $S_n := \sum_{i=1}^{n} X_i \, , n \geq 1$.

(a) Show that $\{S_n/n\}$ is ui.

(b) The *Strong Law of Large Numbers (SLLN)* says that $S_n/n \to \mu$ a.s. We shall prove this in Chapter16. Using (a) and the SLLN, show that $n^{-1}S_n \to \mu$ in L^1.

Exercise 14.5.45 Let $\{X_i\}$ be iid standard Gaussian rvs.
(a) Show that for every $\epsilon > 0$,

$$P\{\omega : \limsup_{n\to\infty} \frac{X_n}{\sqrt{\log n}} \geq \sqrt{2} + \epsilon\} = 0. \tag{14.9}$$

Hint: Use Mills ratio (Exercise 11.4.3) and First Borel-Cantelli lemma.
(b) Show that for every small $\epsilon > 0$,

$$P\{\omega : \limsup_{n\to\infty} \frac{X_n(\omega)}{\sqrt{\log n}} > \sqrt{2} - \epsilon\}) = 1. \tag{14.10}$$

Hint: Use Mills ratio and Second Borel-Cantelli lemma.
(c) Using (14.9) and (14.10), show that

$$P\{\omega : \limsup_{n\to\infty} \frac{X_n(\omega)}{\sqrt{\log n}} = \sqrt{2}\} = P\{\omega : \liminf_{n\to\infty} \frac{X_n(\omega)}{\sqrt{\log n}} = -\sqrt{2}\} = 1. \tag{14.11}$$

Chapter 15

$0-1$ laws

We know that A is independent of itself iff $P(A) = 0$ or 1. We now demonstrate classes of sets with this property in the context of independent random variables. These results are usually called $0-1$ laws.

15.1 Kolmogorov's $0-1$ law

Definition 15.1.1. (**Tail σ-field**) Let $\{X_i\}$ be a sequence of rvs, and

$$\mathcal{T}_n := \sigma(X_n, X_{n+1}, \ldots) \text{ and } \mathcal{T}_\infty := \cap_{n=1}^{\infty} \mathcal{T}_n . \tag{15.1}$$

$\mathcal{T}_\infty$ is known as the *tail σ-field* of $\{X_i\}$ and its elements are known as **tail events**, and $f : (\Omega, \mathcal{T}_\infty) \to (\bar{\mathbb{R}}, \mathcal{B}(\bar{\mathbb{R}}))$ is called a **tail function**. ◇

Since $\{\mathcal{T}_n\}$ is a decreasing sequence, $\mathcal{T}_\infty$ is a σ-field. Tail events as defined in (15.1) are with respect to the given sequence of rvs.

Exercise 15.1.1 (a) Verify that the following sets are tail events:
(i) $A_1 = \{\lim X_n \text{ exists}\}$.
(ii) $A_2 = \{\sum_{n=1}^{\infty} X_n \text{ converges}\}$.
(iii) $A_3 = \{\limsup X_n = \liminf X_n\}$.
(iv) $A_4 = \{X_n < 2 \text{ for infinitely many } n\}$.
(b) Show that $\limsup X_n$, $\liminf X_n$ are tail functions.

Definition 15.1.2. (**Trivial σ-field**) Let $(\Omega, \mathcal{A}, P)$ be a probability space. A σ-field $\mathcal{C} \subset \mathcal{A}$ is *trivial* if $P(A) = 0$ or 1 for all $A \in \mathcal{C}$. ◇

A. Bose, *Elements of Measure and Probability*, Texts and Readings in Mathematics 88,
https://doi.org/10.1007/978-981-95-2758-8_15

Exercise 15.1.2 Let $(\Omega, \mathcal{A}, P)$ be a probability space and let $\mathcal{C} \subset \mathcal{A}$ be a trivial σ-field. Show that if $f : (\Omega, \mathcal{C}, P) \to (\mathbb{R}, \mathcal{B}(\mathbb{R}))$ then it is a constant almost surely.

Theorem 15.1.1 (Kolmogorov's zero-one law). *Let $\{X_i\}$ be a sequence of independent random variables on $(\Omega, \mathcal{A}, P)$. Then their tail σ-field $\mathcal{T}_\infty$ is trivial, and hence every real-valued function which is measurable with respect to $\mathcal{T}_\infty$ is a constant almost surely.* ◆

Proof. Fix $A \in \mathcal{T}_\infty$. Note that $\mathcal{T}_\infty \subseteq \mathcal{T}_1$. Hence for some $A^* \in \mathcal{B}(\mathbb{R}^\infty)$,

$$A = \{\omega : (X_1(\omega), X_2(\omega), \ldots) \in A^*\} . \tag{15.2}$$

For any $C^* \in \mathcal{B}(\mathbb{R}^\infty)$, define

$$C := \{\omega : (X_1(\omega), X_2(\omega), \ldots) \in C^*\} .$$

Let

$$\mathcal{C} = \{C^* \in \mathcal{B}(\mathbb{R}^\infty) : C \text{ and } A \text{ are independent}\} .$$

First suppose C^* is a measurable cylinder, as in Definition 10.4.1. Then for some integer n, and $B_n \in \mathcal{B}(\mathbb{R}^n)$,

$$C = \{\omega : (X_1(\omega), \ldots, X_n(\omega) \in B_n\} \in \sigma(X_1, \ldots, X_n) .$$

Note that $A \in \mathcal{T}_{n+1}$. Since $\{X_i\}$ are independent, clearly A and C are independent. Thus $\mathcal{C}$ contains all measurable cylinders.

Clearly $\mathcal{C}$ is a monotone class. For example, suppose $C_n^* \uparrow C^*$, $C_n^* \in \mathcal{C}$ for all n. Then, $C_n \uparrow C$. Now, $P(A)P(C_n) = P(A \cap C_n) \uparrow P(A \cap C)$. But $P(C_n) \uparrow P(C)$ and hence we have $P(A)P(C) = P(A \cap C)$. Thus A and C are independent. Similarly for decreasing limits.

By monotone class Theorem 2.6.1, $\mathcal{C} = \mathcal{B}(\mathbb{R}^\infty)$. Consequentially, $A^* \in \mathcal{B}(\mathbb{R}^\infty)$. That is, A is independent of itself. By Exercise 14.1.2(c), $P(A) = 0$ or 1. The second part follows from Exercise 15.1.2. ■

Example 15.1.1. Suppose $\{X_i\}$ are iid with mean 0. Define

$$A = \{\omega : \frac{S_n}{n} \to 0\} .$$

Let $\mathcal{T}_\infty$ be the tail σ-field of $\{X_i\}$. Then, $A \in \mathcal{T}_\infty$. So by Theorem 15.1.1, $P(A)$ equals 0 or 1. In Section 16.1 we shall see that $P(A) = 1$. ▲

15.2 Hewitt-Savage 0 − 1 law

If $\{X_i\}$ are iid, a larger σ-field is trivial.

Definition 15.2.1. (**Tail symmetric event**) Let $\{X_i\}$ be a sequence of random variables. Any event $A \in \sigma(X_1, X_2, \ldots)$ is said to be *tail symmetric* if the occurrence of A is not affected by a permutation of the positions of any finitely many X_i in the sequence $\{X_1, X_2, \ldots\}$. Let

$$\mathcal{H}_\infty = \{A \in \sigma(X_1, X_2, \ldots) : A \text{ is tail symmetric}\}. \tag{15.3}$$

Then $\mathcal{H}_\infty$ is called the **tail symmetric σ-field.** ◇

Exercise 15.2.1 Show that $\mathcal{H}_\infty$ defined in (15.3) is indeed a σ-field.

Note that tail symmetry is with respect to the specific sequence $\{X_i\}$ on hand. Let $T : \{1, 2, \ldots\} \to \{1, 2, \ldots\}$ denote a typical permutation that permutes only finitely many co-ordinates. Then A is tail symmetric if and only if there exists some $A^* \in \mathcal{B}(\mathbb{R}^\infty)$ such that

$$A = \{\omega : (X_{T(1)}(\omega), X_{T(2)}(\omega), \ldots) \in A^*\} \quad \text{for all} \quad T.$$

Exercise 15.2.2 (a) Check that any tail event is a tail symmetric event, and hence $\mathcal{T}_\infty \subset \mathcal{H}_\infty$.
(b) Check that $\{X_n = 0 \text{ for all } n\}$ is tail symmetric but not a tail event, and hence $\mathcal{T}_\infty \subsetneq \mathcal{H}_\infty$.

Theorem 15.2.1 (**Hewitt-Savage zero-one Law**)**.** *Suppose $\{X_i\}$ are iid. Then $\mathcal{H}_\infty$ as defined in (15.3) is trivial and hence any real-valued function measurable with respect to $\mathcal{H}_\infty$ is a constant a.s.* ◆

Proof. Fix $A \in \mathcal{H}_\infty$. Write $X = (X_1, X_2, \ldots)$. Then A is as in (15.2) for some A^* and

$$P(A) = P_X(A^*).$$

By Theorem 4.5.1, sets in the σ-field can be approximated by sets in the generating field. So we can obtain measurable cylinders C_k^* such that, $P_X(A^* \Delta C_k^*) \to 0$ as $k \to \infty$.

Let $B_k \in \mathcal{B}(\mathbb{R}^{n_k})$ be such that

$$C_k = \{X \in C_k^*\} = \{(X_1, \ldots, X_{n_k}) \in B_k\}.$$

Let T_k be the permutation that interchanges the sets $\{1, \ldots, n_k\}$ and $\{n_k + 1, \ldots, 2n_k\}$. Let

$$X(T_k) = (X_{n_k+1}, \ldots X_{2n_k}, X_1, \ldots, X_{n_k}, \ldots)$$
$$C_k(T_k) = \{X(T_k) \in C_k^*\} = \{(X_{n_k+1}, \ldots, X_{2n_k}) \in B_k\}.$$

Since $\{X_i\}$ are iid, $X \stackrel{D}{=} X(T_k)$. As a consequence,

$$\begin{aligned} P(A\Delta C_k) &= P_X(A^*\Delta C_k^*) \\ &= P_{X(T_k)}(A^*\Delta C_k^*) \\ &= P[\{X(T_k) \in A^*\}\Delta\{X(T_k) \in C_k^*)\}] \\ &= P[\{X \in A^*\}\Delta\{X(T_k) \in C_k^*)\}], \text{ as } A \text{ is tail symmetric} \\ &= P(A\Delta C_k(T_k)). \end{aligned}$$

Hence, as $k \to \infty$, $P(A\Delta C_k) \to 0$, and $P(A\Delta C_k(T_k)) \to 0$. This implies, $P(A\Delta[C_k \cap C_k(T_k)]) \to 0$. Therefore, $P(C_k)$, $P(C_k(T_k))$, $P[C_k \cap C_k(T_k)]$ all converge to $P(A)$.

On the other hand,

$$\begin{aligned} P[C_k \cap C_k(T_k)] &= P[\{(X_1, \ldots X_{n_k}) \in B_k,\ (X_{n_k+1}, \ldots X_{2n_k}) \in B_k\}] \\ &= P(C_k)P(C_k(T_k)) \text{ by independence.} \end{aligned}$$

If we let $k \to \infty$, we get $P(A) = P(A)P(A)$, and hence $P(A) = 0$ or 1. Therefore, the first part of the theorem is proved.

The second part now follows from Exercise 15.1.2. ∎

15.3 Exercises

Exercise 15.3.1 Prove that the following general version of the 0 − 1 law of Kolmogorov holds : Let $(\Omega, \mathcal{A}, P)$ be a probability space. Suppose $\{\mathcal{A}_n\}$ is a sequence of independent sub-σ-fields of $\mathcal{A}$. Show that for any $A \in \cup_{n=1}^{\infty} \cap_{k=n}^{\infty} \mathcal{A}_n$, $P(A)$ is either 0 or 1.

Exercise 15.3.2 Let $\{X_n\}$ be independent rvs. Show that the radius of convergence of $\sum_{n=1}^{\infty} X_n z^n$ is a constant a.s.

Chapter 16

Sums of independent random variables

For any sequence $\{X_i\}$, the sequence $S_k(X) = S_k := X_1 + \cdots + X_k$ is known as the sequence of **partial sums**. We study the behaviour of this and related sequences in this chapter, when $\{X_i\}$ are independent. We first discuss the strong law of large numbers which roughly says that the time average of iid random variables converges a.s. to the space average, as the number of summands goes to ∞. This extends the weak law of large numbers presented in Exercise 14.2.8. Then we study the probabilistic behaviour of $\max_{1\leq k\leq n} S_k$ as compared to that of S_n. These are known as maximal inequalities. Then we move to the study of the behaviour of partial sums as $n \to \infty$. In particular, for independent rvs, S_n converges a.s. if and only if it converges in probability. The celebrated Kolmogorov three series theorem which relates the convergence of partial sums to the convergence of the sum of means, variance and tail probabilities of related rvs is established. We state without proof the law of iterated logarithm which gives the behaviour of the limit points of appropriately scaled partial sums. Finally, we present a few important inequalities, which substantially improve the polynomial upper bound in Markov's inequality Lemma 8.1.1, to exponential bounds. In particular, we cover Hoeffding's inequality and, without proof, an inequality of Talagrand.

A. Bose, *Elements of Measure and Probability*, Texts and Readings in Mathematics 88,
https://doi.org/10.1007/978-981-95-2758-8_16

16.1 Strong law of large numbers

Let $S_n = X_1 + \cdots + X_n$, $n \geq 1$ be the **partial sums**. In Exercise 14.2.8 we saw that if $\{X_i\}$ are iid with finite variance then $S_n/n \xrightarrow{P} \mathrm{E}(X_1)$. In Exercise 14.2.10 we saw that if $\{X_i\}$ are independent with mean 0 and uniformly bounded fourth moment, then $S_n/n \to 0$ a.s. We now proceed to state and prove a much stronger result.

Definition 16.1.1. Random variables $\{X_i\}$ are said to be **pairwise independent** if for every $i \neq j$, X_i and X_j are independent. ◊

Example 16.1.1. Let X_1, X_2, X_3 be iid with pmf $P\{X_1 = \pm 1\} = 1/2$. Then it is easy to check that X_1X_2, X_1X_3, X_2X_3 are pairwise independent but not jointly independent. ▲

Exercise 16.1.1 Show that if $\{X_i\}$ are pairwise independent rvs with finite variances, then $\mathrm{V}(X_1 + \cdots + X_n) = \mathrm{V}(X_1) + \mathrm{V}(X_2) + \cdots + \mathrm{V}(X_n)$.

Many versions of the strong law of large numbers (SLLN) have been discovered over the last 90 years or so. Traditionally, it is stated for iid rvs, and proved using Kolmogorov's maximal inequality Theorem 16.2.1. We present a relatively recent version by Etemadi [1981]. For a more traditional statement, and outline of proof, see Exercise 16.6.1.

Theorem 16.1.1 (Strong Law of large Numbers (SLLN)). *Let $\{X_i\}$ be pairwise independent identically distributed random variables. Then $\mathrm{E}|X_1| < \infty$ implies $S_n/n \to \mathrm{E}(X_1)$ almost surely and in L^1.* ◆

Proof of Theorem 16.1.1. Since $\{X_i^+\}$ and $\{X_i^-\}$ satisfy the hypothesis of the theorem, we may assume that $\{X_i\}$ are non-negative. Define

$$Y_i := X_i I\{X_i \leq i\}, \quad S_n^* = Y_1 + \cdots + Y_n\,.$$

Note that

$$0 \leq Y_i \leq i \text{ for all } i, \text{ and } \mathrm{E}(Y_i) \to \mathrm{E}(X_1) \text{ as } i \to \infty\,.$$

Fix $\epsilon > 0$, $\alpha > 1$, and define

$$k_n := [\alpha^n], \text{ the largest integer not more than } \alpha^n\,.$$

We will prove the claimed a.s. convergence in three steps:

(i) $\dfrac{S^*_{k_n}}{k_n} \stackrel{a.s}{\to} \mathrm{E}(X_1)$ (i.e. SLLN holds for $\{Y_i\}$ along this sub-sequence).

(ii) $\dfrac{S_{k_n}}{k_n} \stackrel{a.s}{\to} \mathrm{E}(X_1)$ (i.e. SLLN holds for $\{X_i\}$ along this sub-sequence).

(iii) $\dfrac{S_n}{n} \stackrel{a.s}{\to} \mathrm{E}(X_1)$.

Proof of Step (i). We have noted that $\mathrm{E}(Y_n) \to \mathrm{E}(X_1)$. We also know that if $a_n \to a$ then $(a_1 + \cdots + a_n)/n \to a$. Hence

$$\frac{\mathrm{E}(S^*_{k_n})}{k_n} = \frac{\mathrm{E}(Y_1) + \cdots + \mathrm{E}(Y_{k_n})}{k_n} \to \mathrm{E}(X_1). \tag{16.1}$$

Below, C is a constant which depends only on ϵ and α. Now,

$$\begin{aligned}
\sum_{n=1}^{\infty} P\Big\{\Big|\frac{S^*_{k_n} - \mathrm{E}(S^*_{k_n})}{k_n}\Big| > \epsilon\Big\}
&\le \frac{1}{\epsilon^2}\sum_{n=1}^{\infty}\frac{Var(S^*_{k_n})}{k_n^2} \text{ (Chebyshev's inequality)}\\
&= \frac{1}{\epsilon^2}\sum_{n=1}^{\infty}\frac{1}{k_n^2}\sum_{i=1}^{k_n}\mathrm{V}(Y_i) \text{ (independence of } \{Y_i\})\\
&\le C\sum_{i=1}^{\infty}\mathrm{E}(Y_i^2)\sum_{n:k_n\ge i}\frac{1}{k_n^2} \text{ (since } \mathrm{V}(Y_i) \le \mathrm{E}(Y_i^2))\\
&\le C\sum_{i=1}^{\infty}\frac{\mathrm{E}(Y_i^2)}{i^2}\\
&= C\sum_{i=1}^{\infty}\frac{1}{i^2}\int_{(0,\ i]} x^2F(dx) \text{ (} F \text{ is the cdf of } X_1)\\
&= C\sum_{i=1}^{\infty}\frac{1}{i^2}\sum_{k=0}^{i-1}\int_{(k,\ k+1]} x^2F(dx)\\
&= C\sum_{k=0}^{\infty}\int_{(k,\ k+1]} x^2F(dx)\sum_{i=k+1}^{\infty}\frac{1}{i^2} \text{ (Fubini's theorem)}\\
&\le C\sum_{k=0}^{\infty}\frac{1}{k+1}\int_{(k,\ k+1]} x^2F(dx)\\
&\le C\sum_{k=0}^{\infty}\int_{(k,\ k+1]} xF(dx) = C\,\mathrm{E}(X_1) < \infty.
\end{aligned}$$

Hence by First Borel-Cantelli Lemma 3.3.1,

$$\frac{S^*_{k_n} - \mathrm{E}(S^*_{k_n})}{k_n} \overset{a.s.}{\to} 0. \tag{16.2}$$

Thus by using (16.2) and (16.1), $\frac{S^*_{k_n}}{k_n} \overset{a.e.}{\to} \mathrm{E}(X_1)$, proving Step (i).

Proof of Step (ii). This is proved by showing that the sequences $\{X_i\}$ and $\{Y_i\}$ differ only at finitely many indices a.s.

To see this, note that,

$$\sum_{n=1}^{\infty} P\{X_n \neq Y_n\} = \sum_{n=1}^{\infty} P\{X_n > n\} \leq \mathrm{E}(X_1) < \infty.$$

Hence, by First Borel-Cantelli Lemma 3.3.1, $X_n \neq Y_n$ only for finitely many n a.s.

Therefore, using Step (i), $S_{k_n}/k_n \overset{a.e.}{\to} \mathrm{E}(X_1)$, proving Step (ii).

Proof of Step (iii). Since $\{X_i\}$ are non-negative, S_n is non-decreasing.

For every n, let s_n be the largest j such that k_j is smaller than n. Note that $k_{s_{n+1}} \geq n$ and

$$\lim_{n\to\infty} \frac{k_{s_{n+1}}}{k_{s_n}} = \alpha.$$

Then

$$\limsup S_n/n \leq \limsup \frac{S_{k_{s_{n+1}}}}{k_{s_{n+1}}} \frac{k_{s_{n+1}}}{k_{s_n}} \frac{k_{s_n}}{n} \leq \mathrm{E}(X_1)\alpha \quad \text{a.s.}$$

Similarly

$$\liminf S_n/n \geq \mathrm{E}(X_1)\frac{1}{\alpha} \quad \text{a.s}\,.$$

Since $\alpha > 1$ is arbitrary, letting $\alpha \to 1$ (through a countable sequence, so that null sets do not pile up into a non-null set),

$$\lim \frac{S_n}{n} = \mathrm{E}(X_1) \quad \text{a.s}\,.$$

This proves the first part.

By Exercise 14.5.44(a), S_n/n is ui. Hence by applying Theorem 13.4.1 $S_n/n \to \mathrm{E}(X_1)$ in L^1, and this completes the proof. ■

16.2 Two maximal inequalities

We state and prove two historically important probability inequalities for $\{S_k\}$ when $\{X_i\}$ are independent rvs. They imply that the probabilistic behaviour of the maximum of the entire path $S_1, \cdots, S_n$ is controlled in terms of the end-point S_n alone. Both inequalities are extremely useful.

16.2.1 Kolmogorov's maximal inequality

The proof of Kolmogorov's inequality contains ideas that are used elsewhere, for example in the theory of martingales. It can also be used to prove the SLLN for iid rvs.

Theorem 16.2.1 (Kolmogorov's maximal inequality). *Let $\{X_i\}$ be independent rvs where each X_i has zero mean and finite variance. For each $k \geq 1$, let $S_k = X_1 + \cdots + X_k$. Then*

$$P\{\max_{1\leq k\leq n} |S_k| \geq t\} \leq \frac{1}{t^2}\,\mathrm{E}(S_n^2) = \frac{1}{t^2}\sum_{k=1}^{n} \mathrm{E}(X_k^2),\ t > 0\,.$$

$$P\{\sup_{k\geq 1} |S_k| \geq t\} \leq \frac{1}{t^2}\sum_{k=1}^{\infty} \mathrm{E}(X_k^2),\ t > 0\,. \qquad \blacklozenge$$

Proof. The second inequality follows from the first when we let $n \to \infty$. To prove the first inequality, let $S_0 := 0$, and for $1 \leq k \leq n$, define,

$$\begin{aligned} E_k &:= \{|S_j| < t, j = 1, \ldots, k-1,\ |S_k| \geq t\}, \\ E &:= \cup_{i=1}^n E_k = \{\max_{1\leq k\leq n} |S_k| \geq t\}\,. \end{aligned}$$

Note that

$$\mathrm{E}(S_n^2) \geq \mathrm{E}\left(S_n^2 \mathbf{1}_E\right) = \sum_{k=1}^{n} \mathrm{E}\left(S_n^2 \mathbf{1}_{E_k}\right). \tag{16.3}$$

Let $Y_k := X_{k+1} + \cdots + X_n$. Then $S_n = S_k + Y_k$, and hence

$$\mathrm{E}\left(S_n^2 \mathbf{1}_{E_k}\right) = \mathrm{E}\left(S_k^2 \mathbf{1}_{E_k}\right) + 2\,\mathrm{E}\left(S_k Y_k \mathbf{1}_{E_k}\right) + \mathrm{E}\left(Y_k^2 \mathbf{1}_{E_k}\right). \tag{16.4}$$

Note that Y_k is independent of $(S_k, \mathbf{1}_{E_k})$ and $\mathrm{E}(Y_k) = 0$. Hence,

$$\mathrm{E}\left(S_k Y_k \mathbf{1}_{E_k}\right) = \mathrm{E}\left(S_k \mathbf{1}_{E_k}\right)\mathrm{E}(Y_k) = 0\,.$$

Hence using (16.4) it follows that

$$\mathrm{E}\left(S_n^2 \mathbf{1}_{E_k}\right) \geq \mathrm{E}\left(S_k^2 \mathbf{1}_{E_k}\right) \geq t^2 P(E_k), \quad \text{by using definition of } E_k\,.$$

Hence, by (16.3),

$$\mathrm{E}(S_n^2) \geq t^2 \sum_{k=1}^{n} P(E_k) = t^2 P(E). \qquad \blacksquare$$

16.2.2 Lévy's maximal inequality

This maximal inequality does not require the variances to be finite.

Theorem 16.2.2 (Lévy's Maximal inequality). *Suppose that $\{X_i\}$ are independent rvs and $S_k := X_1 + \cdots + X_k$, $k \geq 1$. Then*

$$P\{\max_{1\leq k\leq n} |S_k| \geq 3t\} \leq 3 \max_{1\leq k\leq n} P\{|S_k| \geq t\} \quad \textit{for all} \quad t > 0\,. \qquad \blacklozenge$$

Proof. Define $S_0 = 0$ and

$$B_k := \{|S_k| \geq 3t,\ |S_j| < 3t \text{ for } 1 \leq j < k\}\,,\, 1 \leq k \leq n\,.$$

The sets $B_1, \ldots, B_n$ are disjoint, and let

$$B := B_1 \cup \cdots \cup B_n = \{\max_{1\leq k\leq n} |S_k| \geq 3t\}\,.$$

Notice that

$$\begin{aligned}
P(B) &\leq P\{|S_n| \geq t\} + P\left(B \cap \{|S_n| < t\}\right) \\
&= P\{|S_n| \geq t\} + \sum_{k=1}^{n} P\left(B_k \cap \{|S_n| < t\}\right) \\
&\leq P\{|S_n| \geq t\} + \sum_{k=1}^{n} P\left(B_k \cap \{|S_n - S_k| \geq 2t\}\right) \text{(definition of } B_k) \\
&= P\{|S_n| \geq t\} + \sum_{k=1}^{n} P(B_k) P\{|S_n - S_k| \geq 2t\} \\
&\leq P\{|S_n| \geq t\} + \max_{1\leq k\leq n} P\{|S_n - S_k| \geq 2t\} \\
&\leq P\{|S_n| \geq t\} + \max_{1\leq k\leq n} P\{|S_k| \geq t\} + P\{|S_n| \geq t\} \\
&\leq 3 \max_{1\leq k\leq n} P\{|S_k| \geq t\}\,.
\end{aligned} \qquad \blacksquare$$

16.3 Infinite sum of independent rvs

16.3.1 Equivalence of probability and a.s. convergence

While convergence in probability does not imply a.s. convergence in general, for partial sums of independent rvs this is true.

Theorem 16.3.1. *If $\{X_i\}$ are independent rvs then $\sum_{n=1}^{\infty} X_n$ converges a.s. if and only if it converges in probability.* ◆

The proof of Theorem 16.3.1 is facilitated by the following lemma.

Lemma 16.3.1. A sequence of rvs $\{Z_n\}$ converges to 0 a.s. if

$$\lim_{N\to\infty} \sup_{n\geq 1} P\{\max_{0\leq j\leq n} |Z_{N+j}| > \varepsilon\} = 0 \text{ for all } \varepsilon > 0. \tag{16.5}$$ ◆

Proof. By the given condition, for all $\varepsilon, \delta > 0$, there exists N such that $P\{\sup_{j\geq 0} |Z_{N+j}| > \varepsilon\} \leq \delta$. So there exists $1 \leq n_1 < n_2 < \cdots$ such that

$$P\{\sup_{j\geq 0} |Z_{n_k+j}| > 2^{-k}\} \leq 2^{-k}, \text{ for all } k \geq 1. \tag{16.6}$$

Define

$$A := \limsup_{k\to\infty} \{\sup_{j\geq 0} |Z_{n_k+j}| > 2^{-k}\}.$$

Using (16.6), by the First Borel-Cantelli Lemma 3.3.1, $P(A) = 0$. We claim that for all $\omega \in A^c$, $Z_n(\omega) \to 0$.

To see this, fix $\omega \in A^c$. Then there exists K such that for all $k \geq K$, $\sup_{j\geq 0} |Z_{n_k+j}(\omega)| \leq 2^{-k}$. This implies that

$$\sup_{j\geq n} |Z_j(\omega)| \to 0 \text{ as } n \to \infty.$$

This completes the proof. ∎

Proof of Theorem 16.3.1. Assume that $S_n := \sum_{j=1}^{n} X_j \xrightarrow{P} S$ (say). We shall use Lemma 16.3.1.

Fix $\varepsilon, \delta > 0$ and let N be such that,

$$P\{(|S_m - S| > \varepsilon/12\} \leq \delta \text{ for all } m \geq N.$$

For $n \geq 1$, observe that,

$$
\begin{aligned}
P\{\max_{0\le j\le n}|S_{N+j}-S|>\varepsilon\}
&\le P\{(|S_N-S|>\varepsilon/2\}+P\{\max_{0\le j\le n}|S_{N+j}-S_N|>\varepsilon/2\}\\
&\le \delta+P\{\max_{0\le j\le n}|S_{N+j}-S_N|>\varepsilon/2\}\\
&= \delta+P\{\max_{1\le j\le n}|\sum_{i=N+1}^{N+j}X_i|>\varepsilon/2\}\\
&\le \delta+3\max_{1\le j\le n}P\{|\sum_{i=N+1}^{N+j}X_i|>\varepsilon/6\},\ \text{by Theorem 16.2.2}\\
&= \delta+3\max_{1\le j\le n}P\{|S_{N+j}-S_N|>\varepsilon/6\}\\
&\le \delta+3P\{|S_N-S|>\varepsilon/12\}+3\max_{1\le j\le n}P\{|S_{N+j}-S|>\varepsilon/12\}\\
&\le 7\delta\,.
\end{aligned}
$$

As δ, n are arbitrary, (16.5) holds for $\{S_n-S\}$, completing the proof. ■

16.3.2 Kolmogorov's three series theorem

We now develop the celebrated three series theorem of Kolmogorov. First, an easy corollary of Theorem 16.3.1.

Corollary 16.3.1. *Let $\{X_i\}$ be independent mean zero rvs such that $\sum_{n=1}^{\infty}\mathrm{E}(X_n^2)<\infty$. Then S_n converges a.s. and in L^2.* ◆

Proof. For all $m<n$, using the fact that $\{X_k\}$ have mean 0 and are independent, $\mathrm{E}\left((S_n-S_m)^2\right)=\sum_{k=m+1}^{n}\mathrm{E}(X_k^2)$, which converges to 0 as $n,m\to\infty$, since $\sum_{n=1}^{\infty}\mathrm{E}(X_n^2)<\infty$. In other words, $\{S_n\}$ is Cauchy in L^2. Since L^2 is complete, S_n converges in L^2 to say S_∞. But L^2 convergence implies converges in probability (use Markov's inequality). Now a.s. convergence follows from Theorem 16.3.1. ■

Exercise 16.3.1 Let $\{X_i\}$ be iid Bernoulli rvs with $P(X_1=\pm1)=1/2$. Let $\{a_n\}\subset\mathbb{R}$, $\sum_{n=1}^{\infty}a_n^2<\infty$. Show that $\sum_{n=1}^{\infty}X_na_n$ converges a.s.

A partial converse of Corollary 16.3.1 is the following:

Theorem 16.3.2. *Let $\{X_i\}$ be independent mean zero rvs and there is a finite constant M such that $\sup_n |X_n| \le M$ a.s. Then $\sum_{n=1}^{\infty} X_n$ converges a.s. implies that $\sum_{n=1}^{\infty} \mathrm{E}(X_n^2) < \infty$.* ◆

Proof. Since $\sum_{k=1}^{\infty} X_k$ converges a.s., there exists a $\lambda > 0$ such that

$$P\{\sup_n |\sum_{k=1}^{n} X_k| \le \lambda\} > 0. \tag{16.7}$$

For any such $\lambda > 0$, define the rv τ as

$$\tau = \begin{cases} \text{first } \; n \text{ such that } |\sum_{k=1}^{n} X_k| > \lambda, \\ \infty \; \text{ if no such } n \text{ exists}. \end{cases}$$

τ is an extended real-valued rv. Note that

$$\{\tau = n\} \in \sigma(X_1, \ldots, X_n) \text{ for all } n \ge 1.$$

[Such rvs are called *stopping times* and play a crucial role in the study of stochastic processes].

Due to (16.7), $P(\tau = \infty) > 0$. Using the facts that τ is a stopping time, $\mathrm{E}(X_k) = 0$ for all k, we have for all $N \ge j$,

$$\int_{\{\tau=j\}} (\sum_{k=1}^{N} X_k)^2 dP = \int_{\{\tau=j\}} (\sum_{k=1}^{j} X_k)^2 dP + \int_{\{\tau=j\}} (\sum_{k=j+1}^{N} X_k)^2 dP. \tag{16.8}$$

On the set $\{\tau = j\}$, $|\sum_{k=1}^{j-1} X_k| \le \lambda$, and hence $|\sum_{k=1}^{j} X_k| \le M + \lambda$.

By independence of X_k and $\{\tau = j\}$ for $k > j$,

$$\int_{\{\tau=j\}} X_k^2 dP = P\{\tau = j\}\,\mathrm{E}(X_k^2) \;\text{ for all }\; k > j.$$

Using the above and recalling (16.8), we have

$$\int_{\{\tau=j\}} (\sum_{k=1}^{N} X_k)^2 dP \le (\lambda + M)^2 P\{\tau = j\} + P\{\tau = j\} \sum_{k=1}^{N} \mathrm{E}(X_k^2).$$

Adding the above over $j = 1, \ldots, N$,

$$\int_{\{\tau \le N\}} (\sum_{k=1}^{N} X_k)^2 dP \le [(\lambda + M)^2 + \sum_{k=1}^{N} \mathrm{E}(X_k^2)] P\{\tau \le N\}. \tag{16.9}$$

Further,

$$\int_{\{\tau>N\}} \Big(\sum_{k=1}^{N} X_k\Big)^2 dP \leq \lambda^2 P\{\tau > N\} \leq (\lambda+M)^2 P\{\tau > N\}. \tag{16.10}$$

Adding (16.9) and (16.10),

$$\sum_{k=1}^{N} \mathrm{E}(X_k^2) \leq (\lambda+M)^2 + P\{\tau \leq N\} \sum_{k=1}^{N} \mathrm{E}(X_k^2)$$

which implies

$$P\{\tau > N\} \sum_{k=1}^{N} \mathrm{E}(X_k^2) \leq (\lambda+M)^2 .$$

Letting $N \to \infty$,

$$P\{\tau = \infty\} \sum_{k=1}^{\infty} \mathrm{E}(X_k^2) \leq (\lambda+M)^2.$$

This implies $\sum_{k=1}^{\infty} \mathrm{E}(X_k^2) < \infty$ since we know that $P\{\tau = \infty\} > 0$. ■

The following extension of Theorem 16.3.2 drops the assumption of mean zero, by bringing in the additional series of the expectations.

Theorem 16.3.3. *Let $\{X_n\}$ be uniformly bounded rvs. Then $\sum_{n=1}^{\infty} X_n$ converges a.s. iff $\sum_{n=1}^{\infty} \mathrm{E}(X_n)$ converges and $\sum_{n=1}^{\infty} \mathrm{V}(X_n) < \infty$.* ◆

Proof. Suppose $\sum_{n=1}^{\infty} \mathrm{E}(X_n)$ and $\sum_{n=1}^{\infty} \mathrm{V}(X_n)$ both converge. Then by Theorem 16.3.2, $\sum_{n=1}^{\infty} (X_n - \mathrm{E}(X_n))$ converges a.s. and hence $\sum_{n=1}^{\infty} X_n$ converges a.s.

Now suppose $\sum_{n=1}^{\infty} X_n$ converges. We use a *symmetrization* technique, which is very useful in probability. On a suitable probability space, construct sequences $\{Y_n\}$ and $\{Z_n\}$ which are all independent and $Y_n \stackrel{D}{=} Z_n \stackrel{D}{=} X_n$ for every $n \geq 1$. Then $\{Y_n\}$ and $\{Z_n\}$ inherit all the probabilistic properties of $\{X_n\}$. Let $\tilde{X}_n = Y_n - Z_n$ for all $n \geq 1$. Now, for all n, $\mathrm{E}(\tilde{X}_n) = 0$. We leave it to the reader to show that $\sum_{n=1}^{\infty} X_n$ converges a.s. implies $\sum_{n=1}^{\infty} \tilde{X}_n$ converges a.s. By Theorem 16.3.2, $2\sum_{n=1}^{\infty} \mathrm{V}(X_n) = \sum_{n=1}^{\infty} \mathrm{V}(\tilde{X}_n) < \infty$.

Theorem 16.3.2 implies that $\sum_{n=1}^{\infty}(X_n - \mathrm{E}(X_n))$ converges a.s. Since $\sum_{n=1}^{\infty} X_n$ converges a.s., we get $\sum_{n=1}^{\infty} \mathrm{E}(X_n)$ converges. ■

We are now ready to state and prove the three series theorem.

Theorem 16.3.4 (**Kolmogorov's Three Series**). *Suppose $\{X_n\}$ are independent rvs. Let*

$$X_n^{(c)} := X_n \mathbf{1}_{\{|X_n| \le c\}}, \quad n \ge 1, c > 0,$$

and consider the three series

$$\sum_{n=1}^{\infty} P\{|X_n| > c\}, \ \sum_{n=1}^{\infty} E(X_n^{(c)}), \ \sum_{n=1}^{\infty} \mathrm{Var}(X_n^{(c)}). \tag{16.11}$$

Then the following are equivalent.

(i) The series $\sum X_n$ converges almost surely.

(ii) The series $\sum X_n$ converges in probability.

(iii) For every $c > 0$, the three series in (16.11) *converge.*

(iv) For some $c > 0$, the three series in (16.11) *converge.* ◆

Proof. **Equivalence of (i) and (ii)**: This follows from Theorem 16.3.1.

(iv) implies (i): Note that $\{X_n^{(c)} - \mathrm{E}(X_n^{(c)})\}$ is a uniformly bounded sequence of mean zero rvs. Since $\sum_{n=1}^{\infty} \mathrm{Var}(X_n^{(c)}) < \infty$, by Corollary 16.3.1 $\sum (X_n^{(c)} - \mathrm{E}(X_n^{(c)})$ converges a.s. Since by assumption, $\sum \mathrm{E}(X_n^{(c)})$ converges, we have $\sum X_n^{(c)}$ converges a.s. On the other hand,

$$\sum_{n=1}^{\infty} P\{X_n \neq X_n^{(c)}\} = \sum_{n=1}^{\infty} P\{|X_n| > c\} < \infty.$$

By First Borel-Cantelli Lemma 3.3.1, $P\{X_n \neq X_n^{(c)}$ for infinite n's$\} = 0$. Hence $\sum_{n=1}^{\infty} X_n$ converges if and only if $\sum_{n=1}^{\infty} X_n^{(c)}$ converges a.s. Thus we conclude that $\sum_{n=1}^{\infty} X_n$ converges a.s. Hence (iv) implies (i).

(i) implies (iii): Since $\sum_{n=1}^{\infty} X_n$ converges a.s. we have $X_n \to 0$ a.s. This implies that almost surely, $X_n \neq X_n^{(c)}$ for only finitely many n. In other words $P(\limsup\{X_n \neq X_n^{(c)}\}) = 0$. The Second Borel-Cantelli Lemma 14.1.1 implies that $\sum_{n=1}^{\infty} P\{X_n \neq X_n^{(c)}\} < \infty$. We also conclude that $\sum_{n=1}^{\infty} X_n^{(c)}$ converges a.s.

By Theorem 16.3.3 this implies that the series $\sum_{n=1}^{\infty} \mathrm{V}(X_n^{(c)})$ and $\sum_{n=1}^{\infty} \mathrm{E}(X_n^{(c)})$ converge. This completes the proof of (i) implies (iii).

Finally, trivially (iii) implies (iv). This completes the proof. ■

16.4 Law of iterated logarithm

By SLLN, if $\{X_i\}$ are iid with mean 0 mean, then $n^{-1}S_n \to$ a.s. It is not difficult to show that if further $\mathrm{E}(X_1^2) < \infty$, then $S_n/(n(\log n)^{\delta}) \to 0$ for every $\delta > 1$ (Exercise 16.6.17). The law of iterated logarithm identifies the scaling $\{a_n\}$ so that the set of limit points of $a_n^{-1}S_n$ is a finite non-trivial set. Unfortunately, its proof is beyond the scope of this book.

Theorem 16.4.1. ***(Khinchine's law of iterated logarithm).*** *Let $\{X_i\}$ be iid rvs with zero mean and variance 1. Then,*

$$-\liminf_n \frac{S_n}{\sqrt{2n\log\log n}} = \limsup_n \frac{S_n}{\sqrt{2n\log\log n}} = 1 \ \text{a.s.},$$

and the set of all limit points of the sequence is the interval $[-1,\ 1]$. ◆

16.5 Exponential inequalities

The upper bound in Markov's inequality Lemma 8.1.1, decreases polynomially. In many situations the bounds are exponentially decreasing. Such exponential bounds are sought after highly in advanced probability theory due to a variety of applications.

Historically, the earliest such inequalities were proved in the 1920's and 1930's for bounded random variables (Bernstein [1924, 1927, 1937]), and later by other authors. These are collectively called Bernstein inequalities. There have been many refinements and extensions to different probability models and these results now come under the umbrella of *exponential inequalities* and *concentration inequalities.* We shall state and prove Hoeffding's inequality. We shall state Talagrand's inequality but without proof. For other exponential inequalities, we refer the reader to Bose et al. [2024], McDiarmid [1989]. Bercu et al. [2015] is an excellent resource for concentration inequalities for sums and martingales.

16.5.1 Hoeffding's inequality

Lemma 16.5.1. (**Hoeffding's lemma**) Let X be any real-valued random variable such that $P(a \le X \le b) = 1$. Then, for all $\lambda \in \mathbb{R}$,

$$\mathrm{E}\left(e^{\lambda(X-\mathrm{E}[X])}\right) \le \exp\left(\frac{\lambda^2(b-a)^2}{8}\right). \tag{16.12}$$

◆

Proof. Without loss, we can assume that $\mathrm{E}(X) = 0$, since the inequality does not change if we substitute $Y = X - \mathrm{E}(X)$ for X. Moreover, there is nothing to prove if $P\{X = 0\} = 1$. For the other cases, it follows that $a < 0 < b$. Since $e^{\lambda x}$ is a convex function of x, for all $x \in [a, b]$,

$$e^{\lambda x} \leq \frac{b - x}{b - a} e^{\lambda a} + \frac{x - a}{b - a} e^{\lambda b}.$$

Hence, taking expectations on both sides,

$$\begin{aligned}
\mathrm{E}\left(e^{\lambda X}\right) &\leq \frac{b - \mathrm{E}(X)}{b - a} e^{\lambda a} + \frac{\mathrm{E}(X) - a}{b - a} e^{\lambda b} \\
&= \frac{b}{b - a} e^{\lambda a} + \frac{-a}{b - a} e^{\lambda b} \\
&= e^{\lambda a}\Big[1 + \frac{a - a e^{\lambda(b-a)}}{b - a}\Big] = e^{L(\lambda(b-a))}, \qquad (16.13)
\end{aligned}$$

$$L(h) = \frac{ha}{b - a} + \ln\Big(1 + \frac{a - e^h a}{b - a}\Big). \qquad (16.14)$$

By computing derivatives, we find

$$L(0) = L'(0) = 0, \;\; L''(h) = -\frac{abe^h}{(b - ae^h)^2}.$$

Note that $L''(h) \leq \frac{1}{4}$ for all h. This is easy to see by noting that $4x_1x_2 \leq (x_1 - x_2)^2$ for all x_1, x_2 and using it with $x_1 = b, x_2 = ae^h$.

Hence by Taylor expansion, for some $0 \leq \theta \leq 1$,

$$L(h) = L(0) + hL'(0) + \frac{1}{2}h^2 L''(h\theta) \leq \frac{1}{8}h^2. \qquad (16.15)$$

Now (16.12) follows using $h = \lambda(b - a)$ in (16.15), and noting (16.13). ■

Theorem 16.5.1. ***(Hoeffding's inequality)*** *Let $\{X_i\}$ be independent rvs and for all $1 \leq i \leq n$, $P\{a_i \leq X_i \leq b_i\} = 1$. Let $S_n = X_1 + \cdots + X_n$. Then*

$$P\{S_n - \mathrm{E}[S_n] \geq t\} \leq \exp\Big(-\frac{2t^2}{\sum_{i=1}^n (b_i - a_i)^2}\Big), \; t > 0 \qquad (16.16)$$

$$P\{|S_n - \mathrm{E}[S_n]| \geq t\} \leq 2\exp\Big(-\frac{2t^2}{\sum_{i=1}^n (b_i - a_i)^2}\Big), \; t > 0 \quad ◆$$

Proof. It is enough to establish (16.16).

For $s,t > 0$, by Markov's inequality Lemma 8.1.1, the independence of $\{X_i\}$, and Hoeffding's Lemma 16.5.1

$$\begin{aligned}
P\{S_n - \mathrm{E}[S_n] \geq t\} &= P\{\exp(s(S_n - \mathrm{E}[S_n])) \geq \exp(st)\} \\
&\leq \exp(-st)\,\mathrm{E}[\exp(s(S_n - \mathrm{E}[S_n]))] \\
&= \exp(-st)\prod_{i=1}^{n} \mathrm{E}[\exp(s(X_i - \mathrm{E}[X_i]))] \\
&\leq \exp(-st)\prod_{i=1}^{n} \exp\left(\frac{s^2(b_i - a_i)^2}{8}\right) \\
&= \exp\left(-st + \tfrac{1}{8}s^2\sum_{i=1}^{n}(b_i - a_i)^2\right).
\end{aligned}$$

Since this bound holds for every s, we minimise it with respect to s, which is easily checked to be attained for $s = (4t)/\sum_{i=1}^{n}(b_i - a_i)^2$. Substituting this value of s, we get (16.16). ■

16.5.2 Talagrand's inequality

There are a host of inequalities, each referred to as Talagrand's inequality. We state one of these here and an interesting corollary which gives an exponential probability bound. The proofs are beyond the scope of this book. Interested reader may consult Alon and Spencer [2008], Ledoux [2001] and Steele [1997].

Recall the Euclidean distance in $\mathbb{R}^d$ between points $x = (x_1, \ldots, x_d)'$, $y = (y_1, \ldots, y_d)'$, and between a set V and a points x:

$$d_E(x,y) = \left[\sum_{i=1}^{d}(x_i - y_i)^2\right]^{1/2}, \quad d_E(x,V) = \inf_{y \in V} d(x,y). \tag{16.17}$$

Definition 16.5.1. (Convex distance) Let $\Omega := \Omega_1 \times \cdots \times \Omega_d$ and let $A \subset \Omega$ and $x \in \Omega$. Let

$$U_A(x) = \{s \in \{0,1\}^d : \exists y \in A \text{ with } y_i = x_i \text{ whenever } s_i = 0\}.$$

Let $V_A(x) \subset \mathbb{R}^n$ be the convex hull of $U_A(x)$. Then the *convex distance* from x to A is defined as $d_c(x;A) := d_E(0,\ V_A(x))$, where d_E is the Euclidean distance defined in (16.17). ◇

If $s \in U_A(x)$, then starting at x, to get to A it is sufficient to vary only the coordinates i for which s_i is 1.

Theorem 16.5.2. *Let* $(\Omega_i, \mathcal{A}_i, \mathcal{P}_i)$, $1 \le i \le n$ *be probability spaces and let* $(\Omega, \mathcal{A}, P)$ *be the product probability space. Then for all* $A \in \mathcal{A}$,

$$\int_\Omega e^{d_c(x,A)^2/4} P(dx) \le 1/P(A)\,,$$

$$P\{\omega \in \Omega : d_c(x,A) > t\} \le \frac{1}{P(A)} e^{-t^2/4}\,. \qquad \blacklozenge$$

Exercise 16.5.1 Show that the second inequality follows from the first.

Definition 16.5.2. A function $f : \mathbb{R}^d \to \mathbb{R}$ is called **1-Lipschitz** if there is a constant c such that for all $x, y \in \mathbb{R}^d$, $|f(x) - f(y)| \le c|x - y|$. $\diamond$

Definition 16.5.3. Suppose X is a rv with distribution F. Then $M(X) = M(F) := F^{-1}(1/2)$ is said to be the **median** of X or F. $\diamond$

Corollary 16.5.1. *Let* $X = (X_1, \dots, X_d)$ *be a random vector with independent components, taking values in* $[0,\ 1]$. *Let* $F : \mathbb{R}^d \to \mathbb{R}$ *be a convex* 1*-Lipschitz function. Then*

$$P\{|F(X) - M(F(X))| \ge t\} \le 4e^{-t^2/4}, \ \textit{for all} \ \ t > 0\,. \qquad \blacklozenge$$

16.6 Exercises

The first exercise gives a more traditional statement and proof of the SLLN. We shall need the following lemma. We omit its proof.

Lemma 16.6.1. (**Kronecker's lemma**) Suppose $\{a_n\}$ is a positive sequence $\uparrow \infty$. Then $\sum_{n=1}^\infty x_n/a_n$ converges $\Rightarrow a_n^{-1}\sum_{i=1}^n x_i \to 0$. $\blacklozenge$

Exercise 16.6.1 (Proof of iid SLLN, using Corollary 16.3.1 and Kronecker's lemma). Suppose $\{X_i\}$ are iid rvs where $\mathrm{E}(X_1)$ exists. Then $S_n/n \to \mathrm{E}(X_1)$ almost surely. Conversely if $S_n/n \to m$ almost surely for some finite m, then $\mathrm{E}(X_1) = m$.

Hint: First assume $\mathrm{E}(X_1)$ is finite. Then without loss, $\mathrm{E}(X_1) = 0$. Define $Y_n = X_n\mathbf{1}_{\{|X_n| \le n\}} - \mathrm{E}(X_n\mathbf{1}_{\{|X_n|\le n\}})$. Then prove the following claims:

(i) $\sum_{n=1}^\infty n^{-2}\,\mathrm{V}(Y_n) \le 2\,\mathrm{E}(1 + |X|) < \infty$.

(ii) $n^{-1}\sum_{i=1}^n Y_i \to 0$ a.s. (use Corollary 16.3.1 and Kronecker's lemma).

(iii) $n^{-1}\sum_{i=1}^n X_i \mathbf{1}_{\{|X_i|\le i\}} \to 0$ a.s.

(iv) $\mathrm{E}\,|X_1| < \infty \Leftrightarrow \sum P(|X_n| > n\epsilon) < \infty$ for every $\epsilon > 0 \Leftrightarrow X_n/n \to 0$ a.s. (apply First Borel-Cantelli Lemma 3.3.1).

(v) $n^{-1}\sum_{i=1}^n X_i \mathbf{1}_{\{|X_i|> i\}} \to 0$ a.s.

(vi) If $\mathrm{E}(X_1) = \infty$, then $n^{-1}\sum_{i=1}^n \min(X_i, C) \to \mathrm{E}(\min(X_i, C))$ a.s. for any finite C. Let $C \to \infty$ and note that $S_n \ge \sum_{i=1}^n \min(X_i, C)$. Similar argument when $\mathrm{E}(X_1) = -\infty$.

(vii) Conversely, if $n^{-1}S_n \to m$ a.s. where m is finite then $X_n/n \to 0$ a.s and $\mathrm{E}\,|X_1| < \infty$ (use (iv)).

Exercise 16.6.2 Let $\{X_i\}$ be mean zero independent rvs such that for some constant C and a rv X with $\mathrm{E}(|X|) < \infty$,

$$\sup_n P\{|X_n| > x\} \le CP\{|X| > x\}\,.$$

Show that $S_n/n \to \mathrm{E}(X_1)$ a.s. and in L^1.

Exercise 16.6.3 Let $\{X_i\}$ be independent mean zero rvs such that $\sup \mathrm{E}(X_k^2) < C < \infty$. Show that $S_n/n \to 0$ a.s.
Hint: Use the following steps:

(i) First Borel-Cantelli Lemma 3.3.1 implies that $S_{n^2}/n^2 \to 0$ a.s.

(ii) Let

$$T_m = S_{n^2+m} - S_{n^2}\,, \quad D_n = \max_{n^2\le k<(n+1)^2} |S_k - S_{n^2}|\,.$$

Then $D_n^2 = \max_{1\le m\le 2n} T_m^2 \le \sum_{m=1}^{2n} T_m^2$. Hence $\mathrm{E}(D_n^2) \le 4n^2C$. Then First Borel-Cantelli Lemma 3.3.1 implies that $D_n/n^2 \to 0$ a.s.

(iii) Finally use the fact that

$$|S_k|/k \le |S_{n^2}|/n^2 + D_n/n^2, \text{ for all } n^2 \le k < (n+1)^2\,.$$

Exercise 16.6.4 Suppose that $\{U_n\}$ are iid $U(0,\ 1)$ rvs. Show that $P_n = \prod_{i=1}^n U_i^{1/n}$ converges a.s. and identify the limit.

Exercise 16.6.5 (a) Let $\{X_i\}$ be iid rvs with mean m. Show that $n^{-1}\sum_{i=1}^n X_i X_{i+1} \to m^2$ a.s.

Exercise 16.6.6 (a) Let $\{X_i\}$ be iid non-negative rvs, and $\mathrm{E}(X_1) = \infty$. Using the SLLN, show that $S_n/n \overset{a.e.}{\to} \infty$.

(b) Prove (a) when the non-negativity assumption is dropped.

Exercise 16.6.7 Let $\{X_n\}$ be iid with $\mathrm{E}|X_1| < \infty$. Let $\{c_n\}$ be a bounded sequence of real numbers. Show that $n^{-1}\sum_{k=1}^{n} c_k X_k \to 0$ a.s.

Exercise 16.6.8 Let $\{X_n\}$ be iid standard Gaussian rvs. Show that

$$\liminf n^{-1/2}\sum_{i=1}^{n} X_i = -\infty \quad \text{and} \quad \limsup n^{-1/2}\sum_{i=1}^{n} X_i = \infty, \text{ a.s.}$$

Exercise 16.6.9 (**Denseness of polynomials**). Let $f \in \mathcal{C}[0,\ 1]$, the space of real-valued continuous functions on $[0,\ 1]$ equipped with the sup-norm metric. Let $\{X_i\}$ be iid Bernoulli rvs with probability of success p $(0 \le p \le 1)$. Show that for every p, $\mathrm{E}[f(X_1 + \cdots X_n/n)] \to f(p)$, and the convergence is uniform. Hence the polynomials are dense in $\mathcal{C}[0,\ 1]$.

Exercise 16.6.10 Let $1 \le p < 2$ and $\{X_i\}$ be iid. Then show that $(S_n - \mathrm{E}(S_n))/n^{1/p} \to 0$ a.s. if and only if $\mathrm{E}(|X_1|^p) < \infty$.

Exercise 16.6.11 Let $0 < p < 1$ and $\{X_i\}$ be iid. Then show that $S_n/n^{1/p} \to 0$ a.s. if and only if $\mathrm{E}(|X_1|^p) < \infty$.

Exercise 16.6.12 Let $\{X_n\}$ be iid such that $\mathrm{E}|X_1| = \infty$. Let a_n be a sequence of positive reals such that a_n/n is increasing. Show that

$$\limsup |S_n|/a_n = \begin{cases} 0 & \text{if } \sum_{n=1}^{\infty} P\{|X_1| \ge a_n\} < \infty, \\ \infty & \text{if } \sum_{n=1}^{\infty} P\{|X_1| \ge a_n\} = \infty\,. \end{cases}$$

Exercise 16.6.13 Let Y be a rv uniformly distributed on the interval $[0,\ 2\pi]$. Let $X_k = \sin(kY)$. Show that $(X_1 + \cdots + X_n)/n \to 0$ a.s.

Exercise 16.6.14 Let $\{X_n\}$ be iid $N(0,1)$ rvs. Show that for any t, $\sum X_n \sin(n\pi t)/n$ converges a.s.

Exercise 16.6.15 Let $\{X_n\}$ be non-negative independent rvs. Show that the following are equivalent:

(i) $\sum_{n=1}^{\infty} X_n < \infty$ a.s.

(ii) $\sum_{n=1}^{\infty} \left[P\{X_n > 1\} + \mathrm{E}(X_n I_{\{X_n \le 1\}})\right] < \infty$

(iii) $\sum_{n=1}^{\infty} \left[P\{X_n > 1\} + \mathrm{E}(X_n/(1 + X_n))\right] < \infty$.

Exercise 16.6.16 Let $\{X_i\}$ be iid standard Gaussian rvs and define $S_n = X_1 + \cdots + X_n$, $n \geq 1$. Show that for any $\epsilon > 0$,

$$P\{\omega : \limsup_{n\to\infty} \frac{|S_n(\omega)|}{n^{1/2+\epsilon}} < \infty\} = 1. \tag{16.18}$$

Hint: Use Markov's inequality and first Borel-Cantelli Lemma 3.3.1.

Exercise 16.6.17 Let $\{X_i\}$ be iid with mean 0 and finite variance. Show that for any $\delta > 1$,

$$\frac{S_n}{\sqrt{n(\log n)^\delta}} \to 0 \text{ a.s.}$$

Hint: Use Corollary 16.3.1 and Kronecker's lemma.

Exercise 16.6.18 Let $\{X_n\}$ be rvs such that $\mathrm{E}(X_n) = 0$ for all n, $\sup_n \mathrm{E}(X_n^2) \leq C < \infty$ and $Cov(X_i, X_j) = 0$ for all $i \neq j$. Show that $n^{-1}(X_1 + \cdots + X_n) \to 0$ in probability and in L^2.

Exercise 16.6.19 Let $\{X_n\}$ be iid with $P(X_n = 0) = P(X_n = 2) = 1/2$.

(a) Show that $\sum_{n=1}^{\infty} X_n/3^n$ converges a.s. (to say X).

(b) Show that X has the Cantor distribution given in Exercise 5.3.17.

Exercise 16.6.20 Suppose $\{X_n\}$ are independent gamma rvs where $X_n \sim Gamma(\gamma_n, 1)$, $n \geq 1$. Find necessary and sufficient conditions on $\{\gamma_n\}$ for the a.s. convergence of $\sum_{n=1}^{\infty} X_n$.

Exercise 16.6.21 Let $\{X_n\}$ be independent rvs, $X_n \sim Exp(\lambda_n)$ for each n. Find necessary and sufficient condition on $\{\lambda_n\}$ for the a.s. convergence of $\sum_{n=1}^{\infty} X_n$.

Chapter 17

Convergence of finite measures

We introduce a concept of *convergence of finite measures*. In particular, this will give us a concept of convergence of rvs in distribution, laying the groundwork for the celebrated central limit theorem in Chapter 18.

17.1 Weak convergence: definition

Example 17.1.1. Let P_n be the measure which puts mass $1/n$ on each element of $A_n = \{1/n, 2/n, \ldots, 1\}$. Let P be the uniform probability measure (Lebesgue measure) on $[0,\ 1]$. Then for any reasonable notion of convergence of measures, P_n must converge to P. Let $A = \cup_{n=1}^{\infty} A_n$. Then $P_n(A) = 1$ for all n, but $P(A) = 0$. Thus, $P_n(A) \to P(A)$ for every A cannot serve as a general definition for convergence of measures. ▲

Example 17.1.2. Suppose P_n puts probability $1/2$ each on the points 0 and n. In this case, a mass of $1/2$ escapes to ∞ and so P_n cannot converge to a *probability measure* on $\mathbb{R}$. ▲

Definition 17.1.1. (**Weak convergence on** $\mathbb{R}$) Measures $\{\mu_n\}$ on $\mathcal{B}(\mathbb{R})$ are said to converge *weakly* to μ if for every bounded continuous function $f : \mathbb{R} \to \mathbb{R}$, $\int f d\mu_n \to \int f d\mu$. We write $\mu_n \stackrel{w}{\to} \mu$ or $\mu_n \Rightarrow \mu$. ◇

Weak convergence of measures on $\mathbb{R}^d$ shall be discussed later. Note that $\mu_n \Rightarrow \mu$ implies $\mu_n(\mathbb{R}) \to \mu(\mathbb{R})$, preserving the total mass. In particular, if $\{\mu_n\}$ are probability measures, then the limit μ is also a probability measure.

A. Bose, *Elements of Measure and Probability*, Texts and Readings in Mathematics 88,
https://doi.org/10.1007/978-981-95-2758-8_17

Example 17.1.3. (**Discrete and continuous uniform**) In Example 17.1.1, the measures $\{P_n\}$ are defined on $\mathcal{B}[0,\ 1]$. $P_n \stackrel{w}{\to} \lambda$, the Lebesgue measure on $[0,\ 1]$. To see this, let f be a bounded continuous function on $[0,\ 1]$. By convergence of Riemann sums,

$$\int f dP_n = \frac{1}{n}\sum_{i=1}^{n} f(i/n) \to \int_{(0,\ 1)} f(x)\lambda(dx)\,. \qquad \blacktriangle$$

Exercise 17.1.1 Let P_n be the probability measure that puts mass $1/n$ at each element of $\{1, 2, \ldots, n\}$. Show that P_n does not converge weakly.

Exercise 17.1.2 Suppose P_n puts masses $1/3$ and $2/3$ at $1/n$ and $-1/n$ respectively. Show that $P_n \Rightarrow \delta_{\{0\}}$ which gives measure 1 to $\{0\}$.

Exercise 17.1.3 Let P_n be the measure, $P_n\{k\} := p_{n,k}$, $k = 0, 1, \ldots$, $\sum_{k=0}^{\infty} p_{n,k} = 1$. If $\lim_{n\to\infty} p_{n,k} = p_k$ for all k, such that $\sum_{k=0}^{\infty} p_k = 1$, then show that $P_n \Rightarrow P$ with mass function $P\{k\} = p_k, k = 0, 1, \ldots$

Exercise 17.1.4 (**Binomial to Poisson**) Let P_n be the binomial distribution with parameters n and p_n (defined in Eqn. (14.8)). Let, as $n \to \infty$, $np_n \to \lambda$ $(0 < \lambda < \infty)$. Show that $\{P_n\}$ converges weakly to the Poisson measure (defined in Eqn. (3.3)) with parameter λ.

Exercise 17.1.5 Let $\{P_n\}$ be probability measures on $\mathcal{B}(\mathbb{R})$ with pdfs $\{f_n\}$. Suppose $f_n \to f$ a.e λ, where $\int_{\mathbb{R}} f(x)\lambda(dx) = 1$. Show that $\{P_n\}$ converges weakly to P which has density f.

17.2 Portmanteau theorem

Definition 17.2.1. (**μ-continuity set**) Let μ be a measure on $\mathcal{B}(\mathbb{R})$ (more generally on $\mathcal{B}(\Omega)$ where Ω is metric space). Then $A \in \mathcal{B}(\mathbb{R})$ is called a *μ-continuity set* if $\mu(\partial A) = 0$. ◇

In Example 17.1.1, $\partial A = [0,\ 1]$ and $P(\partial A) = 1$. Thus A is not a P-continuity set. [See Definition 1.4.5 for the notion of ∂A.] The *Portmanteau theorem* gives equivalent criteria for weak convergence of finite measures. It is true for general metric spaces but we shall restrict to $\mathbb{R}$ and $\mathbb{R}^d$. Recall lower and upper semi-continuous (lsc and usc) and envelope of functions from Definitions 1.5.4 and 1.5.5.

Theorem 17.2.1 (Portmanteau theorem). *Let μ and $\{\mu_n\}$ be finite measures on $\mathcal{B}(\mathbb{R})$. Then the following are equivalent to $\mu_n \Rightarrow \mu$.*

(a) $\int f d\mu_n \to \int f d\mu$ *for all real bounded continuous functions f.*

(b) $\liminf \int f d\mu_n \geq \int f d\mu$ *for all real bounded lsc functions f.*

(b′) $\limsup \int f d\mu_n \leq \int f d\mu$ *for all real bounded usc functions f.*

(c) $\int f d\mu_n \to \int f d\mu$ *for all f bounded a.e μ continuous functions f.*

(d) $\liminf \mu_n(A) \geq \mu(A)$ *for every open set A, and $\mu_n(\mathbb{R}) \to \mu(\mathbb{R})$.*

(d′) $\limsup \mu_n(A) \leq \mu(A)$ *for every closed set A, and $\mu_n(\mathbb{R}) \to \mu(\mathbb{R})$.*

(e) $\mu_n(A) \to \mu(A)$ *for every μ-continuity set A.* ◆

Proof of Theorem 17.2.1. (a) $\Rightarrow$ (b), (b′): Suppose g is any function such that $g \leq f$ and g is bounded continuous. Then

$$\liminf \int f d\mu_n \geq \liminf \int g d\mu_n = \int g d\mu .$$

Let f be bounded by M. By Exercise 1.5.4, get continuous $\{g_n\}$ so that $|g_n| \leq M$, and $g_n \to f$ point-wise. Since μ is finite, by DCT Theorem 7.2.7, $\int g_n d\mu \to \int f d\mu$. This implies (b). (b′) follows trivially.

(b) $\Rightarrow$ (c): Let $\overline{f}$ and $\underline{f}$ be the upper and lower envelopes of f. By Exercise 1.5.5, $\underline{f}$ is lsc and $\overline{f}$ is usc. Further

$$\underline{f}(x) \leq f(x) \leq \overline{f}(x) \text{ for all } x, \text{ and } \ f(x) = \underline{f}(x) = \overline{f}(x) \text{ a.e. } \mu.$$

Hence

$$\begin{aligned}
\int f d\mu = \int \underline{f} d\mu &\leq \liminf \int \underline{f} d\mu_n, \text{ by (b)} \\
&\leq \liminf \int f d\mu_n \ \text{ since } \ \underline{f} \leq f \\
&\leq \limsup \int f d\mu_n \\
&\leq \limsup \int \overline{f} d\mu_n, \text{ since } \ \overline{f} \geq f \\
&\leq \int \overline{f} d\mu, \text{ by (b′)} \\
&= \int f d\mu.
\end{aligned}$$

This proves that $\lim \int f d\mu_n = \int f d\mu$ if f is bounded continuous.

(c) $\Rightarrow$ (d): Clearly (c) $\Rightarrow$ (a) $\Rightarrow$ (b). Suppose A is an open set. Then $\mathbf{1}_A$ is an lsc function. Hence by (b), $\liminf \mu_n(A) \geq \mu(A)$. The second part follows by using $\pm\mathbf{1}_{\mathbb{R}}$.

(d) $\Leftrightarrow$ (d'): This follows by taking complements of sets.

(d) $\Rightarrow$ (e): Let A be a μ-continuity set. Let $\bar{A}$ be its closure. Then

$$\begin{aligned}\limsup \mu_n(A) \leq \limsup \mu_n(\bar{A}) &\leq \mu(\bar{A}), \text{ by (d')} \\ &= \mu(A), \text{ since } \ \mu(\partial A) = 0\,.\end{aligned}$$

Similarly, by considering the interior A^o,

$$\liminf \mu_n(A) \geq \mu(A)\,.$$

This proves that $\lim \mu_n(A) = \mu(A)$.

(e) $\Rightarrow$ (a). Let f be a continuous function on $\mathbb{R}$ bounded by M. Let

$$A = \{c \in \mathbb{R} : \mu(f^{-1}\{c\}) \neq 0\}\,.$$

Since μ is finite, by Exercise 5.3.1, A is countable. If $\pm M \in A$, increase M so that they do not belong to A. This is possible since A is countable. Let $t_0 = -M < t_1 < \cdots < t_j = M$ be a partition of $[-M,\ M]$, where none of the t_i belong to A. Let

$$B_i = \{x : t_i \leq f(x) < t_{i+1}\},\ 0 \leq i \leq j-1.$$

Note that $f^{-1}\{(t_i, t_{i+1})\}$ is open and $\partial f^{-1}\{[t_i, t_{i+1})\} \subseteq f^{-1}\{t_i, t_{i+1}\}$. Moreover, $\mu(f^{-1}\{t_i, t_{i+1}\}) = 0$ since $\{t_i\} \notin A$. Now

$$\begin{aligned}&|\int f d\mu_n - \int f d\mu| \\ &\quad\leq |\int f d\mu_n - \sum_{i=0}^{j-1} t_i \mu_n(B_i)| + |\int f d\mu - \sum_{i=0}^{j-1} t_i \mu(B_i)| \\ &\quad\quad + |\sum_{i=0}^{j-1} t_i \mu_n(B_i) - \sum_{i=0}^{j-1} t_i \mu(B_i)| \\ &\quad = T_1 + T_2 + T_3 \ \text{ say.}\end{aligned}$$

Then $T_1 = |\sum_{i=0}^{j-1} \int (f(x) - t_i)\mu_n(dx)| \leq \max_i (t_{i+1} - t_i)\mu_n(\mathbb{R})$, which can be made arbitrarily small by refining the partition (partition points never belonging to A), and noting that $\mu_n(\mathbb{R}) \to \mu(\mathbb{R}) < \infty$ by (e).

Similarly, we have $T_2 \le \max_i(t_{i+1} - t_i)\mu(\mathbb{R})$, which can also be made arbitrarily small. Finally $T_3 \to 0$ since each B_i is a μ-continuity set. ■

Exercise 17.2.1 Let $\{\mu_n\}$, μ be finite measures on $(\Omega, \mathcal{A})$. Show that if $\int f d\mu_n \to \int f d\mu$ for all real bounded *uniformly continuous* functions f, then $\mu_n \overset{w}{\to} \mu$. Hint: Use Theorem 1.5.1.

Exercise 17.2.2 Show that $\mu_n \overset{w}{\to} \mu$, if $\mu_n(A) \to \mu(A)$ for every *open μ-continuity set A*.

17.3 Weak limit via distribution function

Weak convergence of finite measures on $\mathbb{R}$ can be expressed in terms of their distribution functions. Recall that if F is a distribution function on $\mathbb{R}$, then $F(\pm\infty)$ are defined as:

$$F(\pm\infty) = \lim_{y \to \pm\infty} F(y). \tag{17.1}$$

Moreover,

$$F(-\infty) = 0, \; F(\infty) = 1, \text{ if } F \text{ is a probability distribution function}. \tag{17.2}$$

Definition 17.3.1. (**Continuity set**) Define the *continuity set* for any distribution function F on $\mathbb{R}$ as

$$C_F := \{x \in \mathbb{R} : x \text{ is a continuity point of } F\} \cup \{-\infty\} \cup \{\infty\}. \qquad \diamond$$

Theorem 17.3.1. *Let μ and $\{\mu_n\}$ be finite measures on $\mathcal{B}(\mathbb{R})$ with distribution functions F and $\{F_n\}$. Then the following are equivalent.*

(a) $\mu_n \overset{w}{\to} \mu$.

(b) $\mu_n(a,\ b] = F_n(b) - F_n(a) \to \mu(a,\ b] = F(b) - F(a)$ for all $a, b \in C_F$.
Moreover, if all the distributions are 0 at $-\infty$, then (b) is equivalent to

$$F_n(x) \to F(x) \;\; \textit{for all} \;\; x \in C_F.$$

[This convergence is denoted by $F_n \overset{w}{\to} F$ or by $F_n \Rightarrow F$]. ◆

Remark 17.3.1. The notion of **vague convergence** does not stipulate preservation of mass. Measures $\{\mu_n\}$ are said to converge to a measure

μ *vaguely* if $\mu_n(a,\ b] \to \mu(a,\ b]$ for all $a, b \in C_F \setminus \{-\infty, \infty\}$. We shall not draw upon this idea explicitly in this book. The interested reader may consult other probability books, such as Billingsley [1995]. ●

Proof of Theorem 17.3.1. (a) ⇒ (b): Let $A = (a, b]$, a, b finite. Note that $\partial A = \{a, b\}$. If $a, b \in C_F$ then $(a,\ b]$ is a μ-continuity set. Hence $\mu_n(a,\ b] \to \mu(a,\ b] = F(b) - F(a)$. If $a = -\infty$, then the argument is the same. If $b = \infty$, then apply the earlier argument for the set $(a,\ \infty)$.

(b) ⇒ (a): Let A be an open set. Then by Theorem 1.4.1(c) $A = \cup_{i=1}^{\infty} I_i$ where $\{I_i\}$ are disjoint open intervals. Then

$$\begin{aligned} \liminf \mu_n(A) &= \liminf \sum_k \mu_n(I_k) \\ &\geq \sum_k \liminf \mu_n(I_k) \text{ by Fatou's Lemma 7.2.1.} \end{aligned} \tag{17.3}$$

Fix $\epsilon > 0$. Since F has only countably many discontinuity points, for each k, get J_k right semi-closed sub-interval of I_k, such that the end points of J_k are continuity points of F, and

$$\mu(J_k) \geq \mu(I_k) - \epsilon 2^{-k}. \tag{17.4}$$

Then

$$\liminf \mu_n(I_k) \geq \liminf \mu_n(J_k) = \mu(J_k) \text{ by (b),} \tag{17.5}$$

$$\begin{aligned} \liminf \mu_n(A) &\geq \sum_k \mu(J_k) \quad \text{by (17.3) and (17.5)} \\ &\geq \sum_k \mu(I_k) - \epsilon, \quad \text{by (17.4)} \\ &= \mu(A) - \epsilon. \end{aligned}$$

Since ϵ is arbitrary, we have $\liminf \mu_n(A) \geq \mu(A)$. Since A was an arbitrary open set, $\mu_n \stackrel{w}{\to} \mu$ by Theorem 17.2.1(d). ■

17.4 Weak convergence of random variables

Definition 17.4.1. If X and Y are rvs, possibly on different probability spaces, with the same cdf, then we say that X and Y are **equal in distribution**. We write $X \stackrel{D}{=} Y$. ◊

Exercise 17.4.1 (a) If $X \sim N(0, \sigma^2)$, then show that $X \stackrel{D}{=} -X$.

(b) If $X \sim U(0,\ 1)$, then show that $X \stackrel{D}{=} 1 - X$.

(c) If $X \sim Bin(n, p)$ (see (14.8)), when does $X \stackrel{D}{=} n - X$ hold?

Definition 17.4.2. Suppose $\{X_n\}$, X are real-valued rvs, possibly on different probability spaces, with probability distribution functions $\{F_n\}$ and F. Then we say X_n converges to X **in distribution** if $F_n \Rightarrow F$. We write $X_n \Rightarrow X$. This is equivalent to $F_n(x) \to F(x)$ for all $x \in C_F$. ◇

If $X_n \Rightarrow X$, and $X \stackrel{D}{=} Y$, then $X_n \Rightarrow Y$ also holds. Thus *only probability distributions are being identified in the limit.* So, a statement such as "$X_n \Rightarrow X$ and $Y_n \Rightarrow Y$ implies $X_n + Y_n \Rightarrow X + Y$" is meaningless.

Exercise 17.4.2 Show that

(a) for all bounded continuous functions $f : \mathbb{R} \to \mathbb{R}$, $\mathrm{E}[f(X_n)] \to \mathrm{E}[f(X)]$ if and only if, $X_n \Rightarrow X$.

(b) $X_n \Rightarrow X$ if and only if for all Borel sets A for which $P\{X \in \partial A\} = 0$, we have $P\{X_n \in A\} \to P\{X \in A\}$.

Exercise 17.4.3 Let $X_n \Rightarrow X$ and $g : \mathbb{R} \to \mathbb{R}$ be continuous. Show that $g(X_n) \Rightarrow g(X)$. In particular, if c is a constant, then $X_n + c \Rightarrow X + c$.

Exercise 17.4.4 (a) If $X_n \Rightarrow c$ (onstant), then show that $X_n \xrightarrow{P} c$.

(b) If $X_n \xrightarrow{P} X$ then show that $X_n \Rightarrow X$.

(c) Given an example of a sequence $\{X_n\}$ and X, all defined on the same probability space, such that $X_n \Rightarrow X$, but $X_n \xrightarrow{P} X$ does not hold.

Exercise 17.4.5 (**Slutsky's theorem**) Let $\{X_n\}$, $\{Y_n\}$ be real-valued rvs defined on the same probability space where $X_n \Rightarrow X$ and $Y_n \xrightarrow{P} c$ and c is a constant. Show that

(a) $X_n + Y_n \Rightarrow X + c$;

(b) $X_n Y_n \Rightarrow cX$;

(c) $X_n / Y_n \Rightarrow X/c$ whenever $c \neq 0$.

Exercise 17.4.6 Suppose for each n, X_n and Y_n are independent. Suppose $X_n \Rightarrow X$ and $Y_n \Rightarrow Y$. Show that for any continuous function $f : \mathbb{R} \to \mathbb{R}$, $f(X_n + Y_n) \Rightarrow f(X^* + Y^*)$ where the rvs X^* and Y^* are independent and $X^* \stackrel{D}{=} X$ and $Y^* \stackrel{D}{=} Y$.

17.5 Skorokhod's embedding in $\mathbb{R}$

We know that convergence in probability does not imply a.s. convergence. Moreover, when a sequence of rvs converge in distribution, they need not have been defined on the same probability space, hence the question of their convergence in probability or a.s. does not even arise. However, the following result shows how convergence in distribution can be lifted to a.s. convergence on a common probability space.

Theorem 17.5.1. *(**Skorokhod's embedding**) Let $\{X_n\}, X$ be rvs with distribution function $\{F_n\}$ and F. Suppose F_n converges weakly to F. Then there exists $Y_n, Y : (\Omega := (0,\ 1), \mathcal{B}(\Omega), \lambda) \to (\mathbb{R}, \mathcal{B}(\mathbb{R}))$ such that*

(a) $Y_n \stackrel{D}{=} X_n$ and $Y \stackrel{D}{=} X$.

(b) $Y_n(\omega) \to Y(\omega)$ for all $\omega \in (0,\ 1)$. ◆

Proof. For any function g, let C_g be the set of continuity points of g. To begin with, define $Y_n, Y : (\Omega := (0,\ 1), \mathcal{B}(\Omega), \lambda) \to (\mathbb{R}, \mathcal{B}(\mathbb{R}))$ as follows. We shall slightly modify this definition later.

$$Y_n(\omega) = F_n^{-1}(\omega), \quad Y(\omega) = F^{-1}(\omega)\,,\ \omega \in \Omega\,.$$

Then by definition

$$Y(\omega) \le x \ \text{ if and only if } \ F(x) \ge \omega\,. \tag{17.6}$$

Hence $\lambda\{\omega : Y(\omega) \le x\} = F(x)$, proving that $Y \stackrel{D}{=} X$. The same argument yields $Y_n \stackrel{D}{=} X_n$.

(b) (i) We first prove that

$$\liminf_{n\to\infty} Y_n(\omega) \ge Y(\omega) \ \text{ for all } \ \omega \in (0,\ 1)\,. \tag{17.7}$$

Fix $\omega \in (0,\ 1)$, and $\epsilon > 0$. Since C_F is a dense subset of $\mathbb{R}$, choose $x \in C_F, -\epsilon < x < Y(\omega)$. By (17.6), $Y(\omega) > x$ implies that $F(x) < \omega$. Since $F_n(x) \to F(x)$, we have $F_n(x) < \omega$ for all large n.

Again using (17.6) for Y_n and F_n, this means $Y_n(\omega) > x$ for all large n. Hence $\liminf_{n\to\infty} Y_n(\omega) \ge x$. Since ϵ is arbitrary, and C_F is dense, (17.7) is proved.

(ii) Let C_Y be the set of continuity points of Y. Now we show that

$$\limsup_{n\to\infty} Y_n(\omega) \le Y(\omega) \quad \text{for all} \quad \omega \in C_Y\,. \tag{17.8}$$

Let $\omega, \omega_1, \in (0,\ 1)$, $\omega < \omega_1$, and $\epsilon > 0$. Choose $y \in C_F$ such that $Y(\omega') < y < Y(\omega') + \epsilon$. By (17.6), $Y(\omega') < y$ implies $F(Y(\omega')) \le F(y)$.

Since $Y(\omega')$ is the first point where F reaches the height ω', we have $F(Y(\omega')) \ge \omega'$. Thus $\omega < \omega' \le F(Y(\omega')) \le F(y)$. But $F_n(y) \to F(y)$ since $y \in C_F$. Hence for all large n, $F_n(y) \ge \omega$. Now using (17.6), $Y_n(\omega) \le y \le Y(\omega') + \epsilon$. This establishes (17.8).

By (17.7) and (17.8), $Y_n(\omega) \to Y(\omega)$ for all $\omega \in C_Y$. If the cdf F is continuous everywhere, then $C_Y = (0,\ 1)$ and the proof is complete.

Otherwise, we modify our definitions of Y_n and Y slightly. $Y(\cdot)$ is non-decreasing (everywhere), and C_Y^c is at most countable and hence $\lambda(C_Y^c) = 0$. Re-define $Y(\omega) = Y_n(\omega) = 0$ for $\omega \in C_Y^c$. Then we have $Y_n(\omega) \to Y(\omega)$ for all $\omega \in (0,\ 1)$. This does not change Property (a) as the modifications have been done on a set of (Lebesgue) measure 0. ■

Exercise 17.5.1 Let $X_n \Rightarrow X$. Using Skorokhod embedding show that,

(a) if real sequences $a_n \to a$ and $b_n \to b$, then $a_n X_n + b_n \Rightarrow aX + b$;

(b) if $g : \mathbb{R} \to \mathbb{R}$ is a continuous function, then $g(X_n) \Rightarrow g(X)$.

Remark 17.5.1. Proof of Theorem 17.5.1 used the ordering of the real numbers crucially. Weak convergence in many other spaces can also be lifted to a.s. convergence. Proof of these embedding theorems are more difficult since no ordering is available. For a proof of Skorokhod's embedding theorem in $\mathbb{R}^d$, see Billingsley [1995]. ●

17.6 Helly's selection principle in $\mathbb{R}$

Suppose we have a sequence of distribution functions $\{F_n\}$ on $\mathbb{R}$ which is not necessarily weakly convergent. Then can we extract a sub-sequence which converges weakly to some distribution function F? If $\{F_n\}$ are probability distribution functions, then can we guarantee that F is also a probability distribution function? We begin with the following result.

Theorem 17.6.1 (Helly's Selection Principle). *Suppose $\{F_n\}$ is a sequence of distribution functions on $\mathbb{R}$.*

(a) If for all n, $F_n(-\infty) = 0$ and $F_n(\infty) \leq M < \infty$, then there is a distribution function F, and a sub-sequence $\{n_k\}$ so that $F_{n_k}(x) \to F(x)$ at all continuity points $x \in \mathbb{R}$ of F. Further, $F(\infty) \leq M$.

(b) If $\sup_n (F_n(\infty) - F_n(-\infty)) \leq M < \infty$, then there is a distribution function F and a sub-sequence $\{n_k\}$ so that at all continuity points $a, b \in \mathbb{R}$ of F, $F_{n_k}(b) - F_{n_k}(a) \to F(b) - F(a)$ and $F(\infty) - F(-\infty) \leq M$. ◆

Proof of Theorem 17.6.1. (a) Incidentally, if $M = 0$, then there is nothing to prove. So we assume $M > 0$. We use a diagonalisation argument. Let $D = \{x_1, x_2, \ldots\}$ be a countable dense subset of $\mathbb{R}$. The sequence $\{F_n(x_1)\}$ is bounded by M. There is a sub-sequence, $\{F_{1,j}\}$ of $\{F_n\}$ such that $\{F_{1,j}(x_1)\}$ converges to say y_1. Now from $\{F_{1,j}\}$, extract a sub-sequence $\{F_{2,j}\}$ such that $\{F_{2,j}(x_2)\}$ converges to say y_2. Continuing, for every m, there is a sub-sequence $\{F_{m,j}\}$ of $\{F_{m-1,j}\}$ such that $\{F_{m,j}(x_m)\}$ converges to say y_m. Note that $0 \leq y_m \leq M$ for every m. We note in passing that it is possible that $y_m = 0$ for all m.

Let $F_{n_k} = F_{k,k}$, $k \geq 1$ be the "diagonal sequence". Define the function $F_D : D \to \mathbb{R}$ as $F_D(x_j) = y_j$, $j \geq 1$. Note that by construction,

$$F_{n_k}(x) \to F_D(x) \text{ for all } x \in D.$$

Again, note that it is possible that $F_D(x_j) = 0$ for all j. Since $\{F_n\}$ are all non-decreasing functions, it follows that $F_D(\cdot)$ is also non-decreasing on D. Now using the denseness of D, we construct a distribution function out of $F_D(\cdot)$ in the obvious way:

$$F(x) = \inf\{F_D(y) : y \in D,\ y > x\},\ x \in \mathbb{R}. \tag{17.9}$$

By Exercise 5.3.4, F is a distribution function.

It remains to show $F_{n_k}(x) \to F(x)$ at all continuity points x of F. Suppose $x < y \in D$. Then

$$\limsup_{k\to\infty} F_{n_k}(x) \leq \limsup_{k\to\infty} F_{n_k}(y) = F_D(y).$$

Then taking infimum over $y \in D, y > x$, using (17.9), we get

$$\limsup_{k\to\infty} F_{n_k}(x) \leq F(x). \tag{17.10}$$

Now consider $x^* < y < x,\ y \in D$. Then

$$F(x^*) \le F_D(y) = \lim F_{n_k}(y) = \liminf F_{n_k}(y) \le \liminf F_{n_k}(x)\,.$$

Let $x^* \to x$ to obtain

$$F(x-) \le \liminf F_{n_k}(x). \tag{17.11}$$

Now since x is a continuity point of F, $F(x-) = F(x)$. Hence using (17.10) and (17.11), $\lim F_{n_k}(x) = F(x)$.

(b) Consider $G_n(x) = F_n(x) - F_n(-\infty)$, and apply Part (a). ■

In Theorem 17.6.1, we have not claimed that $F_{n_k} \Rightarrow F$. For this we would additionally need $F_{n_k}(\infty) - F_{n_k}(-\infty) \to F(\infty) - F(-\infty)$, and this cannot be guaranteed in general.

Example 17.6.1. Let P_n be the uniform probability measure on the set $\{1, 2, \ldots, n\}$, and F_n its probability distribution function. Then $F_n(x) \to 0$ for all $x \in \mathbb{R}$. Hence the limit is $F(x) \equiv 0$ on $\mathbb{R}$. Note that $F(\infty) = 0$ and $F_n(\infty) = 1$ for all n. So, F_n does not converge to F weakly. ▲

Example 17.6.2. As seen in the above example, the limit F in Helly's theorem can be identically 0. If all F_n's are probability distribution functions, the limit F is not necessarily so. Let

$$F_n(x) = \begin{cases} 0 & \text{if } \ x < 0, \\ 1/2 & \text{if } \ 0 \le x < n, \\ 1 & \text{if } \ x \ge n\,, \end{cases}$$

and let F be the distribution function given by

$$F(x) = \begin{cases} 0 & \text{if } \ x < 0, \\ 1/2 & \text{if } \ x \ge 0. \end{cases}$$

Then F has only one discontinuity (at $x = 0$) and it is easy to see that $F_n(x) \to F(x)$ for all $x \in \mathbb{R}$. But F_n does not converge weakly to F. ▲

17.7 Tightness and Prokhorov's theorem in $\mathbb{R}$

In Example 17.6.2, a mass of 1/2 from F_n escaped to ∞. The concept of "tightness" prevents this situation.

Definition 17.7.1 (Tightness). Finite measures $\{\mu_i, i \in I\}$ on $\mathcal{B}(\mathbb{R})$ is called *tight* if, given $\epsilon > 0$, there is a compact set K such that $\mu_i(K^c) < \epsilon$ for all $i \in I$. Tightness of probability distribution functions $\{F_i, i \in I\}$ or rvs $\{X_i, i \in I\}$ is the same as tightness of their probability measures. ◇

Since any compact set is closed and bounded, the compact set in Definition 17.7.1 can be taken to be $[-M, \ M]$, for some $M > 0$. Note that a finite set of finite measures is always tight.

Exercise 17.7.1 Let $\{X_i\}$ be iid rvs, $S_n = X_1 + \cdots + X_n$. If $\mathrm{E}(X_1) = 0$, $V(X_1) = \sigma^2 < \infty$, show that $\{S_n/\sqrt{n}\}$ is tight. The central limit theorem of Chapter 19 will show that $S_n/\sqrt{n} \Rightarrow N(0, \sigma^2)$.

Definition 17.7.2 (Relatively/sequentially compact). Finite measures $\{\mu_i, i \in I\}$ on $\mathcal{B}(\mathbb{R})$ are called *relatively compact* if given any sequence from $\{\mu_i\}$, there is a further sub-sequence which converges weakly. This definition extends to probability distribution functions $\{F_i, i \in I\}$ on $\mathbb{R}$, or to real-valued rvs $\{X_i, i \in I\}$, in an obvious way. ◇

Theorem 17.7.1 (Prokhorov's theorem). *Let $\{\mu_i, i \in I\}$ be measures on $\mathbb{R}$ with distributions $\{F_i\}$ and $\sup_i(F_i(\infty) - F_i(-\infty)) \leq M < \infty$. Then this family is tight if and only if it is relatively compact.* ◆

Proof. First assume that $\{F_i, i \in I\}$ is tight. Pick a sequence $\{F_n\}$. Then by Helly's Theorem 17.6.1(b), there is a sub-sequence $\{F_{n_k}\}$ and a distribution function F such that $F_{n_k}(b) - F_{n_k}(a) \to F(b) - F(a)$ at all continuity points $a, b \in \mathbb{R}$ of F. Moreover, $F(\infty) - F(-\infty) \leq M$.

Fix $\epsilon > 0$. By tightness, the above convergence, and by the fact that F is a bounded distribution function, we can choose a and b continuity points of F such that for all n_k,

$$\mu_{n_k}(\mathbb{R} \setminus (a, \ b]) < \epsilon \ \text{ and } \ \mu(\mathbb{R} \setminus (a, \ b]) < \epsilon. \tag{17.12}$$

Now if $x \in \mathbb{R}$ is a continuity point of F, then

$$\begin{aligned} F_{n_k}(\infty) - F_{n_k}(x) - [F(\infty) - F(x)] \\ = [F_{n_k}(\infty) - F_{n_k}(b)] - [F(\infty) - F(b)] \\ + [F_{n_k}(b) - F_{n_k}(x)] - [F(b) - F(x)]. \end{aligned}$$

By (17.12), the first two factors on the right side are smaller than ϵ for all n. The third factor on the right side converges to $F(b) - F(x)$ since b and x are continuity points of F. Since ϵ is arbitrary, it follows that

$$F_{n_k}(\infty) - F_{n_k}(x) \to F(\infty) - F(x)\,.$$

Similarly, $F_{n_k}(x) - F_{n_k}(-\infty) \to F(x) - F(-\infty)$. Hence we have shown that $F_{n_k} \stackrel{w}{\to} F$ and thus $\{F_i\}$ is relatively compact.

Now suppose that $\{F_i, i \in I\}$ is relatively compact but not tight. Then there exists an $\epsilon > 0$, such that for each n, we have an F_n such that $\mu_n(\mathbb{R} - (-n\ n)) \geq \epsilon$. By relative compactness, we can get a sub-sequence, say $\{F_{n_k}\}$, of $\{F_n\}$ which converges weakly to a (bounded) distribution function say F with the corresponding finite measure μ. Now note that $\mathbb{R} \setminus (-n,\ n)$ is closed, and hence by Portmanteau Theorem 17.2.1(d),

$$\limsup_{k\to\infty} \mu_{n_k}(\mathbb{R} \setminus (-n,\ n)) \leq \mu(\mathbb{R} \setminus (-n,\ n))\,.$$

Thus $\mu(\mathbb{R} \setminus (-n,\ n)) \geq \epsilon$ for all n, and letting $n \to \infty$, we obtain $\epsilon \leq 0$ which is a contradiction. This completes the proof. ■

Note that Prokhorov's theorem implies that any family of probability measures is tight if and only if it is relatively compact.

Exercise 17.7.2 Let $\{F_n\}$ be a bounded sequence of tight distribution functions on $\mathbb{R}$. If every weakly convergent sub-sequence of $\{F_n\}$ converges to the same distribution function F, then show that $F_n \stackrel{w}{\to} F$.

Exercise 17.7.3 Let $\{X_n\}$ be a tight sequence of rvs. Show that there is a rv X and a sub-sequence $\{n_k\}$ such that $X_{n_k} \Rightarrow X$.

17.8 A metric for weak convergence in $\mathbb{R}$

We discuss a metric which belongs to the class of Mallows' metric (also known as Wasserstein's metric or Kantorovitch-Rubinstein metric).

Consider the space of probability distributions with finite second moment. Then the metric on this space is defined as follows: Let F and G be two distribution functions on $\mathbb{R}$ with finite second moment. Then the W_2 distance between them is defined as

$$W_2(F,G) = \left[\inf_{(X\sim F, Y\sim G)} E(X-Y)^2 \right]^{\frac{1}{2}}.$$

Here $(X \sim F, Y \sim G)$ means that the marginal distributions of the random vector (X,Y) are F and G.

We leave it to the reader to complete the following exercise.

Exercise 17.8.1 Show that W_2 is a metric and is complete.

Theorem 17.8.1. *Let $\{F_n\}$ and F be probability distribution functions on $\mathbb{R}$ with finite second moments. Then $W_2(F_n, F) \to 0$ if and only if $F_n \xrightarrow{D} F$ and $\int_{\mathbb{R}} x^2 F_n(dx) \to \int_{\mathbb{R}} x^2 F(dx)$.* ◆

Proof. First suppose, $W_2(F_n, F) \to 0$. By definition of W_2, get (X_n, Y_n) such that $X_n \sim F_n$, $Y_n \sim F$, and $\mathrm{E}(X_n - Y_n)^2 \to 0$. Further, it can be so arranged that all these rvs are defined on the same probability space.

By Chebyshev's inequality Lemma 8.1.2, $X_n - Y_n \xrightarrow{P} 0$. Moreover, since all Y_n have the same distribution F, by applying Slutsky's Theorem (Exercise 17.4.5(a)), $F_n \xrightarrow{D} F$.

It remains to show that, $\mathrm{E}(X_n^2) \to \mathrm{E}(X^2)$ where X has distribution F. For this, it is enough to show that $\mathrm{E}(X_n^2) - \mathrm{E}(Y_n^2) \to 0$. Note that $\{\mathrm{E}(X_n^2)\}$ is a bounded sequence. Hence,

$$\begin{aligned}[\mathrm{E}(X_n^2) - \mathrm{E}(Y_n^2)]^2 &\le \mathrm{E}(X_n - Y_n)^2)\,\mathrm{E}(X_n + Y_n)^2 \\ &\le [\mathrm{E}(X_n - Y_n)^2)][2\,\mathrm{E}(X_n)^2 + 2\,\mathrm{E}(Y_n^2)] \to 0.\end{aligned}$$

Conversely, suppose $F_n \xrightarrow{D} F$ and $\int_{\mathbb{R}} x^2 F_n(dx) \to \int_{\mathbb{R}} x^2 F(dx)$. We have to show that $W_2(F_n, F) \to 0$. By Skorokhod's theorem 17.5.1, first get $\{Y_n\}$, Y on the same probability space such that $Y_n \to Y$ a.s., $Y_n \sim F_n$ and $Y \sim F$. Fix a large real number k. Define

$$\bar{Y}_{n,k} \begin{cases} = Y_n \text{ if } |Y_n| < k, \\ = k \text{ if } Y_n \ge k, \\ = -k \text{ if } Y_n \le -k. \end{cases}$$

Define $\bar{Y}_k$ similarly using Y instead of Y_n. Let $G_{n,k}$ and G_k denote the distributions of $\bar{Y}_{n,k}$ and $\bar{Y}_k$ respectively. Note that as $n \to \infty$, $\bar{Y}_{n,k} \to \bar{Y}_k$ almost surely. By DCT Theorem 7.2.7, as $n \to \infty$ $\mathrm{E}[\bar{Y}_{n,k} - \bar{Y}_k]^2 \to 0$. Hence $W_2(G_{n,k}, G_k) \to 0$ as $n \to \infty$. Now by triangle inequality,

$$W_2(F_n, F) \leq W_2(F_n, G_{n,k}) + W_2(G_{n,k}, G_k) + W_2(G_k, F)\,.$$

By DCT Theorem 7.2.7,

$$\begin{aligned} W_2^2(G_k, F) &\leq \mathrm{E}[\bar{Y}_k - Y]^2 \leq \mathrm{E}[|Y|^2 \mathbf{1}_{\{|Y| \geq k)\}}] \to 0 \quad \text{as} \quad k \to \infty, \\ W_2^2(G_{n,k}, F_n) &\leq \mathrm{E}[\bar{Y}_{n,k} - Y_n]^2 \leq \mathrm{E}[|Y_n|^2 \mathbf{1}_{\{|Y_n| \geq k\}}]. \end{aligned}$$

Choose k (large) so that $\pm k$ are continuity point of F. This is possible as the set of discontinuity points of F is countable (Exercise 5.3.1). Now

$$\mathrm{E}[|Y_n|^2 \mathbf{1}_{\{|Y_n| \geq k\}}] + \mathrm{E}[|Y_n|^2 \mathbf{1}_{\{|Y_n| < k\}}] = \mathrm{E}(Y_n^2) = \mathrm{E}(Y^2). \tag{17.13}$$

By DCT Theorem 7.2.7, as $n \to \infty$, the second term on the left side of (17.13) converges to $\mathrm{E}[|Y|^2 \mathbf{1}_{\{|Y| < k\}}]$. Hence, the first term converges to $\mathrm{E}[|Y|^2 \mathbf{1}_{\{|Y| \geq k\}}]$. This in turn converges to 0 as $k \to \infty$.

Hence combining all the above, we get $\limsup W_2(F_n, F) \to 0$. This completes the proof. ■

17.9 Weak convergence in $\mathbb{R}^d$

We show how the concept of weak convergence can be extended to finite measures on $\mathbb{R}^d$ for any $d \geq 2$. We skip most of the details.

Definition 17.9.1. Suppose $\{\mu_i\}$ and μ are finite measures on $\mathbb{R}^d$. Then we say that μ_n converges to μ weakly, and write $\mu_n \Rightarrow \mu$ or $\mu_n \stackrel{w}{\to} \mu$, if $\lim_{n\to\infty} \int_{\mathbb{R}^d} f\, d\mu_n = \int_{\mathbb{R}^d} f\, d\mu$ for all bounded continuous $f : \mathbb{R}^d \to \mathbb{R}$. ◊

Exercise 17.9.1 (Continuous mapping theorem) Let μ and $\{\mu_n\}$ be finite measures on $\mathbb{R}^{d_1}$ and $\mu_n \Rightarrow \mu$. If $f : \mathbb{R}^{d_1} \to \mathbb{R}^{d_2}$ is a continuous function, then show that, $\mu_n \circ f^{-1} \Rightarrow \mu \circ f^{-1}$ on $\mathbb{R}^{d_2}$.

The Portmanteau theorem and Helly's Selection Principle remain valid in $\mathbb{R}^d$. We skip the details. The concepts of tightness and relative compactness in $\mathbb{R}^d$ are straightforward extensions of the corresponding concepts in $\mathbb{R}$.

Definition 17.9.2. Let $\{\mu_\alpha : \alpha \in I\}$ be probability measures on $\mathbb{R}^d$.

(a) This collection is said to be tight, if given any $\epsilon > 0$, there is a compact set $K \subset \mathbb{R}^d$ such that $\mu_\alpha(K^c) < \epsilon$ for all $\alpha \in I$.

(b) This collection is said to be relatively compact if any sub-sequence from it has a further sub-sequence which converges weakly. ◇

Prokhorov's theorem also remains valid in $\mathbb{R}^d$. The proof is very similar to the proof in $\mathbb{R}$. The details are left as an exercise.

Theorem 17.9.1. *Let $\{\mu_n\}$ be a sequence of finite measures on $\mathbb{R}^d$ such that $\sup_n \mu_n(\mathbb{R}^d) < \infty$. Then $\{\mu_n\}$ is tight iff it is relatively compact.* ◆

Before we link weak convergence to convergence of distribution functions in $\mathbb{R}^d$, we first note that, distribution functions on $\mathbb{R}^d$ can have uncountably many discontinuity points. A trivial example is the point mass at $(0,0)$ in $\mathbb{R}^2$. Its distribution function is discontinuous at $(0,y)$ and $(x,0)$ for all $x, y \geq 0$. The following result is useful.

Theorem 17.9.2. *Let $X = (X_1, \ldots, X_d)'$ be a d-dimensional random vector. Then, its cdf F is continuous at $x = (x_1, \ldots, x_d) \in \mathbb{R}^d$, iff*

$$P\{\max_{1\leq i\leq d}(X_i - x_i) = 0\} = 0\,. \tag{17.14}$$ ◆

Proof. First assume F is continuous at $x = (x_1, \ldots, x_d)$. Note that

$$P\{X_1 \leq x_1 - \frac{1}{n}, \ldots, X_d \leq x_d - \frac{1}{n}\} = F(x_1 - \frac{1}{n}, \ldots, x_d - \frac{1}{n}), , n \geq 1\,.$$

Since $\{X_1 \leq x_1 - \frac{1}{n}, \ldots, X_d \leq x_d - \frac{1}{n}\} \uparrow \{X_1 < x_1, \ldots, X_d < x_d\}$,

$$\begin{aligned}
P\{X_1 < x_1, \ldots, X_d < x_d\} &= \lim_{n\to\infty} P\{X_1 \leq x_1 - \frac{1}{n}, \ldots, X_d \leq x_d - \frac{1}{n}\} \\
&= \lim_{n\to\infty} F(x_1 - \frac{1}{n}, \ldots, x_d - \frac{1}{n}) \\
&= F(x), \text{ since } F \text{ is continuous at } x \\
&= P\{X_1 \leq x_1, \ldots, X_d \leq x_d\}\,.
\end{aligned}$$

Hence

$$\begin{aligned}
P\{\max_{1\leq i\leq d}(X_i - x_i) = 0\} &= P\{\max_{1\leq i\leq d}(X_i - x_i) \leq 0\} - P\{\max_{1\leq i\leq d}(X_i - x_i) < 0\} \\
&= P\{X_1 - x_1 \leq 0, \ldots, X_d - x_d \leq 0\} \\
&\quad - P\{X_1 - x_1 < 0, \ldots, X_d - x_d < 0\} = 0\,.
\end{aligned}$$

Therefore, (17.14) is implied by continuity of F at x.

Now assume that (17.14) holds. Fix $\varepsilon > 0$. Let $Y = \max_{1\le i\le d}(X_i - x_i)$, Then (17.14) implies $P\{Y = 0\} = 0$. Hence, the cdf of Y is continuous at 0. This implies that there exists $\delta > 0$ such that $P\{-\delta < Y \le \delta\} \le \varepsilon$. In other words,

$$\begin{aligned}\varepsilon &\ge P\{-\delta < Y \le \delta\} \\ &= P\{Y \le \delta) - P(Y \le -\delta\} \\ &= F(x_1 + \delta, \ldots, x_d + \delta) - F(x_1 - \delta, \ldots, x_d - \delta).\end{aligned}$$

If $y = (y_1, \ldots, y_d) \in \mathbb{R}^d$ where $\max_{1\le i\le d} |y_i - x_i| < \delta$, then

$$F(x_1 - \delta, \ldots, x_d - \delta) \le F(y_1, \ldots, y_d) \le F(x_1 + \delta, \ldots, x_d + \delta).$$

Thus, both $F(x)$ and $F(y)$ belong to the interval

$$[F(x_1 - \delta, \ldots, x_d - \delta),\ F(x_1 + \delta, \ldots, x_d + \delta)]$$

whose length is at most ε, showing that,

$$|F(y) - F(x)| \le \varepsilon, \text{ if } \ y \in \mathbb{R}^d,\ \max_{1\le i\le d} |y_i - x_i| \le \delta.$$

Therefore, F is continuous at x. This completes the proof. ∎

The following is a useful consequences of Theorem 17.9.2.

Corollary 17.9.1. *Suppose F is the cdf of $(X_1, \ldots, X_d)$. Let*

$$C_i = \{z \in \mathbb{R} : P\{X_i = z\} = 0\}, i = 1, \ldots, d.$$

Then,

$$\begin{aligned}&C_1 \times \cdots \times C_d \\ &\qquad \subset \left\{x \in \mathbb{R}^d : F \textit{ is continuous at } x\right\} \\ &\qquad \subset \{(x_1, \ldots, x_d) \in \mathbb{R}^d : P\{X_1 = x_1, \ldots, X_d = x_d\} = 0\}. \qquad (17.15)\end{aligned}$$

◆

Exercise 17.9.2 Let $X = (X_1, \ldots, X_d)'$ be a random vector where each X_i is continuous. Show that X is also continuous.

Exercise 17.9.3 Suppose $\{X_n\}$ is a sequence of d-dimensional random vectors such that $X_n \Rightarrow X$ and $X_n \Rightarrow Y$. Then show that $X \stackrel{D}{=} Y$.

We can now state the equivalence of weak convergence of probability measures and the convergence of distribution functions in $\mathbb{R}^d$. We omit its proof which essentially is along the lines of the proof of Theorem 17.3.1 in $\mathbb{R}$. However, for the "if" part, recall that in the proof of Theorem 17.3.1 a suitable dense set was chosen from the set of continuity points of F. Now this set has to be chosen more carefully since, unlike when $d = 1$, there may exist $x \in \mathbb{R}^d$ such that F is discontinuous at x, even though the corresponding rv doesn't take the value x.

Theorem 17.9.3. *Let μ, $\{\mu_n\}$ be probability measures on $\mathbb{R}^d$ with cdfs F, $\{F_n\}$ respectively. Then, $\mu_n \Rightarrow \mu$ iff $\lim_{n\to\infty} F_n(x) = F(x)$, for every $x \in \mathbb{R}^d$ at which F is continuous.* ◆

17.10 Moments, cumulants and weak convergence

Lemma 17.10.1. (**Weak convergence via moments and cumulants**) (a) Let $\{Y_n\}$ be rvs with distribution functions $\{G_n\}$ such that,

$$\lim_{n\to\infty} \mathrm{E}(Y_n^k) = m_k \text{ (finite), for all } k = 1, 2, \ldots.$$

Further, suppose that there is a unique distribution function G whose k-th moment is m_k for every k. Then, $G_n \Rightarrow G$.

(b) Let $\{Y_n\}$ be rvs with distribution functions $\{G_n\}$ such that,

$$\lim_{n\to\infty} c_k(Y_n) = c_k \text{ (finite), for all } k = 1, 2, \ldots.$$

Further suppose that, there is a unique distribution function G whose k-th cumulant is c_k for all $k \geq 1$. Then, $G_n \Rightarrow G$. ◆

Proof. (a) Since all moments converge, by Exercise 13.5.10 all powers of $\{Y_n\}$ are ui. Further, $\{G_n\}$ is tight. Consider any sub-sequence of $\{G_n\}$. By tightness, there is a further sub-sequence which converges weakly to, G, say. Then, by uniform integrability, all moments of G exist and

$$\int x^k G(dx) = m_k \text{ for all } k\,.$$

Now by our assumption, $\{m_k\}$ determines G uniquely. That is, the limit does not depend on the chosen sub-sequence. Hence, the original sequence $\{G_n\}$ converges weakly to G. This completes the proof of (a). Proof of Part (b) can be done easily by using Lemma 12.4.1 in conjunction with the discussion at the end of Section 12.5.2 and applying Part (a). This is left as an exercise. ∎

Exercise 17.10.1 Let $\{X_n\}$ be rvs and $c_j(X_n) \to c_j$ (finite) for all $j \geq 1$.
(a) If $c_j = 0$ for all $j \geq 3$, then show that $X_n \Rightarrow N(c_1, c_2)$.
(b) If $c_j = \lambda$ for all j, then show that X_n converges weakly to $Poi(\lambda)$.

17.11 Pólya's theorem

$G_n \Rightarrow G$ implies point-wise convergence at continuity points of G. Next result shows that, if G is continuous, then G_n converges to G uniformly.

Theorem 17.11.1. *(**Pólya**) Let $\{G_n\}, G$ be bounded non-decreasing functions on $\mathbb{R}$, where G is continuous.*
(a) if there is a dense $D \subset \mathbb{R}$ so that

$$G_n(x) \to G(x) \quad \textit{for all} \quad x \in D \cup \{-\infty\} \cup \{\infty\}. \tag{17.16}$$

Then

$$\sup_{x \in \mathbb{R}} |G_n(x) - G(x)| \to 0. \tag{17.17}$$

(b) In particular, if $\{G_n\}$ and G are cdfs, $G_n \Rightarrow G$, and G is continuous, then (17.17) holds. ◆

Proof. Fix $\epsilon > 0$. Since G is bounded, by definition of $G(\infty)$ and $G(-\infty)$, there exists $C_1 < C_2$ from D such that

$$G(C_1) - G(-\infty) + G(\infty) - G(C_2) < \epsilon. \tag{17.18}$$

On $[C_1,\ C_2]$, G is (uniformly) continuous. So, we can get an integer k, and points $x_1 = C_1 < x_2 < \cdots < x_k = C_2$, from D such that

$$G(x_{i+1}) - G(x_i) < \epsilon \ \text{ for all } \ 1 \leq i \leq k-1. \tag{17.19}$$

Let $x_0 = -\infty, x_{k+1} = \infty$ and $\Delta_n = \max_{0\le i\le k+1} |G_n(x_i) - G(x_i)|$. Then by (17.16), $\Delta_n \to 0$ as $n \to \infty$. Since G_n and G are non-decreasing, for any $x \in [x_i,\ x_{i+1}]$, $1 \le i \le k-1$, using (17.19),

$$\begin{aligned} G_n(x) - G(x) &\le G_n(x_{i+1}) - G(x_i) \\ &\le G_n(x_{i+1}) - G(x_{i+1}) + G(x_{i+1}) - G(x_i) \le \Delta_n + \epsilon. \end{aligned}$$

Likewise $G_n(x) - G(x) \ge -\Delta_n - \epsilon$. Hence, we have

$$\sup_{C_1\le x\le C_2} |G_n(x) - G(x)| \le \Delta_n + \epsilon\,.$$

We now consider any $x \le C_1$. Then using (17.18),

$$\begin{aligned} |G_n(x) - G(x)| &\le |G_n(x) - G_n(-\infty)| + |G_n(-\infty) - G(-\infty)| \\ &\quad + |G(-\infty) - G(x)| \\ &\le |G_n(C_1) - G_n(-\infty)| + |G_n(-\infty) - G(-\infty)| + \epsilon \\ &\le |G_n(C_1) - G(C_1)| + |G(C_1) - G(-\infty)| \\ &\quad + 2|G_n(-\infty) - G(-\infty)| + \epsilon \\ &\le 3\Delta_n + 2\epsilon\,. \end{aligned}$$

The same bound holds for all $x \ge C_2$. Since ϵ is arbitrary and $\Delta_n \to 0$, the proof of Part (a) is complete. Part (b) follows immediately. ■

17.12 Glivenko-Cantelli lemma

There is another special situation where uniform convergence holds.

Definition 17.12.1. (Empirical distribution function) Let $\{X_i\}$ be rvs on $(\Omega, \mathcal{A}, P)$. The *empirical distribution function* (edf) is defined as

$$F_n(x) := \sum_{i=1}^{n} \frac{\mathbf{1}_{\{X_i\le x\}}}{n}, \quad x \in \mathbb{R}\,.$$

◊

So, for every n, F_n is the cdf of the discrete distribution that puts equal mass $1/n$ at each $X_i, 1 \le i \le n$. This is a **random distribution** since it depends on the values of $\{X_i\}$.

For every x, $\{Y_i(x) := \mathbf{1}_{\{X_i \le x\}}\}$ are $0-1$ valued rvs. Hence, if $\{X_i\}$ are iid with a common distribution function F, then by the strong law of large numbers, for every $x \in \mathbb{R}$, $F_n(x) \to \mathrm{E}\, Y_1(x) = F(x)$ a.s. The next result shows that this convergence is uniform a.s.

The rv $\Delta_n = \sup_{x\in\mathbb{R}} |F_n(x) - F(x)|$ is known as *Kolmogorov-Smirnov statistic* and is of immense importance in statistical inference. To see that this supremum over an uncountable set is a rv, observe that

$$\Delta_n = \sup_{x\in\mathbb{Q}} |F_n(x) - F(x)| \quad (\mathbb{Q} \text{ is the set of rationals}). \tag{17.20}$$

Lemma 17.12.1. **(Glivenko-Cantelli)** Let $\{X_i\}$ be iid rvs with a common cdf F. Then $\sup_{x\in\mathbb{R}} |F_n(x) - F(x)| \to 0$ a.s. ♦

Proof of Lemma 17.12.1, needs an extension of Scheffe's Lemma 7.3.9.

Exercise 17.12.1 Let $\{f_n\}$, f be non-negative integrable functions on $(\Omega, \mathcal{A}, \mu)$, $\int f_n d\mu \to \int f d\mu$, and $f_n \to f$ a.s. Show that $\int |f_n - f| d\mu \to 0$.

Proof of Lemma 17.12.1. Let D_F be the (countable) set of discontinuity points of F, enumerated as $\{x_j\}$ with the corresponding jumps of F as $\{p_j\}$. We split F and the edf F_n as follows: Define

$$F_d(x) := \sum_{x_j\in D_F} p_j \mathbf{1}_{\{x_j \le x\}}, \; F_c(x) := F(x) - F_d(x),$$

$$F_{nd}(x) := n^{-1} \sum_{i=1}^{n} \mathbf{1}_{\{X_i \le x, X_i \in D_F\}}, \; F_{nc}(x) := F_n(x) - F_{nd}(x).$$

Define

$$p_{nj} := n^{-1} \sum_{i=1}^{n} \mathbf{1}_{\{X_i = x_j\}},$$

$$p_n := \sum_{x_j\in D_F} p_{nj} = n^{-1} \sum_{x_j\in D_F} \sum_{i=1}^{n} \mathbf{1}_{\{X_i = x_j\}} = n^{-1} \sum_{i=1}^{n} \mathbf{1}_{\{X_i \in D_F\}}.$$

As D_F is countable, by SLLN, outside a single null set, say N, $p_{nj} \to p_j$ for all j, and $p_n \to p := P\{X_1 \in D_F\} = \sum_{x_j\in D_F} p_j$. Hence by Exercise 17.12.1 (using the counting measure on D_F),

$$\sum_{j\in D_F} |p_{nj} - p_j| \to 0 \;\text{ outside }\; N. \tag{17.21}$$

As a consequence,

$$\sup_{x\in\mathbb{R}} |F_{nd}(x) - F_d(x)| \le \sum_{j\in D_F} |p_{nj} - p_j| \to 0 \ \text{ outside } \ N\,. \tag{17.22}$$

Now we focus on $F_{nc}(\cdot)$ and $F_c(\cdot)$. By construction, F_c is non-decreasing and continuous everywhere with $F_c(-\infty) = 0$ and $F_c(\infty) = 1-p$. Again by SLLN, there is a null set N_1 such that

$$F_{nc}(x) \to F_c(x) \ \text{ for all } \ x \in Q \ \text{ outside } \ N_1. \tag{17.23}$$

Moreover, $F_{nc}(-\infty) = 0$ and $F_{nc}(\infty) \equiv 1-p_n \to 1-p$ outside N. So, by Exercise 17.12.1, using (17.23), outside $N \cup N_1$, $\sup|F_{nc}(x)-F(x)| \to 0$. Combining this with (17.22), the result follows. ∎

17.13 Exercises

Exercise 17.13.1 Consider the functions $F_n(\cdot)$ defined by

$$F_n(x) = \begin{cases} 1 - \left(1 - \dfrac{x}{n}\right)^n & \text{if } 0 < x < n, \\ 0 \ \text{ if } \ x \le 0, \\ 1 \ \text{ if } \ x \ge 1. \end{cases}$$

Show that $\{F_n\}$ are continuous cdfs. Identify F such that $F_n \Rightarrow F$.

Exercise 17.13.2 Let $\{X_n\}$ be a sequence of rvs, and for every integer $k \ge 1$, $\mathrm{E}(X_n^k) \to \mu_k$ for some finite μ_k. Show that there is a rv X with some probability distribution F such that $\mathrm{E}(X^k) = \mu_k$. (Note that F need not be unique).

Exercise 17.13.3 Let $(\Omega, \mathcal{A}, P)$ be a probability space where $X_n \xrightarrow{P} X$ and $Y_n \xrightarrow{P} Y$. Show that $X_n + Y_n \xrightarrow{P} X + Y$ and hence in distribution. Show by an example that, convergence in probability in the hypothesis cannot be replaced by convergence in distribution.

Exercise 17.13.4 Suppose $X_n \Rightarrow X$. Let $g(\cdot)$ be a non-negative continuous function. Show that $\liminf \mathrm{E}[g(X_n)] \ge \mathrm{E}[g(X)]$. As a consequence,

$$\liminf \mathrm{E}\,|X_n| \ge \mathrm{E}\,|X| \ \text{ (both sides may equal } \infty).$$

Exercise 17.13.5 Let $S_n \sim Bin(n, p_n)$, $n \geq 1$. If S_n converges in distribution, then show that $np_n \to \lambda$ for some $0 \leq \lambda < \infty$.

Exercise 17.13.6 Let $X_{n,m}$, $1 \leq m \leq n$ be independent non-negative integer valued rvs. Let $P\{X_{n,m} = 1\} = p_{n,m}$, $P\{X_{n,m} \geq 2\} = \epsilon_{n,m}$. Suppose that

(i) $\sum_{m=1}^{n} p_{n,m} \to \lambda$, $0 < \lambda < \infty$,

(ii) $\max_{1 \leq m \leq n} p_{n,m} \to 0$ and

(iii) $\sum_{m=1}^{n} \epsilon_{n,m} \to 0$.

Show that $\sum_{m=1}^{n} X_{n,m}$ converges weakly to $Poi(\lambda)$.

Exercise 17.13.7 For continuous rvs $\{X_1, \ldots, X_n\}$, their *maximum order statistic* (or first maximum) is defined as $X_{(n)} = \max\{X_1, \ldots, X_n\}$. Let $\{X_i\}$ be iid $U(0, 1)$. Show that $n(1 - X_{(n)}) \Rightarrow Exp(1)$. Formulate and prove a similar result for the second maximum, and then for the kth maximum for any fixed k.

Exercise 17.13.8 Suppose $\{X_i\}$ are iid standard Cauchy rvs. Show that $n^{-1} \max_{1 \leq i \leq n} X_i$ converges weakly, and identify the limit.

Exercise 17.13.9 Let $\{X_i\}$ be iid $Exp(1)$ rvs. Let $X_{1n} < \cdots < X_{nn}$ be the ordered values of $X_1, \ldots, X_n$. [Note that a.s. all values are different, hence a strict ordering is possible.]

(a) Show that for every fixed k, $X_{kn} \Rightarrow Y_k$ where $Y_k \sim Gamma(1, k)$.

(b) Show that $X_{nn} - \log n$ converges weakly to a non-degenerate distribution (which is called the Gumbel distribution).

Exercise 17.13.10 Recall the definition of the inverse of a distribution function from Exercise 5.3.19. Let $\{F_n\}$ and F be probability distribution functions. Show that F_n converges weakly to F if and only if $F_n^{-1}(t)$ converges to $F^{-1}(t)$ at each continuity point $t \in (0,\ 1)$ of F^{-1}.

Exercise 17.13.11 Show that if $\{X_n\}$ satisfies (17.24), then it is tight.

$$\sup_{n \geq 1} \mathrm{E}\left(|X_n|^{\alpha}\right) < \infty \text{ for some } \alpha > 0. \tag{17.24}$$

Exercise 17.13.12 Prove the Portmanteau Theorem, Helly's Theorem, and Prokhorov's Theorem in $\mathbb{R}^d$.

Exercise 17.13.13 Show that, Glivenko-Cantelli Lemma 17.12.1 holds if $\{X_i\}$ are pair-wise independent identically distributed rvs.

Exercise 17.13.14 Let P_1 and P_2 be absolutely continuous probability measures on $\mathbb{R}$ with densities f_1 and f_2. Show that

$$\sup_{B \in \mathcal{B}(\mathbb{R})} |P_1(B) - P_2(B)| = \frac{1}{2} \int_{\mathbb{R}} |f_1(x) - f_2(x)| \lambda(dx) .$$

Exercise 17.13.15 Let F_n be the distribution with density

$$f_n(x) = 1 - \cos(2n\pi x), \ 0 \leq x \leq 1 .$$

(a) Show that F_n converges weakly to the $U(0,\ 1)$ distribution.

(a) Show that f_n does not converge to the uniform density.

Exercise 17.13.16 Let $\{U_n\}$ be iid $U(0,\ 1)$ rvs. Let $P_n = \prod_{i=1}^{n} U_i^{1/n}$. Find real sequences $\{a_n\}$ and $\{b_n\}$ such that $b_n(P_n - a_n)$ converges weakly to a non-degenerate distribution.

Exercise 17.13.17 (Lévy metric for probability distributions) [Compare with the Lévy metric on the space of rvs in Exercise 13.5.3.]

For any probability distribution functions F_1 and F_2 on $\mathbb{R}$, define

$$d_L(F_1, F_2) := \inf\{\epsilon > 0 : F_1(x - \epsilon) - \epsilon \leq F_2(x) \leq F_1(x + \epsilon) + \epsilon\} .$$

(a) Show that d_L defines a metric (called the Lévy metric) on the space of all probability distributions on $\mathbb{R}$.

(b) Show that $F_n \Rightarrow F$ if and only if $d_L(F_n, F) \to 0$.

Chapter 18

Characteristic function

Characteristic function is a very important tool in probability. It is directly connected to the Fourier transform, a fundamental object in many areas of science. It uniquely identifies a probability distribution, and is especially useful for handling sums of independent rvs. We shall show its connection to weak convergence and use it to prove the central limit theorem. For more information on characteristic functions, refer to Lukacs [1970], Lukacs and Laha [1964], and Ramachandran [1967].

18.1 Characteristic function

Definition 18.1.1. (**Characteristic function**) If μ is a **finite measure** on $\mathcal{B}(\mathbb{R}^d)$, its *characteristic function* (cf) is defined as

$$\begin{aligned}\hat{\mu}(t) &:= \int_{\mathbb{R}^d} e^{\iota t'x}\mu(dx),\ t \in \mathbb{R}^d \\ &= \int_{\mathbb{R}^d} \cos(t'x)\mu(dx) + \iota \int_{\mathbb{R}^d} \sin(t'x)\mu(dx),\ t \in \mathbb{R}^d.\end{aligned}$$ ◇

We shall mostly deal with the case where $d = 1$, and μ is a probability measure. Note that

$$\hat{\mu}(t) = \begin{cases} \int_{\mathbb{R}^d} e^{\iota t'x} f(x)\lambda(dx) \text{ if } \mu \text{ has density } f(\cdot), \\ \sum_x e^{\iota t'x} p(x) \text{ if } \mu \text{ has the mass function } p(\cdot). \end{cases}$$

We shall use $\phi_\mu(\cdot)$ to denote the cf of the probability measure μ. If X is a d-dimensional random vector with distribution function F, then the

A. Bose, *Elements of Measure and Probability*, Texts and Readings in Mathematics 88,
https://doi.org/10.1007/978-981-95-2758-8_18

characteristic function of X (or of F) is defined in the obvious way:

$$\begin{aligned}\phi_X(t) := \phi_F(t) &= \mathrm{E}\left(e^{\iota t'X}\right)\\ &= \int_{\mathbb{R}^d} e^{\iota t'x} F(dx), \; t \in \mathbb{R}^d.\end{aligned}$$

We will soon see that a cf determines a probability distribution uniquely.

18.1.1 Fourier transform

Fourier transform is an important object in mathematics and has numerous applications. It is closely related to the characteristic function.

Definition 18.1.2. Let $f \in L^1(\mathbb{R}^d, \lambda_d)$. Then its Fourier transform is

$$\hat{f}(t) = \frac{1}{(\sqrt{2\pi})^d} \int_{\mathbb{R}^d} e^{\iota t'x} f(x) \lambda_d(dx), \; t \in \mathbb{R}^d. \tag{18.1}$$ ◇

We note that in the literature, variants of Fourier transforms use a negative sign in front of ι, and/or do not use the constant $1/(\sqrt{2\pi})^d$ in (18.1). Note that $\hat{f}$ is also defined for all $f \in L^1(\lambda)$. If an absolutely continuous finite measure μ has a density f, then its cf satisfies $\hat{\mu}(t) = (\sqrt{2\pi})^d \hat{f}(t)$. We present one basic but famous result on Fourier transforms. For more details on Fourier transforms, see Folland [2009].

Lemma 18.1.1. (Riemann-Lebesgue) Let $f \in L^1(\mathbb{R}^d, \lambda_d)$. Then,

$$\lim_{|t| \to \infty} \hat{f}(t) = \lim_{|t| \to \infty} \int_{\mathbb{R}^d} e^{\iota t'x} f(x) \lambda_d(dx) = 0.$$ ♦

Proof. We prove the result only for the case $d = 1$. First, suppose that, f is continuous and compactly supported. For $t \neq 0$, the substitution $x \to x + \frac{\pi}{t}$ leads to

$$\begin{aligned}\sqrt{2\pi}\hat{f}(t) = \int_{\mathbb{R}} f(x) e^{\iota xt} \lambda(dx) &= \int_{\mathbb{R}} f(x + \frac{\pi}{t}) e^{\iota xt} e^{\iota \pi} \lambda(dx)\\ &= -\int_{\mathbb{R}} f(x + \frac{\pi}{t}) e^{\iota xt} \lambda(dx).\end{aligned}$$

This gives another expression for $\hat{f}(t)$. Using the mean of the two expressions, we get

$$|\hat{f}(t)| \leq \frac{1}{2} \int_{\mathbb{R}} \left| f(x) - f(x + \frac{\pi}{t}) \right| \lambda(dx).$$

Since f is continuous, $\lim_{|t|\to\infty} |f(x) - f(x + \frac{\pi}{t})| = 0$ for all $x \in \mathbb{R}$. Since f is compactly supported, using DCT $\lim_{|t|\to\infty} |\hat{f}(t)| = 0$.

Now suppose $f \in L^1$ is arbitrary. Fix $\epsilon > 0$. By Theorem 9.1.5 and Exercise 9.1.4, pick a compactly supported continuous function g, such that $\|f - g\|_{L^1} \le \epsilon$. Then

$$|\hat{f}(t)| \le \Big| \int (f(x) - g(x)) e^{\iota xt} \, \lambda(dx) \Big| + \Big| \int g(x) e^{\iota xt} \, \lambda(dx) \Big| .$$

The first term is bounded by ϵ by the choice of g. The second term goes to 0 as $|t| \to \infty$ since g is continuous and compactly supported. Since ϵ was arbitrary, the proof is complete. ■

Exercise 18.1.1 (a) (Alternate proof of Theorem 18.1.1 for $d = 1$). Fix $\epsilon > 0$. Using Theorem 9.1.3, get simple functions $s_k = \sum_{i=1}^{k} x_i \mathbf{1}_{B_i}$ such that $\|s_k - f\|_{L^1} < \epsilon$. Then approximate $\{B_i\}$ by finite disjoint union of intervals $\{A_i\}$ so that $\lambda(A_i \Delta B_i) \le \epsilon/(k|x_i|)$. Then directly check that the Fourier transform of $g = \sum_{i=1}^{k} x_i \mathbf{1}_{A_i}$ converges to 0 as $|t| \to \infty$.

(b) Prove Theorem 18.1.1 for general d.

18.1.2 Basic properties of cf

Exercise 18.1.2 (a) Show that

$$|e^{\iota a} - e^{\iota b}| \le \min\{2, |a - b|\}, \text{ for all } a, b \in \mathbb{R} .$$

Hint: Write the left side as the absolute value of an integral from a to b.

(b) Show that $\lim_{x\to 0} \dfrac{e^{\iota x} - 1}{\iota x} = 1$.

Theorem 18.1.1 (Basic properties of characteristic function). *Suppose μ is a finite measure on $\mathbb{R}^d$ with cf $\hat{\mu}$. Then*

(a) $|\hat{\mu}(t)| \le \hat{\mu}(0) = \mu(\mathbb{R}^d)$ for all $t \in \mathbb{R}^d$.

(b) $\hat{\mu}(\cdot)$ is a uniformly continuous function.

(c) For all $t \in \mathbb{R}^d$, $\hat{\mu}(-t) = \overline{\hat{\mu}(t)}$ ($\bar{z}$ is the complex conjugate of z). ◆

The proof of the theorem is left as an exercise.

Definition 18.1.3. (Symmetric distribution) A random variable X or its distribution is said to be symmetric (about 0) if for all $B \in \mathcal{B}(\mathbb{R})$, $P\{X \in B\} = P\{-X \in B\}$. ◇

This means that $X \stackrel{D}{=} -X$. The Gaussian distribution on $\mathbb{R}$ with mean 0 is symmetric about 0.

Theorem 18.1.2. *(a) A rv X has a symmetric distribution if and only if its cf $\phi(\cdot)$ is real-valued.*
(b) Let X be a rv. If $\mathrm{E}(|X|^n) < \infty$ for some positive integer n, then the n-th derivative of $\phi_X(\cdot)$ exists and is continuous on $\mathbb{R}$ and

$$\phi_X^{(n)}(t) = \int_{\mathbb{R}} (\iota x)^n e^{\iota t x} F(dx).$$

In particular,

$$\iota^n \mathrm{E}(X^n) = \phi_X^{(n)}(0). \qquad \blacklozenge$$

The proof of the theorem is left as an exercise.

Exercise 18.1.3 (a) If X is a real-valued rv, show that for any real constants a and b, $\phi_{aX+b}(t) = e^{\iota t b}\phi_X(at)$, for all $t \in \mathbb{R}$.

Exercise 18.1.4 If X is a random vector, state and prove an analogue of Theorem 18.1.2(b) for the mixed moments of its components.

Exercise 18.1.5 Suppose $\{X_i\}$ are real-valued independent rvs, and $S_n = X_1 + \cdots + X_n$. Show that

$$\phi_{S_n}(t) = \prod_{i=1}^{n} \phi_{X_i}(t). \tag{18.2}$$

On the left of (18.2), we have a rv S_n whose distribution may not be easily obtainable. However, its cf is the product of the individual cfs. We use this later to prove the central limit theorem in Chapter 19.

Exercise 18.1.6 (a) Let $X \sim Bin(n, p)$. Find the cf of X.
(b) Let $X \sim Poi(\lambda)$. Show that its cf is $\phi_X(t) = \exp\{\lambda(\exp(\iota t) - 1)\}$.

Exercise 18.1.7 (Characteristic function of the Gaussian distribution)
(a) Let $X \sim N(0,\ 1)$. Since the distribution of X is symmetric, $\phi_X(\cdot)$ is a real-valued function. Show that $\phi'_X(t) = -t\phi_X(t)$. Deduce that $\phi_X(t) = e^{-t^2/2}$. See Exercise 18.12.9 for another proof.
(b) Show that if $Y \sim N(m, \sigma^2)$ then

$$\phi_Y(t) = e^{\iota t m} e^{-t^2\sigma^2/2}. \tag{18.3}$$

18.2 Inversion formula in $\mathbb{R}$

The following theorem shows how to recover a distribution function from a cf. We will need the facts from the following exercise in its proof.

Exercise 18.2.1 Show that

(a) If g is Lebesgue integrable, then $\int_{(a,\ b)} g(x)\lambda(dx)$ is a continuous function of a and b.

(b) If g is a continuous function, then $\int_{(a,\ b)} g(x)\lambda(dx)$ is a differentiable function of both b and a.

(c) If g is a continuous function such that for a distribution function F,

$$F(b) - F(a) = \int_{(a,\ b)} g(x)\lambda(dx) \text{ for all } a < b\,,$$

then g is non-negative.
Note: This exercise can be solved directly, or it can be solved using the developments from Chapter 22.

Theorem 18.2.1 (Inversion formula in $\mathbb{R}$). *Let X be a real-valued rvs with distribution function F and cf $\phi_F(\cdot)$.*
(a) Then for all $a, b \in \mathbb{R}, a < b$,

$$\frac{F(b) + F(b-)}{2} - \frac{F(a) + F(a-)}{2} = \lim_{c\to\infty} \frac{1}{2\pi} \int_{(-c,\ c)} \frac{e^{-\iota ta} - e^{-\iota tb}}{\iota t} \phi_F(t)\lambda(dt)\,.$$

In particular if a, b are continuity points of F, then

$$F(b) - F(a) = \lim_{c\to\infty} \frac{1}{2\pi} \int_{(-c,\ c)} \frac{e^{-\iota ta} - e^{-\iota tb}}{\iota t} \phi_F(t)\lambda(dt)\,.$$

(b) If $\phi_F(\cdot)$ is Lebesgue integrable, then F is absolutely continuous and a density for F is given by

$$f(x) = \frac{1}{2\pi} \int_{\mathbb{R}} e^{-\iota tx} \phi_F(t)\lambda(dt)\,. \tag{18.4}$$

(c) The mass of X at any point can be recovered as

$$P\{X = a\} = \lim_{T\to\infty} \frac{1}{2T} \int_{(-T,\ T)} e^{-\iota ta} \phi_F(t)\lambda(dt)\,.$$

◆

Exercise 18.2.2 (**Uniqueness property**) Let F_1, F_2 be distribution functions with cfs ϕ_1 and ϕ_2. Show that $F_1 \equiv F_2$ if and only if $\phi_1 \equiv \phi_2$.

Remark 18.2.1. (a) Suppose f is a pdf. Recall $\hat{f}$ defined in (18.1). Theorem 18.2.1(b) is known as the *Fourier Inversion Theorem.* Using this, the map $f \to \hat{f}$ can be extended to an isometric map from $L^2(\mathbb{R}, \lambda)$ onto itself. This is one of the fundamental results in Fourier Analysis.

(b) It is also interesting to ask whether a given function is the cf of some distribution function. We shall address this question in Section 18.6. ●

Proof of Theorem 18.2.1. For $a < b$, let

$$\begin{aligned} I_c &= \frac{1}{2\pi}\int_{(-c,\ c)} \frac{e^{-\iota ta} - e^{-\iota tb}}{\iota t}\phi_F(t)\lambda(dt) \\ &= \frac{1}{2\pi}\int_{(-c,\ c)} \frac{e^{-\iota ta} - e^{-\iota tb}}{\iota t}\Big[\int_{\mathbb{R}} e^{\iota tx} F(dx)\Big]\lambda(dt) \end{aligned}$$

Note that by Exercise 18.1.2,

$$\Big|\frac{e^{-\iota ta} - e^{-\iota tb}}{\iota t} e^{\iota tx}\Big| \le b - a,$$

and

$$\int_{(-c,\ c)}\int_{\mathbb{R}} (b-a)F(dx)\lambda(dt) = 2c(b-a) < \infty .$$

Hence by applying Fubini's theorem,

$$I_c = \frac{1}{2\pi}\int_{\mathbb{R}}\int_{(-c,\ c)} \frac{e^{-\iota ta} - e^{-\iota tb}}{\iota t} e^{\iota tx}\lambda(dt)F(dx) = \int_{\mathbb{R}} J_c(x)F(dx),$$

where

$$\begin{aligned} J_c(x) &= \frac{1}{2\pi}\int_{-c}^{c} \frac{\sin t(x-a) - \sin t(x-b)}{t}\lambda(dt) \quad (\text{cos integral is } 0) \\ &= \frac{1}{2\pi}\int_{-c(x-a)}^{c(x-a)} \frac{\sin v}{v}\lambda(dv) - \frac{1}{2\pi}\int_{-c(x-b)}^{c(x-b)} \frac{\sin w}{w}\lambda(dw). \qquad (18.5) \end{aligned}$$

By Exercise 10.5.7, $\int_{(r,\ s)} \frac{\sin t}{t}\lambda(dt) \to \pi$ as $s \to \infty$ and $r \to -\infty$. Using this, and (18.5) it follows that

(i) There is an $M < \infty$ such that $\sup_{c,x} |J_c(x)| \le M$.

(ii)

$$J(x) := \lim_{c\to\infty} J_c(x) = \begin{cases} 0 \ \text{ if } \ x < a \ \text{ or } \ x > b, \\ 1 \ \text{ if } \ a < x < b, \\ \dfrac{1}{2} \ \text{ if } \ x = a \ \text{ or } \ x = b. \end{cases}$$

Hence by DCT Theorem 7.2.7,

$$\begin{aligned} \lim_{c\to\infty} I_c &= \lim_{c\to\infty} \int_{\mathbb{R}} J_c(x) F(dx) \\ &= \int_{\mathbb{R}} J(x) F(dx) \\ &= \int_{\{x<a\}} 0 F(dx) + \frac{1}{2} \int_{\{x=a\}} F(dx) + \int_{\{a<x<b\}} F(dx) \\ &\quad + \frac{1}{2} \int_{\{x=b\}} F(dx) + \int_{\{x>b\}} 0 F(dx) \\ &= \frac{1}{2}[F(a) - F(a-)] + [F(b-) - F(a)] + \frac{1}{2}[F(b) - F(b-)] \\ &= \frac{F(b) + F(b-)}{2} - \frac{F(a) + F(a-)}{2}. \end{aligned}$$

(b) Now suppose $\phi_F(\cdot)$ is integrable. Then f as presented in (18.4) is well-defined. We must show that f is a density for F.

By using DCT, it is easy to see that f is continuous. By Fubini's Theorem 10.3.1,

$$\begin{aligned} \int_a^b f(x)\lambda(dx) &= \frac{1}{2\pi} \int_{\mathbb{R}} \phi_F(t) [\int_{(a,\, b)} e^{-\iota t x} \lambda(dx)] \lambda(dt) \\ &= \lim_{c\to\infty} \frac{1}{2\pi} \int_{(-c,\, c)} \phi_F(t) \Big(\int_a^b e^{-\iota t x} \lambda(dx) \Big) \lambda(dt) \text{ by DCT} \\ &= \lim_{c\to\infty} \frac{1}{2\pi} \int_{(-c,\, c)} \frac{e^{-\iota t a} - e^{-\iota t b}}{\iota t} \phi_F(t) \lambda(dt) \\ &= F(b) - F(a) \ \ \text{(by (a) if, } a, b \text{ continuity points of } F). \end{aligned}$$

Thus for all continuity points $a < b$ of F,

$$F(b) - F(a) = \int_{(a,\, b)} f(x)\lambda(dx).$$

We claim this holds for all $a < b$. Since f is continuous, by Exercise 18.2.1(a), the integral is a continuous function of a and b. Since the continuity points of F are dense in $\mathbb{R}$, the above holds for all a and b. Since f is continuous everywhere, by Exercise 18.2.1(b), F is differentiable everywhere, and moreover $F'(x) = f(x)$ for all x. Since F is non-decreasing, f must be non-negative. Thus f is a density of F.

(c) This can be proved using (a). We leave it as an exercise. ■

Exercise 18.2.3 (a) Let $X_i \sim Bin(n_i, p)$, $i = 1, 2$ be independent rvs. Show that $X_1 + X_2$ is a binomial rv and identify the parameters.

(b) Let $\{X_i\}$ be independent Poisson rvs with parameters $\{\lambda_i\}$. Show that $X_1 + \cdots + X_n \sim Poi(\lambda_1 + \cdots + \lambda_n)$.

18.3 Moments and uniqueness

We have seen in Exercise 12.2.6 that different probability distributions on $\mathbb{R}$ can have identical moments of all order. But then when does a sequence of moments determine a distribution uniquely? There are several sufficient conditions for this. We establish two that use the uniqueness property of the cf. We shall find the following result very useful.

Exercise 18.3.1 Show that for all $x \in \mathbb{R}$,

(a) $|e^{\iota x} - (1 + \iota x)| \leq \dfrac{x^2}{2}$.

(b) $|e^{\iota x} - (1 + \iota x - x^2/2)| \leq \min\{\dfrac{|x|^3}{6}, x^2\}$.

Hint. For (a), take $g_2(t) = e^{\iota t} - 1$ and use Exercise 18.1.2(a) and integrate. For (b), take $g_3(t) = e^{\iota t} - (1 + \iota t)$ and use (a).

(c) Generalize (b) by showing that for any non-negative integer n,

$$|\exp(\iota x) - (1 + \iota x + \frac{(\iota x)^2}{2!} + \ldots + (-1)^n \frac{(\iota x)^n}{n!})| \leq \frac{|x|^{n+1}}{(n+1)!}.$$

18.3.1 Uniqueness via mgf

Theorem 18.3.1. *Let μ be a probability measure on $\mathbb{R}$ with finite moments $\{m_k\}$. If $\sum_{k=1}^{\infty} m_k x^k/k!$ is finite for x in a neighbourhood of 0, then μ is the only measure with the moments $\{m_k\}$.* ◆

Proof. Let $\beta_k = \int_{\mathbb{R}} |x|^k \mu(dx)$ be the *absolute moments*. Then, for an $r > 0$,

$$\frac{\beta_k r^k}{k!} \to \text{ as } k \to \infty\,. \tag{18.6}$$

To see this, first, by the hypothesis, there is $0 < s < 1$ such that $m_k s^k/k! \to 0$ as $k \to \infty$. Choose $0 < r < s$. Then $2kr^{2k-1} < s^{2k}$ for all large k. Since $|x|^{2k-1} \le 1 + |x|^{2k}$,

$$\frac{\beta_{2k-1} r^{2k-1}}{(2k-1)!} \le \frac{r^{2k-1}}{(2k-1)!} + \frac{\beta_{2k} s^{2k}}{(2k)!}\,.$$

Hence (18.6) holds as $k \to \infty$ through odd values. Since $\beta_k = m_k$ for even k, (18.6) holds.

By Exercise 18.3.1(c),

$$\big|e^{\iota tx}\big(e^{\iota hx} - \sum_{k=0}^{n} \frac{(\iota hx)^k}{k!}\big)\big| \le \frac{|hx|^{n+1}}{(n+1)!}.$$

Integrating the quantity within the absolute sign with respect to μ, the cf ϕ of μ satisfies

$$\big|\phi(t+h) - \sum_{k=0}^{n} \frac{h^k}{k!} \int_{\mathbb{R}} (\iota x)^k) e^{\iota tx} \mu(dx)\big| \le \frac{|h|^{n+1} \beta_{n+1}}{(n+1)!}$$

By Theorem 18.1.2(b), above integral equals $\phi^{(k)}(t)$. Hence using (18.6),

$$\phi(t+h) = \sum_{k=0}^{\infty} \frac{\phi^{(k)}(t)}{k!} h^k,\ |h| \le r,\ t > 0\,. \tag{18.7}$$

Now suppose that ν is another probability measure with moments $\{m_k\}$ and cf $\psi(\cdot)$. Then the same arguments as above lead to

$$\psi(t+h) = \sum_{k=0}^{\infty} \frac{\psi^{(k)}(t)}{k!} h^k,\ |h| \le r,\ t > 0\,. \tag{18.8}$$

By Theorem 18.1.2(b), we also know that $\phi^{(k)}(0) = \psi^{(k)}(0) = \iota^k m_k$ for all k. Hence

$$\phi(h) = \psi(h) \text{ for all } |h| \le r\,.$$

As a consequence, they have identical derivatives in this domain. Taking $t = r - \epsilon$ and $t = -r + \epsilon$ in (18.7) and (18.8) shows that φ and ψ also agree in $(-2r + \epsilon, 2r - \epsilon)$, and hence in $(-2r, 2r)$. But then by the same argument, they must agree in $(-3r, 3r)$ as well, and so on. Thus $\phi \equiv \psi$ and hence $\nu \equiv \mu$. ■

18.3.2 Uniquess via Riesz's condition

We now establish a second sufficient condition which is useful in applications. We need the well-known **Stirling's approximation** for factorials. It implies that for some $C_1, C_2 > 0$,

$$C_1(2\pi)^{\frac{1}{2}}e^{-n}n^{n+\frac{1}{2}} \leq n! \leq C_2(2\pi)^{\frac{1}{2}}e^{-n}n^{n+\frac{1}{2}}, \text{ for all } n \geq 1. \tag{18.9}$$

For a probabilistic proof of Stirling's approximation, see Exercise 19.5.8.

Theorem 18.3.2. *Let $\{m_k\}$ be the sequence of moments of the distribution function F. Then F is the unique distribution with these moments if the following condition given by Riesz [1923]) holds:*

$$\liminf_{k\to\infty} \frac{1}{k} m_{2k}^{\frac{1}{2k}} < \infty \quad \text{(Riesz's condition)}. \tag{18.10}$$ ◆

Proof. Suppose F and G are two distributions with the common moments $\{m_k\}$. Let the cfs of F and G be

$$\phi_F(t) = \int_{\mathbb{R}} e^{\iota tx} F(dx) \quad \text{and} \quad \phi_G(t) = \int_{\mathbb{R}} e^{\iota tx} G(dx).$$

By the uniqueness property of cf (Exercise 18.2.2), it is enough to show $\phi_F(t) = \phi_G(t)$ for all t. We focus on $t \geq 0$, the other case is similar.

Since F and G have common moments, we have, for all $j = 0, 1, \ldots,$ $\phi_F^{(j)}(0) = \phi_G^{(j)}(0) = \iota^j m_j$. Define

$$t_0 = \sup\{t \geq 0 : \ \phi_G^{(j)}(s) = \phi_F^{(j)}(s), \ 0 \leq s \leq t, \ j \geq 1\}.$$

The result will follow if we can show $t_0 = \infty$ for all $j = 0, 1, \ldots$. Let, if possible, $t_0 < \infty$. By the continuity property (Theorem 18.1.2(b)),

$$\int_{\mathbb{R}} x^j e^{\iota t_0 x}[F(ds) - G(ds)] = 0.$$

By condition (18.10), there is a constant $M > 0$ such that

$$m_{2k} \leq (Mk)^{2k} \text{ for infinitely many } k.$$

By Exercise 18.3.1(c),

$$|e^{\iota a} - 1 - \iota a \ldots - (\iota a)^k/k!| \leq |a|^{k+1}/(k+1)!, \quad \text{for all} \quad k \geq 1.$$

Choose $s \in (0, 1/(eM))$. Use the above inequality and the relation $k! > (k/e)^k$, to obtain that, for any fixed $j \geq 0$,

$$\begin{aligned}
\left|\phi_F^{(j)}(t_0+s) - \phi_G^{(j)}(t_0+s)\right| &= \left|\int_{\mathbb{R}} x^j e^{\iota(t_0+s)x}[F(dx) - G(dx)]\right| \\
&= \left|\int_{\mathbb{R}} x^j e^{\iota t_0 x}\left[e^{\iota sx} - 1 - \iota sx - \ldots - \frac{(\iota sx)^{2k-j-1}}{(2k-j-1)!}\right][F(dx) - G(dx)]\right| \\
&\leq 2\frac{s^{2k-j}m_{2k}}{(2k-j)!} \leq 2\frac{(sMk)^{2k}}{s^j(2k-j)!} \\
&\leq 2(esMk/(2k-j))^{2k}(2k/s)^j \to 0,
\end{aligned}$$

as $k \to \infty$ along those k such that $m_{2k} \leq (Mk)^{2k}$. The last inequality follows by Stirling's approximation. This violates the definition of t_0. ■

Exercise 18.3.2 (a) Show that the moments of bounded rvs satisfy Riesz's condition (18.10).

(b) Show that the moments given in (12.1) satisfy Riesz's condition. Hence, they identify uniquely the standard Gaussian distribution.

(c) Show that moments $\{m_k\}$ satisfy Riesz's condition if,

$$m_{2k} \leq \frac{(2k)!}{k!2^k}\Delta^k, \quad k = 0, 1, \ldots, \text{ for some } 0 < \Delta < \infty.$$

Remark 18.3.1. (**Carleman's condition**) A more general condition than Riesz which is quite popular is that of Carleman [1926]).

Let $\{m_k\}$ be a moment sequence of some probability measure μ. Then μ is the unique probability measure with these moments if

$$\sum_{i=1}^{\infty} m_{2k}^{-\frac{1}{2k}} = \infty \quad (\text{Carleman's condition}).$$

For a proof see, Bai and Silverstein [2010]. ●

18.4 Tightness, cf, and weak convergence in $\mathbb{R}$

Theorem 18.4.1. *Let $\{F_n\}$ be tight probability distributions on $\mathbb{R}$.*

(a) If every weakly convergent subsequence of $\{F_n\}$ converges weakly to the same distribution function F, then $F_n \stackrel{w}{\to} F$.

(b) $F_n \stackrel{w}{\to} F$ if and only if $\phi_{F_n}(t)$ converges to a finite limit $g(t)$ for every t, and then $g(\cdot)$ is the cf of F. ◆

Proof. Part (a) is left as an exercise. For Part (b), let $\phi_{F_n}(t) \to g(t)$ (finite) for every t. Since $\{F_n\}$ is tight, by Helly's Theorem 17.6.1, there is a sub-sequence $\{n_k\}$ such that $F_{n_k} \Rightarrow F$. If F_n does not converge weakly to F, then there is another sub-sequence $F_{m_k}(t)$ that converges to say $G \neq F$. By uniqueness (see Exercise 18.2.2) this is a contradiction since $g(t) = \phi_F(t) = \phi_G(t)$ for all t. Hence $F_n \Rightarrow F$. The reverse implication follows from the Portmanteau Theorem 17.2.1 since $x \to e^{\iota tx}$ is a bounded continuous for every t. ■

18.5 Lévy's continuity theorem in $\mathbb{R}$

This is an extremely useful result to show weak convergence of probability measures. We shall later see a version of this result in $\mathbb{R}^d$.

Theorem 18.5.1 (Lévy's continuity theorem). *Let $\{F_n\}$ and F be probability distribution functions on $\mathbb{R}$.*

(a) If $F_n \stackrel{w}{\to} F$, then $\phi_{F_n}(t) \to \phi_F(t)$ for all t.

(b) If $\phi_{F_n}(t) \to g(t)$ for all t where g is continuous at 0 and $g(0) = 1$, then $F_n \stackrel{w}{\to} F$ where F has the cf g. ◆

To prove the theorem, we need the following technical lemma. Recall that for any $z \in \mathbb{C}$, $\mathcal{R}\mathrm{e}(z)$ denotes its real part.

Lemma 18.5.1 (Truncation Inequality). Let F be a probability distribution function on $\mathbb{R}$ with cf $\phi_F(\cdot)$. Then for some $k > 1$, for all $u > 0$,

$$\int_{\{x:|x|\geq 1/u\}} F(dx) \leq \frac{k}{u} \int_{(0,\ u)} [1 - \mathcal{R}\mathrm{e}(\phi_F(t))]\lambda(dt).$$

◆

Proof Lemma 18.5.1.

$$
\begin{aligned}
\frac{1}{u}\int_{(0,\,u)} &[1-\mathcal{R}\mathrm{e}\ \phi_F(t)]\lambda(dt) \\
&= \frac{1}{u}\int_0^u \int_{\mathbb{R}} (1-\cos tx)F(dx)\lambda(dt) \\
&= \int_{\mathbb{R}} \Big[\frac{1}{u}\int_0^u (1-\cos tx)\lambda(dt)\Big] F(dx) \quad \text{by Fubini's theorem} \\
&= \int_{\mathbb{R}} \Big(1-\frac{\sin ux}{ux}\Big)F(dx) \\
&\geq \inf_{\{t:\ |t|\geq 1\}} \Big(1-\frac{\sin t}{t}\Big) \int_{\{x:\ |ux|\geq 1\}} F(dx) \\
&= \frac{1}{k}\int_{\{x:\ |x|\geq 1/u\}} F(dx).
\end{aligned}
$$

■

Proof of Theorem 18.5.1. (a) follows from the definition of weak convergence since $e^{\iota tx}$ is a bounded continuous function.

(b) We first claim that $\{F_n\}$ is tight.

Using Lemma 18.5.1,

$$
\begin{aligned}
\int_{\{x:\ |x|\geq 1/u\}} F_n(dx) &\leq \frac{k}{u}\int_0^u [1-\mathcal{R}e\ \phi_{F_n}(t)]\lambda(dt) \\
&\to \frac{k}{u}\int_0^u [1-\mathcal{R}e\ g(t)]\lambda(dt) \quad \text{by DCT} \\
&\to 0 \quad \text{as } u\to 0, \quad \text{since } g \text{ is continuous at } 0.
\end{aligned}
$$

Hence $\{F_n\}$ is tight.

By Theorem 18.4.1(b) F_n converges weakly to a distribution function F. As a consequence, $\phi_{F_n}(t) \to \phi_F(t)$ for all t. This implies that, $\phi_F(t) = g(t)$ for all t. Hence g must be the cf of F. ■

Exercise 18.5.1 Let $X_n \sim Bin(n, 1/2)$. Show that as $n \to \infty$, $\dfrac{2X_n - n}{\sqrt{n}}$ converges weakly to a standard Gaussian distribution.

Exercise 18.5.2 Let X_λ be a Poisson rv with parameter λ. Show that as $\lambda \to \infty$, $\dfrac{X_\lambda - \lambda}{\sqrt{\lambda}}$ converges weakly to a standard Gaussian distribution.

Exercise 18.5.3 Let $X_n \sim Bin(n, p_n)$. Show that X_n converges to $Poi(\lambda)$ if and only if $np_n \to \lambda, 0 < \lambda < \infty$.

18.6 Bochner's theorem

Given a function g, when can we say that it is the cf of some (finite) measure? We now give a criterion for this.

Definition 18.6.1. A function $h : \mathbb{R} \to \mathbb{C}$ is said to be **non-negative definite**, if the matrix $((h(u_i - u_j)_{1 \leq i,j \leq n}))$ is non-negative definite, for any $n \geq 1$, $\{u_i\} \subset \mathbb{R}$, and $\{t_i\} \subset \mathbb{C}$. That is, $\sum_{i,j=1}^{n} t_i h(u_i - u_j)\bar{t}_j \geq 0$. ◇

It is easy to show that the cf of any finite measure is non-negative definite and continuous.

The reverse implication is the main contention of Bochner's Theorem 18.6.1. Its proof is beyond the scope of this book. The reader may consult Folland [1999].

Theorem 18.6.1. *Let $h : \mathbb{R} \to \mathbb{C}$ be a function. Then $h(\cdot)$ is the cf of a finite measure on $\mathbb{R}$ if and only if $h(\cdot)$ is continuous at 0 and $h(\cdot)$ is non-negative definite.* ◆

18.7 Multivariate Gaussian distribution

From Definition 11.2.3, $X = (X_1, \ldots, X_d)'$ is multivariate Gaussian if and only if for every real vector $t = (t_1, \ldots, t_d)'$, $t'X$ is real-valued Gaussian. Suppose X is multivariate Gaussian. Let m be the mean vector and Σ be the dispersion matrix of X. Then it is easy to see that, $\mathrm{E}(t'X) = t'm$ and $\mathrm{V}(t'X) = t'\Sigma t$. That is, $t'X \sim N(t'm, t'\Sigma t)$. Hence, utilising the cf formula (18.3) for a real-valued Gaussian variable,

$$\phi_X(t) = e^{\iota t'm - t'\Sigma t/2}. \tag{18.11}$$

Exercise 18.7.1 (a) Using Bochner's theorem, show that $\phi_Y(\cdot)$ in (18.3) is a cf.

(b) If X and Y are Gaussian random vectors of dimension d, then show that $X \stackrel{D}{=} Y$ if and only if $t'X \stackrel{D}{=} t'Y$ for every $t \in \mathbb{R}^d$. What happens if we know that $t'X \stackrel{D}{=} t'Y$ for $t \in D$, where D is a dense set of $\mathbb{R}^d$?

Exercise 18.7.2 Let $X \sim N(\mu, \Sigma)$. Then show that,

(a) its co-ordinates are independent if and only if Σ is diagonal.

(b) any of its sub-vector is also a Gaussian random vector.

(c) X is absolutely continuous if and only if Σ is non-singular. In that case, the density of X is given by

$$f_X(x) = \frac{1}{(2\pi)^{d/2}|\Sigma|^{d/2}} e^{-(x-\mu)'\Sigma^{-1}(x-\mu)/2}, \ x \in \mathbb{R}^d.$$

(d) The mgf of X is given by

$$M_X(t) = e^{t'\mu + t'\Sigma t/2}. \tag{18.12}$$

18.8 Isserlis' formula

Let $(X_1, \ldots, X_d)'$ be Gaussian with mean vector 0 and dispersion matrix $\Sigma = ((\sigma_{ij}))$. Then using (18.12), the cgf of X is $C_X(t) = t'\Sigma t/2$. Thus all cumulants (mixed or pure) of order larger than two are zero, and the second order cumulants are $\{\sigma_{ij} = \mathrm{E}(X_i X_j)\}$. As a consequence, all moments are functions of $((\sigma_{ij}))$.

The following formula, which was discovered by Isserlis [1918], gives all the moments in terms of the covariances. Specializing relation (12.14) to the present situation,

$$\begin{aligned} \mathrm{E}(X_1 X_2 \cdots X_d) &= 0 \quad (\text{if } d \text{ is odd}), \\ \mathrm{E}(X_1 X_2 \cdots X_d) &= \sum \prod \mathrm{E}(X_{i_k} X_{j_k}) \quad (\text{if } d \text{ is even}), \end{aligned}$$

where the sum is over the product of all *pair-partitions* $\{\{i_k, j_k\}\}$ of $\{1, \ldots, d\}$. Note that some or all of the X_i are allowed to be identical in the above formula.

This formula is also referred to as *Wick's formula* after the work of Wick [1950] in particle physics.

18.9 Inversion formula in $\mathbb{R}^d$

The inversion formula and its proof in one dimension can be easily extended to dimension d. We omit the proof. Note that, as in dimension one, the inversion formula implies that the cfs and probability distribution functions are in one-to-one correspondence.

Theorem 18.9.1. *Suppose μ is a probability measure on $\mathbb{R}^d$. Let A be the rectangle $A := (a_1,\ b_1] \times \cdots \times (a_d,\ b_d]$ such that $\mu(\partial A) = 0$. Then*

$$\mu(A) = \lim_{c\to\infty} \frac{1}{(2\pi)^d} \int_{[-c,\ c]^k} \prod_{j=1}^{d} [\frac{e^{-\iota t_j a_j} - e^{-\iota t_j b_j}}{\iota t_j}] \phi_\mu(t)\lambda_d(dt). \qquad \blacklozenge$$

Exercise 18.9.1 (a) Give a proof of Theorem 18.9.1.

(b) Theorem 18.9.1 corresponds to only Part (a) of Theorem 7.2.1. Formulate and prove the analogues of other parts of the latter in $\mathbb{R}^d$.

18.10 Lévy's continuity theorem in $\mathbb{R}^d$

Theorem 18.10.1 (Lévy's continuity theorem in $\mathbb{R}^d$). *For probability measures μ and $\{\mu_i\}$ on $\mathbb{R}^d$, with cfs ϕ, and $\{\phi_i\}$,*

$$\mu_n \Rightarrow \mu\,, \tag{18.13}$$

if and only if

$$\lim_{n\to\infty} \phi_n(t) = \phi(t) \textit{ for all } t \in \mathbb{R}^d\,. \tag{18.14}$$

$\blacklozenge$

Proof. (a) First suppose (18.13) holds. Then $\lim_{n\to\infty} \phi_n(t) = \phi(t)$ for all $t \in \mathbb{R}^d$, since for every t, $e^{\iota t' x}$ is a bounded continuous function of x.

(b) Now suppose (18.14) holds. We shall first show that $\{\mu_n : n \geq 1\}$ is tight. Let $X, X_1, X_2, \ldots$ be rvs defined on some probability space $(\Omega, \mathcal{A}, P)$ such that μ_n and μ are the distributions of X_n and X respectively. Write $X_n = (X_{n1}, \ldots, X_{nd})'$ and $X = (X_l, \ldots, X_d)'$.

Fix $s \in \mathbb{R}$, $1 \leq j \leq d$. Let t be the vector in $\mathbb{R}^d$ whose j-th coordinate is s and the remaining coordinates are zero. Then (18.14) yields

$$\lim_{n\to\infty} \mathrm{E}\left(e^{\iota s X_{nj}}\right) = \mathrm{E}\left(e^{\iota s X_j}\right). \tag{18.15}$$

Since (18.15) holds for every $s \in \mathbb{R}$, by Levy's continuity Theorem 18.5.1 $X_{nj} \Rightarrow Y_j$. In particular, $\{X_{nj} : n \geq 1\}$ is tight. That is,

$$\lim_{T\to\infty} \limsup_{n\to\infty} P\{|X_{nj}| > T\} = 0\,,\ j = 1, \ldots, d\,.$$

Then it is easy to conclude that $\{X_n\}$ itself is also tight.

Let $\{\mu_{n_k}\}$ be any subsequence of $\{\mu_n\}$. Since $\{\mu_n\}$ is tight, so is $\{\mu_{n_k}\}$. By Prokhorov's theorem on $\mathbb{R}^d$, $\{\mu_{n_k}\}$ is relatively compact and hence there is a further subsequence $\{\mu_{n_{k_l}}\}$ such that $\mu_{n_{k_l}} \Rightarrow \nu$, for some probability measure ν on $\mathbb{R}^d$.

Hence by Part (a) of the proof, $\lim_{l\to\infty} \phi_{n_{k_l}}(t) = \phi_\nu(t)$, for all $t \in \mathbb{R}^d$, where ϕ_ν is the cf of ν. Now use (18.14) again to conclude that $\phi_\nu(t) = \phi(t)$, $t \in \mathbb{R}^d$.

This, in conjunction with Theorem 18.9.1 shows that $\nu = \mu$. Thus, every subsequence of μ_n has a further subsequence that converges weakly to μ. Therefore (18.13) follows, and the proof is complete. ■

18.11 Cramér-Wold technique

Recall that in $\mathbb{R}^d$, $X \stackrel{D}{=} Y$ if and only if $t'X \stackrel{D}{=} t'Y$ for all $t \in \mathbb{R}^d$. Thus questions on weak convergence on $\mathbb{R}^d$ can be reduced to those on $\mathbb{R}$ by the following simple but powerful result. We omit the proof, since it follows immediately from Exercise 17.9.1 and Theorem 18.10.1.

Theorem 18.11.1 (Cramér-Wold theorem). *Let $X, X_1, X_2, \ldots$ be $\mathbb{R}^d$-valued rvs. Then $X_n \Rightarrow X$ if and only if $t'X_n \Rightarrow t'X$ for all $t \in \mathbb{R}^d$.* ◆

18.12 Exercises

Exercise 18.12.1 Show that if $\phi(\cdot)$ is a cf then $\mathcal{R}e(\phi(\cdot))$ and $|\phi(\cdot)|^2$ are also so.

Exercise 18.12.2 A rv X is said to be **lattice** if there exist a, and $h > 0$ such that $\sum_{k=-\infty}^{\infty} P\{X = a + kh\} = 1$. Show that

(a) X is a lattice rv, if and only if $|\phi_X(t_0)| = 1$ for some $t_0 \neq 0$;

(b) If X is lattice then $\phi_X(\cdot)$ is a periodic function;

(c) If X is lattice, then the pmf of X can be obtained from its cf as:

$$P\{X = a + hk\} = \frac{h}{2\pi} \int_{(-\pi/h,\ \pi/h)} e^{-\iota t(a+kh)} \phi_X(t) \lambda(dt) .$$

Exercise 18.12.3 Let $\phi(\cdot)$ be a cf, with $|\phi(t)| = 1$ for distinct points $t = u$, and $t = \alpha u$, α irrational. Show that the distribution is degenerate.

Exercise 18.12.4 Let X be a real rv with cf $\phi(\cdot)$. Let $\{x_1, x_2, \ldots\ldots\}$ be the set of jump points of the cdf of X. Show that

(a)

$$\lim_{T\to\infty} \frac{1}{2T} \int_{(-T,\ T)} |\phi(t)|^2 \lambda(dt) = \sum_{k=1}^{\infty} [P\{X = x_k\}]^2;$$

(b) the probability distribution of X has no jumps if $|\phi(t)|^2$ is integrable.

Exercise 18.12.5 Let X be a rv with finite expectation. Show that

$$\mathrm{E}(|X|) = \frac{2}{\pi} \int_{(0,\infty)} \frac{1 - \mathcal{R}e(\varphi_X(t))}{t^2} \lambda(dt)\,.$$

Exercise 18.12.6 Let E_1 and E_2 be iid $Exp(1)$ rv. Show that the cf of $E_1 - E_2$ is,

$$\varphi(t) = 1/(1 + t^2)\,.$$

Exercise 18.12.7 Let X be a standard Cauchy rv with pdf

$$f(x) = \frac{1}{\pi(1 + x^2)}, \quad -\infty < x < \infty\,.$$

(a) Using the inversion formula, show that $\phi_X(t) = e^{-|t|}$ must be the cf of the standard Cauchy distribution.

(b) Using Exercise 18.12.6, show that $\phi_X(t) = e^{-|t|}$.

(c) Using *contour integration*, show that $\phi_X(t) = e^{-|t|},\ t \in \mathbb{R}$.

(d) Let $\{X_i\}$ be iid standard Cauchy rvs. Show that $(X_1 + \cdots + X_n)/n$ is also a standard Cauchy rv for every n.

Exercise 18.12.8 Consider the pdf, $f(x) = \frac{1-\cos x}{\pi x^2}, |x| \le 1$. Show that its cf is given by

$$\phi(t) = \begin{cases} 1 - |t| & \text{if } |t| \le 1, \\ 0 & \text{otherwise}\,. \end{cases}$$

Exercise 18.12.9 (Needs knowledge of analyticity) Let $X \sim N(0, 1)$. From Exercise 12.2.4, its mgf is $M_X(t) = e^{t^2/2},\ t \in \mathbb{R}$.

(a) Define $M_X(z)$ for complex numbers z by replacing t by z in (12.2.4). Show that $M_X(z)$ is a complex analytic function of z.

(b) Using (a) and the mgf, show that $\phi_X(t) = e^{-t^2/2}$.

Exercise 18.12.10 Show that a sequence of Gaussian rvs converges in distribution iff both the sequences of means and variances converge.

Exercise 18.12.11 Let X_1, X_2 be iid with $\mathrm{E}(X_1) = 0$, $\mathrm{V}(X_1) = 1$.

(a) If they are Gaussian, show that $X_1 - X_2$ and $X_1 + X_2$ are iid Gaussian.

(b) If $X_1 - X_2$ and $X_1 + X_2$ are independent, show that X_1 and X_2 are standard Gaussian.

(c) The **Lévy-Cramér theorem** is a significantly stronger result: if X and Y are independent and $X + Y$ is Gaussian, then X and Y are Gaussian. Its proof is beyond the scope of this book.

Exercise 18.12.12 Let $\{X_i\}$ be iid Gaussian. Define the sequence of sample means and sample variances as:

$$\bar{X}_j = \sum_{i=1}^{j} X_i, \quad s_j^2 = (j-1)^{-1} \sum_{i=1}^{j} (X_i - \bar{X}_j)^2, \ 1 \leq j \leq n\,.$$

Show that for every n, $\bar{X}_n$ and the vector $(s_2^2, \ldots s_n^2)'$ are independent. Hint: Linear combinations of independent Gaussians are again jointly Gaussian. If $\{A_i\}$ are independent, then $\{f_i(A_i)\}$ are also independent for all measurable functions $\{f_i\}$. Choose $\{A_i\}$ and $\{f_i\}$ judiciously and relate them to the sample mean and sample variances.

Exercise 18.12.13 (a) Let X be a gamma rv with density given in (11.1). Find the cf of X. Hint: Identify the appropriate values of t for which the mgf is finite, and find a formula for $M_X(t)$. Now argue by analytic continuity as in Exercise 18.12.9.

(b) Suppose $\{X_i\}$ are i.i.d with exponential density $g(x) = \lambda e^{-\lambda x}$, $x > 0$. Using cfs, show that $X_1 + \cdots + X_n$ is a gamma rv with parameters (n, λ).

Exercise 18.12.14 Let $\{X_n\}$ be independent. Show that $\sum_{n=1}^{\infty} X_n$ converges a.s. iff $\Pi_{n=1}^{\infty} \phi_n(t)$ converges for ever $t \geq 0$.

Exercise 18.12.15 If $\{F_n\}$ is tight, then show that $\{\phi_n(\cdot)\}$ is uniformly equi-continuous.

Exercise 18.12.16 If $\{F_n\}$ converges weakly, then show that $\{\phi_n(\cdot)\}$ converge uniformly on bounded sets.

Exercise 18.12.17 (Independence and weak convergence). Let $\{X_n\}$ and $\{Y_n\}$ be independent and weakly convergent sequences. What can you say about the convergence in distribution of the vector $(X_n, Y_n)'$?

Exercise 18.12.18 Show that a probability measure on $\mathbb{R}$ is compactly supported iff there exists $C < \infty$ such that the moments of μ satisfy $|m_k(\mu)| \leq C^k$ for all non-negative integers k. Further, the support of μ is contained in the interval $[-C,\ C]$.

Exercise 18.12.19 Show that Riesz's condition implies Carleman's.

Exercise 18.12.20 Suppose that, for some constant C, a sequence of cumulants $\{c_k\}$ satisfies $|c_k| \leq C^k$ for all $k \in \mathbb{N}$. Show that the moments also satisfy this bound (with a possibly different C).

Exercise 18.12.21 Show by using cumulants that the sum of independent Gaussian rvs is again Gaussian.

Exercise 18.12.22 Let $\{X_i\}$ be iid with finite mean m. Using cf, show that $(X_1 + \cdots + X_n)/n \xrightarrow{P} m$.

Exercise 18.12.23 Let $X_n \sim Bin(n, p_n)$ such that $np_n \to \lambda$ as $n \to \infty$. Using cumulants, show that the distribution of X_n converges to the Poisson distribution with parameter λ.

Chapter 19

Central limit theorem

Central limit theorem is one of the fundamental results in probability theory. Since it has significant applications in statistics, numerous variations of this result have been generated. Two excellent books on the historical development of the CLT are Adams [2009] and Fischer [2010]. We shall prove the basic CLT in two ways–via cumulants and via cfs.

19.1 Central limit theorem

Recall the notation $S_n = X_1 + \cdots + X_n$. The classical central limit theorem (CLT) is the following:

Theorem 19.1.1 (Central Limit Theorem (CLT)). *Suppose $\{X_i\}$ is a sequence of iid rvs with mean μ and variance σ^2. Then $(S_n - n\mu)/\sqrt{n}\sigma$ converges weakly to a standard Gaussian variable.* ◆

19.1.1 Cumulant based proof

As mentioned earlier, the CLT can be proved by many methods. Let us first sketch a proof via cumulants.

Theorem 19.1.2 (CLT via cumulants). *(a) Let $\{Y_n : n \in \mathbb{N}\}$ be a sequence of iid bounded rvs with mean zero and variance one. Then,*

$$n^{-1/2} \sum_{i=1}^{n} Y_i \Rightarrow Y \sim N(0,1).$$

(b) The above remains true without the boundedness condition. ◆

A. Bose, *Elements of Measure and Probability*, Texts and Readings in Mathematics 88,
https://doi.org/10.1007/978-981-95-2758-8_19

Proof. Moments of the standard Gaussian distribution satisfy Riesz's condition, and hence they identify the distribution uniquely.

(a) Define

$$Z_n := n^{-1/2}\sum_{i=1}^{n} Y_i,\ n \ge 1\,.$$

Since the variables are bounded, all moments are finite. Hence by using Exercise 17.10.1(a), it suffices to show that, for all $k \in \mathbb{N}$,

$$\lim_{n\to\infty} c_k(Z_n) = \begin{cases} 1 & \text{if } k = 2\,, \\ 0 & \text{if } k \neq 2\,. \end{cases} \tag{19.1}$$

Recall from Exercise 12.3.1 that for any rv X (with finite mgf), and any real constants a and b, $c_j(aX+b) = a^j c_j(X)$ for all $j \ge 2$. Also recall from Exercise 14.2.4 that for independent rvs X and Y (with finite mgfs), $c_j(X+Y) = c_j(X) + c_j(Y)$ for all $j \ge 1$.

Using these, and the fact that Y_i are identically distributed, with mean 0 and variance 1, it follows that, for any $k \in \mathbb{N}$,

$$\begin{aligned} c_k(Z_n) &= n c_k(n^{-1/2}Y_1) \\ &= n^{1-k/2} c_k(Y_1) \to 0 \text{ if } k \ge 3. \end{aligned}$$

On the other hand, it is easy to check that

$$c_1(Y_1) = \mathrm{E}(Y_1) = 0,\quad c_2(Y_1) = \mathrm{E}(Y_1^2) - (c_1(Y_1))^2 = 1\,.$$

Therefore, $c_1(Z_n) = 0$ and $c_2(Z_n) = 1$ for all n. This establishes (19.1) and the proof in this case is complete by invoking Lemma 17.10.1(b).

(b) We can first truncate the rvs at a fixed level B and invoke Part (a) to show that its standarized sum obeys the CLT. Then we can show that the variance of the difference between the two standardised sums, truncated and untruncated, is upper bounded (across n) by a quantity which depends on B and which tends to 0 as $B \to \infty$. Then the proof is completed by using the Mallows' metric W_2 in a limit argument. The details are left as an exercise. ∎

A proof can also be based on the moments. But that argument is slightly more involved. The reader is invited to try this.

19.1.2 Characteristic function based proof

We now present a proof of the CLT using characteristic functions and Lévy's continuity Theorem 18.5.1. The method of this proof can be used in other situations where the underlying sequence need not be iid and the limit need not be Gaussian. Recall the formula for the cf of the Gaussian distribution from Exercise 18.1.7.

We shall need two lemma. The proof of Lemma 19.1.1 follows easily by induction on n, and is omitted.

Lemma 19.1.1. Suppose $\{a_i\}$ and $\{b_i\}$ are complex numbers such that

$$\max_{1\le i\le n}\{|a_i|, |b_i|\} \le 1\,.$$

Then

$$\Big|\prod_{i=1}^n a_i - \prod_{i=1}^n b_i\Big| \le \sum_{i=1}^n |a_i - b_i|\,.$$

In particular, if $a_i \equiv a, b_i \equiv b$, then $|a^n - b^n| \le n|a-b|$. ♦

Proof of Lemma 19.1.2 follows from Exercise 18.3.1(b) in conjunction with DCT Theorem 7.2.7. Details are omitted.

Lemma 19.1.2. (a) For all $x \in \mathbb{R}$,

$$\Big|e^{\iota x} - (1 + \iota x - \frac{x^2}{2})\Big| \le \min\Big\{\frac{|x|^3}{6}, x^2\Big\}\,.$$

(b) If X is a rv with $\mathrm{E}(X^2) < \infty$, then

$$\phi_X(t) = 1 + \iota t\,\mathrm{E}(X) - \frac{1}{2}t^2\,\mathrm{E}(X^2) + o(t^2), \quad \text{as} \quad t \to 0. \qquad ♦$$

Proof of Theorem 19.1.1. Define $Y_i = (X_i - \mu)/\sigma$. Then $\{Y_i\}$ are iid with mean 0 and variance 1.

$$Z_n := (S_n - n\mu)/\sqrt{n}\sigma = \sum_{i=1}^n Y_i/\sqrt{n}\,.$$

Let $\phi(\cdot)$ denote the cf of Y_1. Noting that $\{Y_i\}$ are iid,

$$\begin{aligned}\phi_{Z_n}(t) &= \mathrm{E}\left(e^{\iota t n^{-1/2}\sum_{i=1}^n Y_i}\right)\\ &= \Big[\phi(\frac{t}{\sqrt{n}})\Big]^n. \qquad (19.2)\end{aligned}$$

From Exercise 18.1.7, the cf of the $N(0,1)$ distribution is $\psi(t) = e^{-t^2/2}$. Hence by Lévy's continuity Theorem 18.5.1, it is enough to show that

$$\left[\phi(\frac{t}{\sqrt{n}})\right]^n \to e^{-t^2/2} \quad \text{for all} \quad t \in \mathbb{R}. \tag{19.3}$$

By using Eqn. (19.2), Lemma 19.1.1, and Lemma 19.1.2(b),

$$\begin{aligned}
|\phi_{Z_n}(t) - e^{-t^2/2}| &= \left|\left[\phi(\frac{t}{\sqrt{n}})\right]^n - \left[e^{-t^2/(2n)}\right]^n\right| \\
&\le n\left|\phi(\frac{t}{\sqrt{n}}) - e^{-t^2/(2n)}\right| \\
&= n\left|(1 - \frac{t^2}{2n} + o(t^2/n)) - (1 - \frac{t^2}{2n} + o(t^2/n))\right| = o(1).
\end{aligned}$$

This proves the theorem. ■

19.2 Lindeberg CLT

This extension of the CLT to independent rvs which are not necessarily identically distributed, is due to Lindeberg. The proof for the iid case via cfs can be modified to give a proof of the following theorem. Details are left as an exercise.

Theorem 19.2.1 (Lindeberg CLT). *Suppose that $r_n \to \infty$, and for every n, $X_{n1}, \ldots, X_{nr_n}$ are independent rvs with*

$$\mathrm{E}(X_{nk}) = 0, \quad \textit{and} \ \sum_{k=1}^{r_n} \mathrm{E}(X_{nk}^2) = 1.$$

Assume that

$$\lim_{n\to\infty} \sum_{k=1}^{r_n} \mathrm{E}\left[X_{nk}^2 \mathbf{1}(|X_{nk}| > \epsilon)\right] = 0 \ \textit{for all} \ \epsilon > 0\,. \tag{19.4}$$

Then, as $n \to \infty$, $\sum_{k=1}^{r_n} X_{nk} \Rightarrow N(0,1)$. ◆

Condition (19.4) is known as **Lindeberg condition**. This condition is indeed necessary when no individual summand is too large, as seen in the following theorem of Feller. This is why Lindeberg CLT is also

known as the Lindeberg-Feller CLT. The proof of the following theorem is beyond the scope of the book.
A triangular array $\{X_{nk}\}$ is **uniformly asymptotically negligible** if

$$\lim_{n\to\infty} \max_{1\le k\le r_n} P\{|X_{nk}| > \epsilon\} \text{ for every } \epsilon > 0. \tag{19.5}$$

Theorem 19.2.2 (Feller). *Let $X_{n1}, \ldots, X_{nr_n}$ be independent rvs for every n, $r_n \to \infty$, and $\{X_{nk}\}$ satisfy condition (19.5). Further,*

$$\mathrm{E}(X_{nk}) = 0, \quad \textit{and} \quad \sum_{k=1}^{r_n} \mathrm{E}(X_{nk}^2) = 1 \textit{ for every } k, n\,.$$

If $\sum_{k=1}^{r_n} X_{nk} \Rightarrow N(0,1)$, then Lindeberg Condition (19.4) holds. ◆

19.3 Multivariate CLT

The Cramér-Wold technique and Lévy's continuity theorem together immediately lead to the following multivariate version of the CLT.

Theorem 19.3.1 (Multivariate CLT). *Suppose $X_1, X_2, \ldots$ are iid $\mathbb{R}^d$-valued random vectors with mean μ and covariance matrix Σ. Then, as $n \to \infty$, $n^{-1/2} \sum_{i=1}^{n} (X_i - \mu) \Rightarrow Z$, where Z has the Gaussian distribution with mean 0 and dispersion matrix Σ.* ◆

Remark 19.3.1 There are other methods to establish convergence to the Gaussian distribution and beyond. These include the "replacement method" and Stein's method. For detailed information, the interested reader may consult Tao [2010] (accessed April 25, 2025).

19.4 Large deviations

The theory of large deviations is a well-developed advanced topic in probability. We give a very short introduction to this theory, restricted to two results on the mean and on the edf of of iid rvs, without proof. Interested reader may consult Dembo and Zeitouni [1998], Ellis [2006], Deuschel and Stroock [1989] den Hollander [2000], and Varadhan [2016] for details of this theory and its applications.

Recall the Portmanteau Theorem 17.2.1 for weak convergence. A sequence of probability measure $\{P_n\}$ on $\mathbb{R}$ converges to P if any of the following two equivalent conditions hold:

$$\liminf_{n\to\infty} P_n(U) \ge P(U) \quad \text{for every open set } U\,.$$

$$\limsup_{n\to\infty} P_n(C) \le P(C) \quad \text{for every closed set } C\,.$$

The CLT says that if $\{X_i\}$ are iid with $\mathrm{E}(X_1) = 0$ and $Var(X_1) = \sigma^2$, then $S_n/\sqrt{n}$ converges weakly to $N(0, \sigma^2)$. On the other hand, we also know that $S_n/n \to 0$ a.s. How small would $P\{n^{-1}S_n \in U\}$ be, where U is a suitable set that is away from 0?

As an example, let $\{X_i\}$ be iid with common distribution $Ber(p)$. Let $S_n = X_1 + \cdots + X_n$. Then using Stirling's approximation (18.9), it can be shown that

$$P\{S_n = [nx]\} = \exp\left(-nI(x) + O(\log n)\right),$$

where

$$I(x) = x \log \frac{x}{p} + (1-x)\log\frac{1-x}{1-p}\,.$$

As a consequence,

$$\lim_{n\to\infty} n^{-1} \log P\{S_n = \lfloor nx \rfloor\} = -I(x) \quad \text{for all } \; 0 \le x \le 1\,.$$

A general formulation of such convergence results is the following:

Definition 19.4.1. (**Large deviation principle (LDP)**) Probability measures $\{P_n\}$ on $\mathbb{R}$ are said to satisfy a large deviation principle (LDP) with a rate function $I(\cdot)$ if the following conditions hold:

(i) For every $a \ge 0$, $\{x : I(x) \le a\}$ is compact.

(ii) For every open set U, $\liminf_{n\to\infty} n^{-1} \log P_n(U) \ge -\inf_{x\in U} I(x)$.

(iii) For every closed C, $\limsup_{n\to\infty} n^{-1} \log P_n(C) \le -\inf_{x\in C} I(x)$. ◊

19.4.1 Cramér's large deviation

We now state the most basic large deviation result for iid rvs. Let $\{X_i\}$ be iid $\mathbb{R}^d$-valued random vectors on $(\Omega, \mathcal{A}, P)$. Let

$$P_n(A) = P\{n^{-1}(X_1 + \cdots + X_n) \in A\}, \; A \in \mathcal{B}(\mathbb{R}^d)\,. \tag{19.6}$$

Theorem 19.4.1. ***(Cramér's LDP)*** *Suppose $\{X_i\}$ are iid $\mathbb{R}^d$-valued rvs with probability distribution P, and the mgf $M(t)$ is finite for every $t \in \mathbb{R}^d$. Then $\{P_n\}$ defined in (19.6) satisfies a large deviation principle with rate function $I(x) = \sup_{t\in\mathbb{R}^d}(x't - \log M(t))$.* ◆

We have seen several results on the convergence of sums of independent rvs: Kolmogorov's three series theorem, the law of iterated logarithm, and the central limit theorem. Theorem 19.4.1 completes the picture for the asymptotic behaviour of sums of iid rvs. For a short proof of the theorem when $\{X_i\}$ are real-valued, see Cerf and Petit [2011].

19.4.2 Sanov's large deviation

The concept of LDP is available for probability measures in more general spaces (such as a completely separable metric space).

Let $\{X_i\}$ be iid $\mathbb{R}^d$-valued rvs. Then the empirical distribution (edf) is a random probability measure on $\mathbb{R}^d$, and an appropriate LDP holds for this sequence. This is known as Sanov's theorem, stated below without proof. For more information see the references listed earlier.

Definition 19.4.2. **(Relative entropy)** The *relative entropy* on the space of probability measures on $\mathbb{R}^d$ is defined as

$$H(Q|P) = \begin{cases} \int_{\mathbb{R}} \big(\log \frac{dQ}{dP}\big) dQ \text{ if } Q \ll P \text{ and } \big(\frac{dQ}{dP}\big)\big(\log \frac{dQ}{dP}\big) \in L^1(P), \\ \infty \text{ otherwise}. \end{cases}$$ ◇

Theorem 19.4.2. ***(Sanov's LDP)*** *Let $\{X_i\}$ be iid random vectors in $\mathbb{R}^d$ with distrbution P. Let F_n be the edf. Then $\{F_n\}$ satisfies the LDP with rate function $I(Q) = H(Q|P)$* ◆

19.5 Exercises

Exercise 19.5.1 Let $\{X_i\}$ be iid with cf $\phi(\cdot)$. Show that

(a) $\phi'(0) = ia$ implies $S_n/n \xrightarrow{P} a$.

(b) if $S_n/n \xrightarrow{P} a$, then $\phi(t/n)^n \to e^{iat}$. Hence the weak law of large numbers holds if and only if $\phi'(0)$ exists.

Exercise 19.5.2 Using Theorem 17.8.1, give a rigorous justification of the steps in the proof of Theorem 19.1.2(b).

Exercise 19.5.3 Suppose $\{X_i\}$ are iid with mean 0 and finite positive variance. Show that

(a) $\limsup S_n/\sqrt{n} = \infty$ a.s.

(b) $S_n/\sqrt{n}$ does not converge in probability.

Exercise 19.5.4 Complete the proof of Lindeberg CLT Theorem 19.2.1.

Exercise 19.5.5 Suppose $\{X_i\}$ are independent rvs. Show that the Lindeberg CLT can be applied to appropriately centered and scaled S_n in the following cases once we define $X_{nk}, 1 \le k \le n, n \ge 1$ suitably.

(i) $\{X_i\}$ are uniformly bounded.

(ii) $\{X_i\}$ is iid with finite variance.

(iii) Let $\mu_k = \mathrm{E}(X_k)$ and $s_n^2 = \sum_{k=1}^n \mathrm{E}[(X_k - \mu_k)^2]$. Then

$$\frac{1}{s_n^{2+\delta}} \sum_{k=1}^{n} \mathrm{E}[|X_k - \mu_k|^{2+\delta}] \to 0 \text{ for some } \delta > 0 \text{ (**Lyapunov's condition**)}.$$

(iv) $\sup_k \mathrm{E}[|X_k|^3] < \infty$ and $s_n^2 \to \infty$.

Exercise 19.5.6 Let $\{X_n\}$ be independent rvs with distributions

$$P\{X_k = \pm 1\} = \frac{1}{2}(1 - \frac{1}{k^2}),\ P\{X_k = \pm k\} = \frac{1}{2k^2}.$$

Show that $n^{-1/2}(X_1 + \cdots + X_n) \Rightarrow N(0,\ 1)$.

Exercise 19.5.7 Let $\{X_n\}$ be iid with density $f(x) = |x|^{-3},\ |x| \ge 1$. Show that $(X_1 + \cdots + X_n)/\sqrt{n \log n} \Rightarrow N(0,1)$. Hint: Verify that Lyapunov's condition holds for $\{Y_n\}$, where $Y_n = X_n \mathbf{1}_{\{|X_n| \le \sqrt{n}\}}$, and that $\sum_{n=1}^{\infty} P(X_n \ne Y_n) < \infty$.

Exercise 19.5.8 (Proof of Stirling's approximation) Consider iid $Poi(1)$ rvs $\{X_i\}$, and let $S_n = X_1 + \cdots + X_n$.

(a) Show that $n^{-1/2}(S_n - n)^-$ is uniformly integrable and hence converges to $\mathrm{E}(N^-) = 1/\sqrt{2\pi}$ where N is a $N(0,\ 1)$ rv.

(b) Hence prove that $n! \sim \sqrt{2\pi} n^{n+1/2} e^{-n}$ as $n \to \infty$.

Exercise 19.5.9 Let $\{X_n\}$ be iid, $\mathrm{E}(X_1) = 0$, $\mathrm{E}(X_1^2) = 1$, $\mathrm{E}(X_1^4) < \infty$, and

$$\bar{X}_n = n^{-1} \sum_{i=1}^{n} X_i,\quad s_n^2 = n^{-1} \sum_{i=1}^{n} (X_i - \bar{X}_n)^2. \tag{19.7}$$

(a) Show that $Z_n := n^{1/2}(s_n^2 - 1)$ converges to the Gaussian distribution.

(b) Does the random vector $(n^{1/2}\bar{X}_n, Z_n)$ converge weakly?

Exercise 19.5.10 Let $\{X_n\}$ be iid with mean m and variance σ^2. Suppose $g : \mathbb{R} \to \mathbb{R}$ is a continuous function and is differentiable at m. Show that $n^{1/2}(g(\bar{X}_n) - g(m))$ converges weakly to a Gaussian distribution. Here $\bar{X}_n$ is as defined in (19.7).

Definition 19.5.1. (a) If a rv X has probability distribution μ, then for real constants a, b, the distribution of $aX + b$ is the **shifted and scaled** version of μ.

(b) If X and Y are independent rvs with probability distributions μ_1 and μ_2, then the distribution of $X + Y$ is called the **additive convolution** of μ_1 and μ_2, and is written as $\mu_1 \star \mu_2$. This convolution can be extended to several probability measures in the natural way. ◇

Exercise 19.5.11 (a) Show that

$$\mu_1 \star \mu_2(B) = \int_{\mathbb{R}} \mu_2(B - x)\mu_1(dx) \text{ for all } B \in \mathcal{B}(\mathbb{R}).$$

(b) Show that the clt (in $\mathbb{R}$) can be expressed as follows: Let μ be a probability measure on $\mathcal{B}(\mathbb{R})$ such $\int_{\mathbb{R}} x^2\mu(dx) < \infty$. Then an appropriately shifted and scaled n-fold convolution of μ converges to the standard Gaussian measure. Find the centering and scaling (which depend on n and the first two moments of μ).

(c) Formulate a version of (b) in $\mathbb{R}^d$.

Chapter 20

Signed measure

We now extend the notion of a measure by dropping the condition of non-negativity. This chapter is devoted to establishing a crucial technical result—the Jordan-Hahn theorem.

This will be a key ingredient in establishing the Radon-Nikodym theorem in Chapter 21 and subsequently the fundamental theorem of calculus in Chapter 22.

20.1 Jordan-Hahn decomposition

Definition 20.1.1. (**Signed measure**). Let $(\Omega, \mathcal{A})$ be a measurable space. Then the set function $\nu : \mathcal{A} \to [-\infty, \infty]$ is said to be a *signed measure* if it is countably additive and $\nu(\emptyset) = 0$. Signed measures on a field $\mathcal{F}$ are defined in the obvious way. ◇

Exercise 20.1.1 Show that,

(a) if ν is a signed measure, then it cannot take both values $\pm\infty$;

(b) if ν_1 and ν_2 are two measures, at least one of which is finite, then $\nu = \nu_1 - \nu_2$ is a signed measure;

(c) if $(\Omega, \mathcal{A}, \mu)$ is a measure space and integral of f exists, then the indefinite integral of f is a signed measure.

Theorem 20.1.1. *Let ν be a signed measure on $(\Omega, \mathcal{A})$. Then there exists sets $C, D \in \mathcal{A}$ such that*

$$\nu(C) = \sup\{\nu(A) : A \in \mathcal{A}\}, \quad \textit{and} \quad \nu(D) = \inf\{\nu(A) : A \in \mathcal{A}\}. \quad \blacklozenge$$

A. Bose, *Elements of Measure and Probability*, Texts and Readings in Mathematics 88,
https://doi.org/10.1007/978-981-95-2758-8_20

Example 20.1.1. (a) Suppose ν is a measure. Then $C = \Omega$, and $D = \emptyset$.
(b) Suppose ν is the indefinite integral of f. Then it can be easily checked that we can take the sets C and D in Theorem 20.1.1 to be

$$C = \{\omega : f(\omega) \geq 0\}, \quad D = \{\omega : f(\omega) < 0\}.$$

The set $\{\omega : f(\omega) = 0\}$ could be taken out of C and included in D. Hence the pair (C, D) in Theorem 20.1.1 is not unique. ▲

We know that if μ is a measure, then it is continuous from below, and is continuous from above if it is finite. We shall need the following extension to signed measures.

Exercise 20.1.2 Let ν be a countably additive set function on a σ-field $\mathcal{A}$, and $\{A_n\}_{n\geq 1} \subset \mathcal{A}$. Show that then the following hold:
(a) If $A_n \uparrow A$, then $\nu(A_n) \to \nu(A)$ (convergence may not be monotone).
(b) If $A_n \downarrow A$, and $\nu(A_i) < \infty$ for some i, then $\nu(A_n) \to \nu(A)$.
(c) The results (a) and (b) hold if, ν is defined on a field $\mathcal{F}$, and we assume that $A \in \mathcal{F}$.

Proof of Theorem 20.1.1. (i) If for some $A_0 \in \mathcal{A}$, $\nu(A_0) = \infty$, then take $C = A_0$.
(ii) Suppose now that $\nu(A) < \infty$ for all $A \in \mathcal{A}$. Let

$$S := \sup\{\nu(A) : A \in \mathcal{A}\}\,. \tag{20.1}$$

Get $A_n \in \mathcal{A}$ such that $\nu(A_n) \to S$. Let $A_0 = \cup_{n=1}^{\infty} A_n$. For every n, consider the 2^n disjoint sets $A_1^* \cap A_2^* \cap \cdots \cap A_n^*$ where each A_i^* is either A_i or $A_0 \setminus A_i$. Some of these sets could be empty. Label them as $A_{nm}, m = 1, 2, \ldots, 2^n$. Let

$$B_n := \cup_m \{A_{nm} : \nu(A_{nm}) \geq 0\}\,.$$

Since each A_n is a finite disjoint union of some sets A_{nm}, and negative-valued sets have been dropped in the definition of B_n, using countable additivity of ν, we have $\nu(A_n) \leq \nu(B_n)$.

There is "nesting" in the sense that if $n_1 > n_2$, then each $A_{n_1 m}$ is either a subset of $A_{n_2 m}$ or disjoint from it. This implies that for $r \geq n$,

$$\cup_{k=n}^{r} B_k = B_n \cup (\cup_j E_j) \text{ where, for all } j,\ E_j \cap B_n = \emptyset, \text{ and } \nu(E_j) \geq 0\,.$$

Hence we have

$$\begin{aligned}\nu(A_n) &\le \nu(B_n)\\ &\le \nu(\cup_{k=n}^{r} B_k), \text{ (additivity, and the above observation)}\\ &\to \nu(\cup_{k=n}^{\infty} B_k) \text{ as } r \to \infty, \text{ (continuity from below, Exc. 20.1.2)}\end{aligned}$$

Then $\cup_{k=n}^{\infty} B_k \downarrow \limsup B_n =: C$. Also $0 \le \nu(\cup_{k=n}^{\infty} B_k) < \infty$ for all n.
Thus

$$\begin{aligned}S &= \lim_{n\to\infty} \nu(A_n)\\ &\le \lim_{n\to\infty} \nu(\cup_{k=n}^{\infty} B_k)\\ &= \nu(C) \text{ (continuity from above, Exercise 20.1.2(b))}\\ &\le S \text{ since } S \text{ is the supremum in (20.1).}\end{aligned}$$

Hence $S = \nu(C)$. The set D can be defined by considering $-\nu$. ∎

Theorem 20.1.2 (Jordan-Hahn decomposition). *Let ν be a signed measure on a measurable space $(\Omega, \mathcal{A})$. Let*

$$\begin{aligned}\nu^+(A) &:= \sup\{\nu(B) : B \in \mathcal{A}, B \subset A\},\\ \nu^-(A) &:= -\inf\{\nu(B) : B \in \mathcal{A}, B \subset A\}.\end{aligned}$$

Then ν^+ and ν^- are measures on $\mathcal{A}$ and $\nu = \nu^+ - \nu^-$. ◆

Proof. Since ν is a signed measure, it does not take both values $\pm\infty$. Without loss, assume that ν does not take the value $-\infty$. Let D be a set with the property described in Theorem 20.1.1. Since $\nu(\emptyset) = 0$, we have $-\infty < \nu(D) \le 0$. Take any set $A \in \mathcal{A}$. Then

$$\nu(D) = \nu(A \cap D) + \nu(A^c \cap D) \text{ since } \nu \text{ is additive.}$$

Since both terms on the right side of the above equation are finite,

$$\begin{aligned}\nu(D) &\le \nu(A^c \cap D) \text{ (since } D \text{ yields the infimum)}\\ &= \nu(D) - \nu(A \cap D).\end{aligned}$$

This implies that

$$\nu(A \cap D) \le 0 \text{ for any } A \in \mathcal{A}. \tag{20.2}$$

On the other hand

$$\begin{aligned}\nu(D) &\leq \nu(D \cup (A \cap D^c)) \quad \text{(since } D \text{ yields the infimum)}\\ &= \nu(D) + \nu(A \cap D^c) \quad \text{(by additivity).}\end{aligned}$$

This implies that

$$\nu(A \cap D^c) \geq 0 \quad \text{for any} \quad A \in \mathcal{A}. \tag{20.3}$$

Take any $B \in \mathcal{A}, B \subset A$. Then

$$\begin{aligned}\nu(B) &= \nu(B \cap D) + \nu(B \cap D^c) \quad \text{(by additivity)}\\ &\leq \nu(B \cap D^c) \quad \text{(by (20.2))}\\ &\leq \nu(B \cap D^c) + \nu((A \setminus B) \cap D^c) \quad \text{(by (20.3))}\\ &= \nu(A \cap D^c) \quad \text{(by additivity).}\end{aligned}$$

Hence

$$\begin{aligned}\nu^+(A) &\leq \nu(A \cap D^c) \quad \text{(taking supremum above over all } B \subseteq A)\\ &\leq \nu^+(A) \quad \text{(by definition of } \nu^+).\end{aligned}$$

This shows that

$$\nu^+(A) = \nu(A \cap D^c) \geq 0. \tag{20.4}$$

Similarly,

$$\begin{aligned}\nu(B) &= \nu(B \cap D) + \nu(B \cap D^c) \quad \text{(by additivity)}\\ &\geq \nu(B \cap D) \quad \text{(by (20.3))}\\ &\geq \nu(B \cap D) + \nu((A \setminus B) \cap D) \quad \text{(by (20.2))}\\ &= \nu(A \cap D) \quad \text{(by additivity).}\end{aligned}$$

Hence by taking infimum over all such B,

$$\begin{aligned}\nu^-(A) &\leq -\nu(A \cap D)\\ &\leq \nu^-(A) \quad \text{(by definition of } \nu^-).\end{aligned}$$

This shows that

$$\nu^-(A) = -\nu(A \cap D) \geq 0\,. \tag{20.5}$$

Hence for all $A \in \mathcal{A}$,

$$\nu^+(A) - \nu^-(A) = \nu(A \cap D^c) + \nu(A \cap D) = \nu(A)\,,$$

and by (20.4) and (20.5) ν^+ and ν^- are measures on $\mathcal{A}$. ■

Remark 20.1.1. Let ν be a signed measure on $\mathcal{A}$ with Jordan-Hahn decomposition (ν^+, ν^-). The statements below follow from the arguments used in the proof of Theorem 20.1.2. Details are left as an exercise.

(a) ν is the difference of two measures, $\nu = \nu^+ - \nu^-$ and at least one of these measures is finite. Moreover, $|\nu| = \nu^+ + \nu^-$ is a measure.

(b) If $|\nu(A)| < \infty$ for all $A \in \mathcal{A}$, then $\sup_{A \in \mathcal{A}} |\nu(A)| < \infty$.

(c) There is a set $D \in \mathcal{A}$ such that

$$\nu(A \cap D) \leq 0 \text{ and } \nu(A \cap D^c) \geq 0 \text{ for all } A \in \mathcal{A}\,. \tag{20.6}$$

(d) If $D \in \mathcal{A}$ is any set that satisfies (20.6), then

$$\nu^+(A) = \nu(A \cap D^c) \text{ and } \nu^-(A) = -\nu(A \cap D)\,.$$

(e) If the conditions in (d) for D are satisfied for any set E, then

$$\nu^+(D \Delta E) + \nu^-(D \Delta E) = 0\,.$$ ●

Definition 20.1.2. (**Upper, lower and total variation**). Measures ν^+, ν^-, $|\nu| := \nu^+ + \nu^-$ are known respectively as the *upper*, *lower*, and *total variations* of ν. ◇

20.2 Exercises

Exercise 20.2.1 Let P be a probability measures on $\mathcal{B}(\mathbb{R})$. Define

$$Q(A) := \begin{cases} 1 & \text{if } 0 \in A, \quad A \in \mathcal{B}(\mathbb{R}), \\ 0 & \text{if } 0 \notin A, \quad A \in \mathcal{B}(\mathbb{R}). \end{cases}$$

Find the Jordan-Hahn decomposition of $\nu = P - Q$.

Exercise 20.2.2 (**Minimality of Jordan-Hahn decomposition**) Let ν be a signed measure on $\mathcal{A}$. If $\nu = \nu_1 - \nu_2$ where ν_1 and ν_2 are measures, then show that

$$\nu_1(A) \geq \nu^+(A), \quad \text{and} \quad \nu_2(A) \geq \nu^-(A), \quad \text{for all} \quad A \in \mathcal{A}.$$

Exercise 20.2.3 Suppose ν is a signed measure on $\mathcal{A}$. Show that for any $A \in \mathcal{A}$, its total variation measure $|\nu|(A)$ is given by

$$|\nu|(A) = \sup\Big\{\sum_{i=1}^{n} |\nu(E_i)| : \{E_i\} \text{ are disjoint measurable subsets of } A\Big\}.$$

Exercise 20.2.4 Suppose ν_1 and ν_2 are signed measures. Then show that $|\nu_1 + \nu_2| \ll |\nu_1| + |\nu_2|$.

Exercise 20.2.5 Give an example where g is finite almost surely μ, $\int g d\mu$ exists, and if ν is defined by $\nu(A) := \int_A g d\mu$, $A \in \mathcal{A}$ then none of the measures $|\nu|$ and ν are σ-finite.

Exercise 20.2.6 (**Jordan-Hahn decomposition for indefinite integral**) Suppose $f : (\Omega, \mathcal{A}, \mu) \to (\mathbb{R}, \mathcal{B}(\mathbb{R}))$, and the integral of f exists. Recall the indefinite integrals ν, ν^+ and ν^- from Definition 7.2.3 and discussion following it. Show that the Jordan-Hahn decomposition of ν is given by

$$\nu^+(A) = \int_A f^+ d\mu, \quad \nu^-(A) = \int_A f^- d\mu, \quad \text{and} \quad |\nu|(A) = \int_A |f| d\mu.$$

Chapter 21

Radon-Nikodym theorem

We have seen that if f is a non-negative measurable function, then its indefinite integral ν (with respect to μ) is a measure. We may visualize the relation between μ, f, and ν as, $d\nu = fd\mu$. That is, the "derivative of ν with respect to μ equals f". We now make this and related ideas precise, by introducing the concept of absolute continuity of measures and prove a crucial result namely, the Radon-Nikodym theorem. The concept of conditional probability and expectation in Chapter 23 is based on this theorem.

We will also define the singularity of measures and establish the Lebesgue decomposition theorem, which in particular splits any probability measure on $\mathcal{B}(\mathbb{R})$ into three unique components in terms of their relation with the Lebesgue measure. This result will also be crucially used to identify links between different properties of functions such as bounded variability, absolute continuity, and differentiability (fundamental theorem of calculus) in Chapter 22.

21.1 Absolute continuity and Radon-Nikodym theorem

Definition 21.1.1. (**Absolutely continuity**) Let μ and ν be a measure and a signed measure respectively, on $(\Omega, \mathcal{A})$. We say that ν is *absolutely continuous* with respect to μ if, for every $A \in \mathcal{A}$, $\mu(A) = 0$ implies $\nu(A) = 0$. We write $\nu \ll \mu$. $\diamond$

A. Bose, *Elements of Measure and Probability*, Texts and Readings in Mathematics 88,
https://doi.org/10.1007/978-981-95-2758-8_21

Exercise 21.1.1 Let μ and ν be a measure and a signed measure respectively. Show that $\nu \ll \mu$ if and only $\nu^+ \ll \mu$ and $\nu^- \ll \mu$.

Example 21.1.1. Let $f : (\Omega, \mathcal{A}, \mu) \to (\bar{\mathbb{R}}, \mathcal{B}(\bar{\mathbb{R}}))$ be such that $\int_\Omega f d\mu$ exists. If ν is the indefinite integral of f, then $\nu \ll \mu$. ▲

The Radon-Nikodym theorem, stated below, provides a converse to Example 21.1.1. It is used to define conditional expectation in Chapter 23, leading to the study of martingales (see Bose et al. [2024]).

Exercise 21.1.2 Suppose $f, g : (\Omega, \mathcal{A}, \mu) \to (\bar{\mathbb{R}}, \mathcal{B}(\bar{\mathbb{R}}))$, and consider the condition

$$\int_A g d\mu \leq \int_A f d\mu, \quad \text{for all} \quad A \in \mathcal{A}. \tag{21.1}$$

Then show the following:

(a) If f and g are integrable and (21.1) holds, then $g \leq f$ almost surely.

(b) If integrals of f and g exist, (21.1) holds, and μ is σ-finite, then $g \leq f$ a.s. *Hint*: Reduce to finite measure case and then use the set,

$$A_n = \{\omega : g(\omega) \geq f(\omega) + \frac{1}{n}, \ |f(\omega)| \leq n\}.$$

Theorem 21.1.1 (Radon-Nikodym theorem). *Let μ and ν be a σ-finite and a signed measures respectively, on $(\Omega, \mathcal{A}, \mu)$, and $\nu \ll \mu$. Then there exists $f : (\Omega, \mathcal{A}, \mu) \to (\bar{\mathbb{R}}, \mathcal{B}(\bar{\mathbb{R}}))$ such that,*

$$\nu(A) = \int_A f d\mu \quad \textit{for all } A \in \mathcal{A}. \tag{21.2}$$

If a function g satisfies (21.2), then $f = g$ a.s. $[\mu]$. ◆

We write (21.2) as $d\nu = f d\mu$, or $d\nu / d\mu = f$ and call f the **Radon-Nikodym derivative** of ν with respect to μ.

Proof of Theorem 21.1.1. We will prove the existence in five steps. Then the uniqueness will follow immediately from Exercise 21.1.2.

Step 1: Suppose that μ and ν are finite measures. Define

$$\mathcal{S} := \{f : f \geq 0, \text{ is } \mu \text{ integrable, and } \int_A f d\mu \leq \nu(A) \ \text{ for all } \ A \in \mathcal{A}\}.$$

Clearly $\mathcal{S} \neq \emptyset$. Let $s := \sup\{\int_\Omega f d\mu : f \in \mathcal{S}\}$. Then $s \leq \nu(\Omega) < \infty$. Partially order $\mathcal{S}$ by declaring that $f \geq g$ if and only $f \geq g$ a.s.$[\mu]$.

Let $f, g \in \mathcal{S}$. Then $h := \max(f, g) \in \mathcal{S}$. This follows by taking the set $B := \{\omega : f(\omega) \leq g(\omega)\}$, and observing that for any $A \in \mathcal{A}$,

$$\begin{aligned}\int_A h d\mu &= \int_{A\cap B} g d\mu + \int_{A\cap B^c} f d\mu \\ &\leq \nu(A\cap B) + \nu(A\cap B^c) = \nu(A)\,.\end{aligned}$$

We identify a *maximal element* of $\mathcal{S}$. Let $\{f_n\}$ be a sequence in $\mathcal{S}$ such that $\int f_n d\mu \to s$. Let $g_n := \max(f_1, \ldots, f_n)$. Then $g_n \in \mathcal{S}$ and g_n is non-decreasing. Let $g := \lim g_n$. By MCT Theorem 7.2.6,

$$\int_\Omega g d\mu = \lim \int_\Omega g_n d\mu \geq \lim \int_\Omega f_n d\mu = s\,.$$

Let $A \in \mathcal{A}$. Then $0 \leq g_n I_A \uparrow g I_A$. Hence, by MCT Theorem 7.2.6,

$$\int_\Omega g I_A d\mu = \lim \int_\Omega g_n I_A d\mu\,.$$

But we know that $\int_\Omega g_n I_A d\mu \leq \nu(A)$ for all n. Hence $g \in \mathcal{S}$. Since $\int_\Omega g d\mu = s$, g is a maximal element of $\mathcal{S}$.

Now consider the set function

$$\nu_1(A) = \nu(A) - \int_\Omega g I_A d\mu, \ A \in \mathcal{A}. \tag{21.3}$$

We claim that ν_1 is a finite measure and $\nu_1 \ll \mu$.

If $\nu_1 \equiv 0$ then we are done. Let if possible, ν_1 is not identically 0. Then $\nu_1(\Omega) > 0$. Hence there exists a $k > 0$, such that

$$\mu(\Omega) - k\nu_1(\Omega) < 0. \tag{21.4}$$

Consider the signed measure $\mu - k\nu_1$. By Remark 20.1.1(c), we obtain a $D \in \mathcal{A}$, such that for all $A \in \mathcal{A}$,

$$\mu(A\cap D) - k\nu_1(A\cap D) \leq 0 \ \text{ and } \ \mu(A\cap D^c) - k\nu_1(A\cap D^c) \geq 0\,. \tag{21.5}$$

Let if possible $\mu(D) = 0$. Then by absolute continuity, $\nu(D) = 0$. Hence using (21.3) we have $\nu_1(D) = 0$.

Using (21.5) with $A = \Omega$, we obtain

$$\begin{aligned} 0 &\le \mu(D^c) - k\nu_1(D^c) \\ &= \mu(\Omega) - k\nu_1(\Omega) \quad (\text{since} \quad \mu(D) = \nu_1(D) = 0) \\ &< 0 \quad (\text{by (21.4)}), \end{aligned}$$

which is a contradiction. Hence $\mu(D) > 0$. Now define

$$h(\omega) := \begin{cases} \dfrac{1}{k} & \text{if} \quad \omega \in D, \\ 0 & \text{if} \quad \omega \notin D. \end{cases}$$

If $A \in \mathcal{A}$, then

$$\begin{aligned} \int_A h d\mu = \frac{1}{k}\mu(A \cap D) &\le \nu_1(A \cap D) \quad (\text{by (21.5)}) \\ &\le \nu_1(A) = \nu(A) - \int_A g d\mu. \end{aligned}$$

This implies that,

$$\int_A (h + g) d\mu \le \nu(A).$$

But then $h + g > g$ on D with $\mu(D) > 0$, contradicting the maximality of g. Thus, $\nu_1 \equiv 0$, and the theorem is proved in this special case.

Step 2: Now let μ and ν be finite and σ-finite measures respectively. Let $\{\Omega_n\}$ be disjoint sets in $\mathcal{A}$ with $\cup_{n=1}^{\infty}\Omega_n = \Omega$, and $\nu(\Omega_n) < \infty$ for all n. Define

$$\nu_n(A) := \nu(A \cap \Omega_n), \quad A \in \mathcal{A}.$$

Then it trivially follows that for every n, ν_n is a finite measure, and $\nu_n \ll \mu$. Hence by Step 1, for every n, there exists a non-negative measurable function g_n such that ν_n is the indefinite integral of g_n with respect to μ. Take $g = \sum_{n=1}^{\infty} g_n$. Then ν is the indefinite integral of g with respect to μ. Note that g may be extended real-valued.

Step 3: Now let μ and ν be finite and arbitrary measures respectively. For any $C \in \mathcal{A}$, let $\mathcal{A}_C := \{A \cap C : A \in \mathcal{A}\}$. Then $\mathcal{A}_C$ is a σ-field of subsets of C. Define

$$\mathcal{C} := \{C \in \mathcal{A} : \nu \text{ restricted to } \mathcal{A}_C \text{ is a } \sigma\text{-finite measure}\}.$$

If for every non-empty set $A \in \mathcal{A}$, $\nu(A) = \infty$, then we can take $f \equiv \infty$. If there is a set with finite positive ν-measure, then $\mathcal{C}$ is not empty. Let

$$s := \sup\{\mu(A) : A \in \mathcal{C}\}.$$

Pick $C_n \in \mathcal{C}$ such that $\mu(C_n) \to s$. Let $C := \cup_{n=1}^{\infty} C_n$. Then $C \in \mathcal{C}$. Since $s \geq \mu(C) \geq \mu(C_n) \to s$, we have $\mu(C) = s$.

Now consider μ and ν restricted to $\mathcal{A}_C$ for the above choice of C. Since μ and ν are finite and σ-finite respectively, on $\mathcal{A}_C$, by Step 2, there exists a non-negative function $f_C : C \to \bar{\mathbb{R}}$, which is measurable with respect to $\mathcal{A}_C$, and ν is the indefinite integral of μ.

In other words,

$$\nu(A \cap C) = \int_{A \cap C} f_C d\mu, \quad \text{for all} \quad A \in \mathcal{A}.$$

Consider any $A \in \mathcal{A}$. Then we have either Case 1 or Case 2 given below:

Case 1. $\mu(A \cap C^c) > 0$. Then, suppose if possible, $\nu(A \cap C^c) < \infty$. But this would imply that $C \cup (A \cap C^c) \in \mathcal{C}$, and

$$s \geq \mu(C \cup (A \cap C^c)) = \mu(C) + \mu(A \cap C^c) > \mu(C) = s\,.$$

This is a contradiction. Hence, we must have $\nu(A \cap C^c) = \infty$.

Case 2. $\mu(A \cap C^c) = 0$. Then, by absolute continuity, $\nu(A \cap C^c) = 0$.

Observe that in either Case 1 or Case 2,

$$\nu(A \cap C^c) = \int_{A \cap C^c} \infty \, d\mu\,.$$

It follows that,

$$\nu(A) = \nu(A \cap C) + \nu(A \cap C^c) = \int_A f d\mu, \text{ for all } \; A \in \mathcal{A}\,,$$

where,

$$f(\omega) = \begin{cases} f_C(\omega) & \text{if } \; \omega \in C, \\ \infty & \text{if } \; \omega \in C^c. \end{cases}$$

Clearly f is a Borel measurable function.

Step 4: Let μ and ν be σ-finite and arbitrary measures respectively. Let $\{\Omega_n\}$ be a measurable partition of $\mathcal{A}$, so that $\mu(\Omega_n) < \infty$ for all n.

By Step 3, for every n, there exists $g_n : \Omega_n \to \bar{\mathbb{R}}$ which is Borel measurable with respect to $\mathcal{A}_{\Omega_n}$, and $\nu(A \cap \Omega_n) = \int g_n d\mu$ for every $A \in \mathcal{A}$.

Extend g_n to all of Ω by defining it as 0 on Ω_n^c. Call this new function f_n. Note that f_n is measurable, and

$$\nu(A \cap \Omega_n) = \int_A f_n d\mu, \; A \in \mathcal{A}.$$

Then for all $A \in \mathcal{A}$,

$$\begin{aligned}\nu(A) &= \sum_{n=1}^{\infty} \nu(A \cap \Omega_n)\\ &= \sum_{n=1}^{\infty} \int_A f_n d\mu = \int_A f d\mu \quad \text{where, } f := \sum_{n=1}^{\infty} f_n.\end{aligned}$$

Step 5: Now assume that μ and ν are σ-finite and signed measures respectively. Write $\nu = \nu^+ - \nu^-$. Without loss, assume that ν^- is a finite measure. Note that, $\nu^+ \ll \mu$ and $\nu^- \ll \mu$. By Step 4, there exist non-negative Borel measurable functions f_1 and f_2, such that, ν^+ and ν^- are the indefinite integrals of f_1 and f_2 with respect to μ. Since ν^- is finite, f_2 is μ-integrable. Hence

$$\begin{aligned}\nu(A) &= \nu^+(A) - \nu^-(A)\\ &= \int_A f_1 d\mu - \int f_2 d\mu\\ &= \int_A (f_1 - f_2) d\mu \quad \text{(by the additivity of integrals).}\end{aligned}$$

■

Remark 21.1.1. The following facts follow from the above theorem and its proof. Details are left as exercises. Suppose $\nu \ll \mu$ where ν is a signed measure and μ is σ-finite.

(a) If ν is a finite measure, then $d\nu/d\mu$ is μ-integrable, and hence is finite almost surely μ.

(b) If $|\nu|$ is σ-finite, then $d\nu/d\mu$ is finite almost surely μ.

(c) If ν is a measure, then $d\nu/d\mu \geq 0$ almost surely μ. ●

21.2 Singularity and Lebesgue decomposition

Definition 21.2.1. (**Singularity**) Measures μ_1 and μ_2 are said to be **mutually singular** if, there is $A \in \mathcal{A}$ such that $\mu_1(A) = \mu_2(A^c) = 0$. We write $\mu_1 \perp \mu_2$. Signed measures ν_1 and ν_2 are said to be mutually singular if $|\nu_1| \perp |\nu_2|$. ◇

Exercise 21.2.1 Let ν be a signed measure with Jordan-Hahn decomposition $\nu = \nu^+ - \nu^-$. Then show that $\nu^+ \perp \nu^-$.

Lemma 21.2.1. Let μ be a measure, and λ_1 and λ_2 be signed measures on $\mathcal{A}$. Then the following hold:

(a) If $\lambda_1 \perp \mu$, $\lambda_2 \perp \mu$, and $\lambda_1 + \lambda_2$ is well defined, then $\lambda_1 + \lambda_2 \perp \mu$;

(b) $\lambda_1 \ll \mu$ if and only if $|\lambda_1| \ll \mu$;

(c) If $\lambda_1 \ll \mu$ and $\lambda_2 \perp \mu$, then $\lambda_1 \perp \lambda_2$;

(d) If $\lambda_1 \ll \mu$ and $\lambda_1 \perp \mu$, then $\lambda_1 \equiv 0$;

(e) If λ_1 is finite, then $\lambda_1 \ll \mu$ if and only if $\lim_{\mu(A)\to 0} \lambda_1(A) = 0$. ♦

Proof. (a) Suppose A and B are such that

$$\mu(A) = |\lambda_1|(A^c) = 0, \quad \text{and} \quad \mu(B) = |\lambda_2|(B^c) = 0.$$

Then $\mu(A \cup B) = 0$. For every $C \subset (A \cup B)^c$, $C \in \mathcal{A}$,

$$\lambda_1(C) = \lambda_2(C) = 0, \text{ and hence } \ |\lambda_1 + \lambda_2|((A \cup B)^c) = 0.$$

(b) Suppose $\lambda_1 \ll \mu$ and $\mu(A) = 0$. If $\lambda_1^+(A) > 0$, then there exists a $B \subset A$ such that $\lambda_1(B) > 0$. Since $\mu(B) = 0$, this is a contradiction to $\lambda_1 \ll \mu$. Hence $\lambda_1^+(A) = 0$. That is $\lambda_1^+ \ll \mu$. Similarly, $\lambda_1^- \ll \mu$, and then it easily follows that $|\lambda_1| \ll \mu$. The converse is easy.

(c) Suppose A is such that $\mu(A) = 0$ and $|\lambda_2|(A^c) = 0$. But since $\lambda_1 \ll \mu$, by Part (b) we know that $|\lambda_1| \ll \mu$. Hence $|\lambda_1|(A) = 0$. That is $\lambda_1 \perp \lambda_2$.

(d) By Part (c) $\lambda_1 \perp \lambda_1$. So, pick A such that $|\lambda_1|(A) = |\lambda_1|(A^c) = 0$. Hence $|\lambda_1|(\Omega) = 0$.

(e) Suppose $\lambda_1 \ll \mu$. Suppose if possible the condition does not hold. Then there exists $\varepsilon > 0$ and sets such that

$$\mu(A_n) < 2^{-n}, \quad |\lambda_1|(A_n) \geq \varepsilon\,.$$

Let $A = \limsup A_n$. By Borel-Cantelli Lemma 3.3.1, $\mu(A) = 0$. But $|\lambda_1|(\cup_{k=n}^{\infty} A_k) \geq |\lambda_1|(A_n) \geq \varepsilon$. So $|\lambda_1|(A)| = \lim_{n\to\infty} |\lambda_1|(\cup_{k=n}^{\infty} A_k) \geq \varepsilon$. This is a contradiction, and the proof is complete. ■

Exercise 21.2.2 Suppose f is integrable with respect to a measure μ.

(a) Show that $\lim_{n\to\infty} \int_{\{\omega:|f(\omega)|\geq n\}} f d\mu = 0$.

(b) By using (a), show that $\lim_{\mu(A)\to 0} \int_A f d\mu = 0$.

Definition 21.2.2. A signed measure ν is called σ-finite if its total variation measure $|\nu|$ is σ-finite. ◇

Theorem 21.2.1 (Lebesgue decomposition). *Let μ and ν on $(\Omega, \mathcal{A})$ be a measure and a σ-finite signed measure respectively. Then ν has a unique decomposition $\nu = \nu_1 + \nu_2$ so that ν_i, $i = 1, 2$ are signed measures, and $\nu_1 \ll \mu$, $\nu_2 \perp \mu$.* ◆

Proof. (a) First suppose ν is a finite measure. Let

$$\mathcal{C} := \{A \in \mathcal{A} : \mu(A) = 0\}, \text{ and } \; s := \sup\{\nu(A) : A \in \mathcal{C}\} \leq \nu(\Omega) < \infty.$$

Suppose $\{A_i\}$ from $\mathcal{C}$ are such that $\nu(A_n) \to s$. Then $C = \cup_{n=1}^{\infty} A_n \in \mathcal{C}$ and $\nu(C) = s$. Suppose $B \in \mathcal{C}$. Note that $C \cup B \in \mathcal{C}$, and hence

$$s \geq \nu(C \cup B) = \nu(C) + \nu(B \setminus C) \geq s\,.$$

It follows that $\nu(B \setminus C) = 0$. Define

$$\nu_1(A) := \nu(A \setminus C), \;\; \nu_2(A) := \nu(A \cap C), \;\; A \in \mathcal{A}.$$

Then $\nu_1 \ll \mu$. To see this, suppose $\mu(B) = 0$. Then $B \in \mathcal{C}$. Hence

$$\nu_1(B) = \nu(B \setminus C) = 0.$$

Note that $\mu(C) = 0$, and $\nu_2(C^c) = \nu(C^c \cap C) = \nu(\emptyset) = 0$. So $\nu_2 \perp \mu$. Finally, $\nu = \nu_1 + \nu_2$. Uniqueness follows by using Lemma 21.2.1(d).

(b) Now suppose ν is a σ-finite measure. Suppose $\{A_i\}$ is a disjoint partition of Ω such that $\nu(A_n) < \infty$ for every n. Define

$$\nu_n(A) := \nu(A \cap A_n), \;\; A \in \mathcal{A}\,.$$

Then by (a), get $\{\nu_{1n}\}$, $\{\nu_{2n}\}$, such that for every n, $\nu_n = \nu_{1n} + \nu_{2n}$, and $\nu_{1n} \ll \mu$, $\nu_{2n} \perp \mu$. Adding over n (these infinite sums make sense), we get $\nu = \sum_{n=1}^{\infty} \nu_{1n} + \sum_{n=1}^{\infty} \nu_{2n} = \nu_1 + \nu_2$, where $\nu_1 \ll \mu$, and $\nu_2 \perp \mu$. Uniqueness follows by using the uniqueness proved in Part (a).

(c) Finally let ν be a σ-finite signed measure. Then use Jordan-Hahn decomposition, and apply Part (b). Uniqueness also follows. ■

Let F be a bounded distribution function with the finite measure μ. Then the Lebesgue decomposition Theorem 21.2.1 splits μ as $\mu = \mu_1 + \mu_2$ where $\mu_1 \ll \lambda$ and $\mu_2 \perp \lambda$. The following exercise splits μ_2 further.

Exercise 21.2.3 Let F be a bounded distribution function on $\mathbb{R}$, and λ be the Lebesgue measure. Show that we can write F, uniquely up to additive constants, as $F = F_1 + F_{21} + F_{22}$ where F_1, F_{21} and F_{22} are bounded distribution functions (with measures μ_1, μ_{21} and μ_{22}), such that:

(a) $\mu_1 \ll \lambda$. So F_1 is absolutely continuous;

(b) μ_{21} is discrete. So F_{21} increases only by jumps and $\mu_{21} \perp \lambda$;

(c) F_{22} is continuous (but not absolutely continuous) and $\mu_{22} \perp \lambda$.

Hint: Define F_{21} using the jumps of $F - F_1$ and $F_{22} = F - F_1 - F_{21}$.

The measure μ_{22} or the corresponding distribution F_{22} is often called **continuous singular**. Exercise 21.3.7 provides a specific example.

21.3 Exercises

Exercise 21.3.1 Give an example to show that the condition μ is σ-finite cannot be dropped from the Radon-Nikodym Theorem 21.1.1.

Exercise 21.3.2 Let ν be a finite signed measure, and μ be a measure. Show that $\nu \ll \mu$ if and only if given any $\varepsilon > 0$ there exists a $\delta > 0$ such that for any $A \in \mathcal{A}$, $\mu(A) < \delta$ implies $|\nu(A)| < \varepsilon$.

Exercise 21.3.3 Let $g : (\Omega, \mathcal{A}, \mu) \to (\bar{\mathbb{R}}^+, \mathcal{B}(\bar{\mathbb{R}}^+))$, and ν is the indefinite integral of g with respect to μ. When is μ an indefinite integral of some function f with respect to ν?

Exercise 21.3.4 Suppose ν_1 and ν_2 are signed measures, and $\nu_1 + \nu_2$ is well-defined. Let μ be a σ-finite measure such that $\nu_i \ll \mu$, $i = 1, 2$. Show that $(\nu_1 + \nu_2) \ll \mu$, and $d(\nu_1 + \nu_2)/d\mu = (d\nu_1/d\mu) + (d\nu_2/d\mu)$.

Exercise 21.3.5 Let μ_1, μ_2, μ_3, be σ-finite measures where $\mu_1 \ll \mu_2$ and $\mu_2 \ll \mu_3$. Show that $\mu_1 \ll \mu_3$, and $d\mu_1/d\mu_3 = (d\mu_1/d\mu_2)(d\mu_2/d\mu_3)$.

Exercise 21.3.6 Let μ and ν be mutually absolutely continuous σ-finite measures. Show that $d\mu/d\nu = (d\nu/d\mu)^{-1}$ a.s. $[\mu]$ (and a.s. $[\nu]$).

Exercise 21.3.7 (Cantor distribution function) Consider the Cantor distribution function F from Exercise 5.3.17.

(a) Show that F is not absolutely continuous.

(b) If μ is the Lebesgue-Stieltjes measure corresponding to F, then show that $\mu \perp \lambda$, and $\mu\{x\} = 0$ for every $x \in \mathbb{R}$.

Exercise 21.3.8 Let μ be a translation invariant measure on $\mathcal{B}(\mathbb{R}^d)$. Show that there exists a constant $c \in [0, \infty)$ such that $\mu(B) = c\lambda_d(B)$ for all Borel subsets B, where λ_d is the d-dimensional Lebesgue measure. [For a solution see Elekes and Keleti [2006].]

Chapter 22

Fundamental theorem of calculus

The fundamental theorem of calculus in its simplest form says that if F is everywhere differentiable with an everywhere continuous derivative f, then F is the Riemann integral of f in the sense that,

$$F(y) - F(x) = \int_x^y f(t)dt \text{ for all } x, y \in \mathbb{R},\ x < y\,. \tag{22.1}$$

Now that we have the concepts of continuity a.e. λ and of the Lebesgue integral, we may ask to what extent (22.1) can be generalised. The primary goal of this chapter is to state and prove the fundamental theorem in a general form, using the Radon-Nikodym derivative. Along the way, we shall also develop and use the concepts of functions of bounded variation, absolutely continuity of functions, and differentiability of measures.

22.1 A covering lemma

The following interesting lemma shall be very useful to us. Let us denote the open ball of radius $r > 0$, centered at $x \in \mathbb{R}^d$ by

$$B(x, r) := \{y \in \mathbb{R}^d : ||x - y|| < r\}\,.$$

Lemma 22.1.1. Let $\mathcal{C}$ be a collection of open balls in $\mathbb{R}^d$ and U be the union of all these balls. For any $c < \lambda_d(U)$, there exists finitely many disjoint $B_1, \ldots, B_k$ from $\mathcal{C}$ such that $c < 3^d \sum_{i=1}^k \lambda_d(B_k)$. ♦

A. Bose, *Elements of Measure and Probability*, Texts and Readings in Mathematics 88,
https://doi.org/10.1007/978-981-95-2758-8_22

Proof. Since U is open, by Theorem 5.2.4 there exists a compact set $K \subset U$ such that $c < \lambda_d(K)$. Since K is compact, its open cover $\mathcal{C}$ has a finite sub-cover, say $\mathcal{C}_0 = \{A_1, \ldots, A_m\}$ of K. Let $B_1 \in \mathcal{C}_0$ be the set with the largest radius. Now let $B_2 \in \mathcal{C}_0$ be the next set with the largest radius amongst the remaining sets of $\mathcal{C}_0$ that are disjoint from B_1. Continuing this process till completion, we get sets $B_1, \ldots B_k$ from $\mathcal{C}_0$ which are disjoint and have non-increasing radius. If $A_i \in \mathcal{C} \backslash \mathcal{C}_0$, then there exists a B_j such that $A_i \cap B_j \neq \emptyset$, and let j be the minimum such j. For any s, let $B_{s,3}$ be the ball with the same centre as B_s with radius three times that of B_s. Then $A_i \subseteq B_{j,3}$. It follows that $K \subseteq \cup_{j=1}^{k} B_{j,3}$ and hence $c < \lambda_d(K) < \sum_{i=1}^{k} \lambda_d(B_{k,3}) = 3^d \sum_{i=1}^{k} \lambda_d(B_k)$. ■

22.2 Maximal function

Definition 22.2.1. (Locally integrable) Let $f : \mathbb{R}^d \to \mathbb{C}$ be a measurable function. It is called *locally integrable* if for every compact set K, $\int_K |f| d\lambda_d < \infty$. We denote the class of all such functions as L^1_{loc}. ◇

For any $f \in L^1_{loc}$, define for $r > 0$ and $x \in \mathbb{R}^d$,

$$A_r f(x) := \frac{1}{\lambda_d(B(x,r))} \int_{B(x,r)} f(y) \lambda_d(dy) = \frac{1}{r^d} \int_{B(x,r)} f(y) \lambda_d(dy).$$

Exercise 22.2.1 Show that $A_r f(x)$ is continuous for $r > 0,\ x \in \mathbb{R}^d$.

Definition 22.2.2. For $f \in L^1_{loc}$, the **Hardy-Littlewood maximal function** is defined as $Hf(x) := \sup_{r>0} A_r |f|(x)$. ◇

Exercise 22.2.2 Show that Hf is a Borel measurable function.

Theorem 22.2.1. *(**Maximal Theorem**) For any $f \in L^1(\lambda_d)$,*

$$\lambda_d\{x : Hf(x) > \alpha\} \leq \frac{3^d}{\alpha} \int_{\mathbb{R}^d} |f| d\lambda_d.$$ ◆

Proof. Let $E_\alpha := \{x : HF(x) > \alpha\}$. Then E_α is open for all α. For each $x \in E_\alpha$ choose r_x such that $A_{r_x}|f|(x) > \alpha$. Then $\{B(x, r_x)\}$, $x \in E_\alpha$ cover E_α. Fix $c < \lambda_d(E_\alpha)$. Using Lemma 22.1.1 obtain $\{B_1, \ldots, B_k\}$, and then use Markov's inequality Lemma 8.1.1 to conclude that

$$c < 3^d \sum_{i=1}^{k} \lambda_d(B_k) \le \frac{3^d}{\alpha} \sum_{i=1}^{k} \int_{B_i} |f(y)| \lambda_d(dy) \le \frac{3^d}{\alpha} \int_{\mathbb{R}^d} |f(y)| \lambda_d(dy) .$$

Now let $c \to \lambda_d(E_\alpha)$ to complete the proof. ■

22.3 Lebesgue differentiation

We now investigate for what functions $f : \mathbb{R}^d \to \mathbb{C}$ and for what values of $x \in \mathbb{R}^d$ we can claim that

$$\lim_{r \to 0} A_r f(x) = f(x).$$

Note that in particular this is also a question about $\lim_{r\to 0} \dfrac{\nu(B(x,r))}{\lambda_d(B(x,r))}$ when $\nu \ll \lambda_d$ with the Random-Nikodym derivative f.

Let $f : \mathbb{R} \to \mathbb{R}$ be measurable. Define the function $\limsup f(\cdot)$ as

$$\limsup f(x) = \lim_{\epsilon \to 0} \sup_{y:|x-y|<\epsilon} f(y) .$$

Exercise 22.3.1 (a) Show that measurability of $f(\cdot)$ implies that of $\limsup f(\cdot)$.

(b) $\lim_{y\to x} f(y) = c$ if and only if $\limsup |f(x) - c| = 0$.

We need the following definition.

Definition 22.3.1. We say $\{E_r\} \subset \mathcal{B}(\mathbb{R}^d)$ **shrink nicely** to $x \in \mathbb{R}^d$ if

(i) $E_r \subseteq B(x,r)$ for each $r > 0$, and

(ii) there is a constant $\alpha > 0$, such that $\lambda_d(E_r) > \alpha \lambda_d(B(x,r))$. ◇

Theorem 22.3.1. *(**Lebesgue differentiation**) Let $f \in L^1_{loc}(\lambda_d)$. Then*

(a) $\lim A_r f(x) = f(x)$ a.e. λ_d. In other words,

$$\lim_{r\to 0} \frac{1}{\lambda_d(B(x,r))} \int_{B(x,r)} (f(y) - f(x)) \lambda_d(dy) = 0 \quad a.e.\ \lambda_d . \tag{22.2}$$

(b) In addition,

$$\lim_{r\to 0} \frac{1}{\lambda_d(B(x,r))} \int_{B(x,r)} |f(y) - f(x)| \lambda_d(dy) = 0 \quad a.e.\ \lambda_d . \tag{22.3}$$

(c) Eqns. (22.2) and (22.3) continue to hold when $B(x, r)$ is replaced by $\{E_r\}$ which shrink nicely to x. ◆

Proof. (a) It is enough to prove the result for $|x| \leq N$ for an arbitrary integer N. Since $r \to 0$, only the values of f in $B(0, N+1)$ are involved and hence we may assume $f \in L^1(\lambda_d)$.

If we further assume that f is continuous, then the proof is left to the reader as an easy exercise.

For general f, we approximate it by a continuous function. Fix $\epsilon > 0$. By Theorem 9.1.5, get a continuous function $g \in L^1(\lambda_d)$ such that $\int |f - g|\lambda_d < \epsilon$. Then

$$\begin{aligned}\limsup |A_r f(x) - f(x)| &\leq \limsup |A_r(f-g)(x)| \\ &\quad + \limsup |A_r g(x) - g(x)| + |f(x) - g(x)| \\ &\leq H(f-g)(x) + 0 + |f-g|(x)\,.\end{aligned}$$

Define the sets

$$C_\alpha := \{x : \limsup |A_r f(x) - f(x)| > \alpha\},\ D_\alpha := \{x : |f(x) - g(x)| > \alpha\}\,.$$

Then by definition of H, it immediately follows that

$$C_\alpha \subseteq D_{\alpha/2} \cup \{x : H(f-g)(x) > \alpha/2\}\,.$$

On the other hand using Markov's inequality Lemma 8.1.1,

$$(\alpha/2)\lambda_d(D_{\alpha/2}) \leq \int_{D_{\alpha/2}} |f(y) - g(y)|\lambda_d(dy) < \epsilon\,.$$

Hence by Maximal Theorem 22.2.1,

$$\lambda_d(C_\alpha) \leq \frac{2\epsilon}{\alpha} + \frac{2C\epsilon}{\alpha}\,.$$

Since ϵ is arbitrary, $\lambda_d(C_\alpha) = 0$ for all $\alpha > 0$. But $\lim A_r f(x) = f(x)$ on $\cup_{n=1}^{\infty} C_{1/n}$ and this proves (a).

(b) Fix $c \in \mathbb{C}$. Apply Part (a) to get a set E_c with $\lambda_d(E_c) = 0$ such that (22.2) holds for the function $g_c(x) = |f(x) - c|$ on E_c^c. Let D be a countable dense set in $\mathbb{C}$ and let $E = \cup_{c \in D} E_c$.

It now suffices to show that (22.3) holds for all $x \notin E$.

Fix $\epsilon > 0$. Choose $c \in D$ such that $|f(x) - c| < \epsilon$. Then

$$\begin{aligned}
&\limsup_{r\to 0} \frac{1}{\lambda_d(B(x,r))} \int_{B(x,r)} |f(y) - f(x)|\lambda_d(dy) \\
&\leq \limsup_{r\to 0} \frac{1}{\lambda_d(B(x,r))} \int_{B(x,r)} |f(y) - c)|\lambda_d(dy) + |f(x) - c| \\
&\leq 2|f(x) - c| \leq 2\epsilon \text{ using (22.2)}.
\end{aligned}$$

This completes the proof of (b). Proof of (c) is left as an exercise. ■

Exercise 22.3.2 Suppose $\mu \ll \lambda_d$ is a signed measure with a Radon-Nikodym derivative f. Show that

$$\lim_{r\to 0} \frac{\mu(B(x,r))}{\lambda_d(B(x,r))} = f(x) \text{ a.e. } \lambda_d. \tag{22.4}$$

What can we say about the existence of the limit in (22.4) for *any* signed measure μ on $\mathcal{B}(\mathbb{R}^d)$? We quickly realise that, this limit cannot exist always, since, for a given x, $\mu(B(x,r))$ may equal ∞ for all $r > 0$. Thus, we need to restrict the class of measures μ.

This class is precisely the class of Lebesgue-Stieltjes measures that were discussed in Chapter 5. These are also known as **regular** measures. A measure ν is regular iff $\nu(K) < \infty$ whenever K is compact. A signed measure μ will be called regular (or Lebesgue-Stieltjes) if $|\mu|$ is so.

Theorem 22.3.2. *Let μ be a Lebesgue-Stietjes signed measure on $\mathbb{R}^d$. Then $\mu = \mu_1 + \mu_2$, $\mu_1 \perp \lambda_d$, $\mu_2 \ll \lambda_d$¿ Let $d\mu_2/d\lambda_d = f$ a.e. λ_d. Then for every class of sets $\{E_r\}$ that shrink nicely to $x \in \mathbb{R}^d$,*

$$\lim_{r\to 0} \frac{\mu(E_r)}{\lambda_d(E_r)} = f(x) \quad a.e.\ \lambda_d\,.$$

◆

Proof. Note that $|\mu| = |\mu_1| + |\mu_2|$. Since μ is regular, μ_1 and μ_2 are regular. Further $d|\mu_2|/d\lambda_d = |f|$ is integrable. Hence by Exercise 22.3.2,

$$\lim_{r\to 0} \frac{\mu_2(E_r)}{\lambda_d(E_r)} = f(x) \text{ a.e. } \lambda_d\,.$$

Thus it suffices to show that for any regular positive measure $\nu \perp \lambda_d$,

$$\lim_{r\to 0} \frac{\nu(E_r)}{\lambda_d(E_r)} = 0 \text{ a.e. } \lambda_d.$$

To show this, without loss of generality, we can assume $E_r = B(x,r)$. Let A be a Borel set such that $\lambda(A^c) = 0$ and $\nu(A) = 0$. Define

$$D_k = \left\{x \in A : \limsup \frac{\nu(B(x,r))}{\lambda_d(B(x,r))} > \frac{1}{k}\right\}.$$

It is enough to show that $\lambda_d(D_k) = 0$ for all k. Fix $\epsilon > 0$. Since ν is regular, by Theorem 5.2.4(c), there is an open set $U_\epsilon \supseteq A$ such that $\nu(U_\epsilon) < \epsilon$. For any $x \in D_k$, let $r_x > 0$ be such that $B(x,r_x) \subseteq U_\epsilon$ and $\nu(B(x,r_x) > k^{-1}\lambda_d(B(x,r_x))$. Let $V_\epsilon = \cup_{x\in D_k} B(x,r_x)$ and $c < \lambda_d(V_\epsilon)$. By Lemma 22.1.1, get disjoint $B(x_1,r_{x_1}),\ldots,B(x_m,r_{x_m})$ such that

$$\begin{aligned} c &< 3^d \sum_{i=1}^{m} \lambda_d(B(x_i,r_{x_i})) \\ &\leq 3^d k \sum_{i=1}^{m} \nu(B(x_i,r_{x_i})) \\ &\leq 3^d k\nu(V_\epsilon) \leq 3^d k\epsilon\,. \end{aligned}$$

Now let $c \to \lambda_d(V_\epsilon)$ to conclude that $\lambda_d(V_\epsilon) \leq 3^d k\epsilon$. Since $D_k \subseteq V_\epsilon$ and ϵ is arbitrary, we have $\lambda_d(D_k) = 0$, and this completes the proof. ■

Example 22.3.1. (a) Let $G : \mathbb{R} \to \mathbb{R}$ be a distribution function. Let μ_G be the corresponding Lebesgue-Stieltjes measure and let $\mu_G = \mu_1 + \mu_2$, $\mu_1 \perp \lambda_1$, $\mu_2 \ll \lambda_1$ be its Lebesgue decomposition and let $f = d\mu_2/d\lambda_1$. Now, both $\{(x-h,\ x]\}$ and $\{(x,\ x+h]\}$ as $h \to 0$ shrink nicely to x. Hence by Theorem 22.3.2 with $d = 1$, $\mu_G(x-h,\ x]/h$ and $\mu_G(x,\ x+h]/h$ converge to $f(x)$ a.e λ. This is the same as saying $G'(x) = f(x)$ a.e λ.

(b) Let μ be a regular measure on a countable set $E = \{x_1, x_2, \ldots\}$. Then $\mu(x-h,\ x]/h$ and $\mu(x,\ x+h]/h$ both go to 0 as $h \to 0$. ▲

22.4 Functions of bounded variation

Now we shall dig further in the case where $d = 1$ and obtain results on the differentiation and integration of functions on $\mathbb{R}$. We begin with a result on the a.e differentiability of non-decreasing functions on $\mathbb{R}$.

Theorem 22.4.1. *Let $F : \mathbb{R} \to \mathbb{R}$ be a non-decreasing function, and let $G(x) = F(x+)$. Then*

(a) the set of discontinuity points of F is countable;

(b) F and G are differentiable a.e. λ, and $F' = G'$ a.e. λ. ◆

Proof. (a) This is left as an exercise. Compare with Exercise 5.3.1.

(b) Note that G is also non-decreasing and in addition it is right continuous. By Part (a), $G(x) = F(x)$ a.e. λ. By Example 22.3.1(a), G is differentiable a.e .λ.

Let $H = G - F$. To complete the proof, it is enough to show that H is differentiable a.e. and $H' = 0$ a.e. λ.

Note that $H(\cdot)$ is a non-negative function. Let $E = \{x_1, x_2, \ldots\}$ be the set of all possible values at which $H > 0$. Define the measure $\mu = \sum_{j=1}^{\infty} H(x_j)\delta_j$ where δ_j is the point mass at x_j. For any integer N, it is easy to check $\sum_{\{j:|x_j|<N\}} H(x_j) < \infty$. That is, μ is a regular measure. Now

$$\left|\frac{H(x+h) - H(x)}{h}\right| \leq \frac{H(x+h) + H(x)}{|h|} \leq 4\frac{\mu(x - 2|h|,\ x + 2|h|)}{4|h|},$$

which tends to 0 a.e. λ_1 by Example 22.3.1(b). Thus $H' = 0$ a.e. λ_1, and the proof is complete. ■

Definition 22.4.1. (**Total variation**) Suppose $F : \mathbb{R} \to \mathbb{R}$. Then

$$V_F(x) := \sup\{\sum_{i=1}^{n} |F(x_i) - F(x_{i-1})|,\ n \geq 1,\ -\infty < x_0 < \cdots < x_n = x\},$$

is called the *total variation function* of F. ◇

Exercise 22.4.1 Show that for any $a < b$,

$$V_F(b) - V_F(a) = \sup\{\sum_{i=1}^{n} |F(x_i) - F(x_{i-1})|,\ n \geq 1,\ a = x_0 < \cdots < x_n = b\},$$

so that $V_F(\cdot)$ is a non-decreasing function.

Define,

$$V_F(-\infty) := \lim_{x \downarrow -\infty} V_F(x), \text{ and } V_F(\infty) := \lim_{x \uparrow \infty} V_F(x).$$

Definition 22.4.2. (**Bounded variation**) For $F : \mathbb{R} \to \mathbb{R}$, $V_F(\infty)$ is called the **total variation** of F on $\mathbb{R}$, and $V_F(b) - V_F(a)$ is called the total variation of F on $[a, b]$. F is said to be of *bounded variation* on $\mathbb{R}$ if $V_F(\infty) < \infty$, and likewise for F restricted to $[a, b]$. The class of bounded variation functions on $\mathbb{R}$ and $[a, b]$ will be denoted by

$$BV := \{F : F \text{ is of bounded variation on } \mathbb{R}\},$$
$$BV[a, b] := \{F : F \text{ is of bounded variation on } [a, b] \qquad \diamond$$

If F is defined on $[a, b]$ then F can be considered as a function on $\mathbb{R}$, by defining $F(x) = F(a)$ for all $x < a$, and $F(x) = F(b)$ for all $x > b$. Further, Definition 22.4.2 remains valid when F is complex-valued.

Example 22.4.1. (a) Suppose F is a monotone function on $[a, b]$. Then $F \in BV[a, b]$, and for any $a < x < b$, $V_F(b) - V_F(x) = |F(b) - F(x)|$.

(b) If $F : \mathbb{R} \to \mathbb{R}$ is bounded and non-decreasing, then $F \in BV$ and $V_F(x) = F(x) - F(-\infty)$.

(c) If $F(x) = \cos x$, $x \in \mathbb{R}$, then $F \notin BV$ but $F \in BV[a, b]$ for every $a, b \in \mathbb{R}$, $a < b$. ▲

Exercise 22.4.2 Let $F : \mathbb{R} \to \mathbb{R}$ be everywhere differentiable with a bounded derivative F'. Then show that $F \in BV[a, b]$ for all $a, b \in \mathbb{R}$.

Lemma 22.4.1 (Structure of functions of bounded variation). Let $F : \mathbb{R} \to \mathbb{R}$.

(a) Then the two functions $V_F \pm F$ are non-decreasing.

(b) If $F = F_1 - F_2$ where F_1 and F_2 are bounded and non-decreasing functions, then $F \in BV$.

(c) If $F \in BV$, then $F = F_1 - F_2$ for some F_1 and F_2 which are bounded non-decreasing functions. Indeed we can choose

$$F_1 = \frac{1}{2}(V_F + F) \text{ and } F_2 = \frac{1}{2}(V_F - F).$$

(d) If $F \in BV$ then the following limits exist.

$$F(x+) := \lim_{y \downarrow x} F(y), \quad F(x-) := \lim_{y \uparrow x} F(y), \text{ for all } x, y \in \mathbb{R}.$$
$$F(-\infty) := \lim_{y \downarrow -\infty} F(y), \quad F(\infty) := \lim_{y \uparrow \infty} F(y).$$

(e) If $F \in BV$, then the set of discontinuity points of F is countable.

(f) If $F \in BV$, and $G(x) = F(x+)$, then F' and G' exist a.e. λ, and are equal a.e. λ. ♦

Proof. (a) Suppose $x < y$ and $\epsilon > 0$. Then since $F \in BV$, there is a partition $\{x = x_1 < x_2 < \cdots < x_n = y\}$ such that

$$\sum_{j=1}^{n} |F(x_j) - F(x_{j-1})| \geq V_F(x) - \epsilon \pm F(x) .$$

Hence

$$\begin{aligned} V_F(y) \pm F(y) &\geq \sum_{j=1}^{n} |F(x_j) - F(x_{j-1}) \pm |F(y) - F(x)| \pm F(x) \\ &\geq V_F(x) - \epsilon \pm F(x). \end{aligned}$$

Since ϵ is arbitrary, (a) is proved.

(b) follows immediately from Example 22.4.1(b).

(c) Observe that

$$F(x) - F(y) \leq V_F(y) - V_F(x) \leq F(y) - F(x), \text{ for all } \; y > x .$$

This implies

$$|F(x) - F(y)| \leq V_F(y) - V_F(x) \leq V_F(\infty) - V_F(-\infty) .$$

Hence $F, V_F \pm F \in BV$. We already know they are non-decreasing. Finally, (d), (e) and (f) follows from (a), (b), and Theorem 22.3.2. ∎

22.5 Absolutely continuous function

Definition 22.5.1. (**Absolutely continuous function**) Any function $F : [a,\ b] \to \mathbb{R}$ is said to be *absolutely continuous* if, given any $\varepsilon > 0$, there is a $\delta > 0$, such that $\sum_{i=1}^{n} |F(b_i) - F(a_i)| < \varepsilon$ for any disjoint sub-intervals $(a_i,\ b_i)$, $1 \leq i \leq n$, of $[a,\ b]$ with $\sum_{i=1}^{n}(b_i - a_i) < \delta$. ◇

Due to triangle inequality, it is clear that in Definition 22.5.1, we can assume that $a = a_1 < b_1 = a_2 < b_2 = a_3 < \cdots < b_{n-2} = a_{n-1} < b_n = b$ without loss. We shall denote this generic choice as π.

Exercise 22.5.1 (a) If F is absolutely continuous on $[a,\ b]$, then it is uniformly continuous on $[a,\ b]$. Give an example of f which is continuous but not absolutely continuous.

(b) Show that if F is everywhere differentiable on $(a,\ b)$ with a uniformly bounded derivative, then it is absolutely continuous.

(c) Show that if F_1 and F_2 are absolutely continuous, then so is $F_1 - F_2$.

Theorem 22.5.1 (Absolutely continuous function and measure). *Let F_1, F_2 be distribution functions on $[a,\ b]$ with finite Lebesgue-Stieltjes measures μ_1 and μ_2. Then $\mu := \mu_1 - \mu_2 \ll \lambda$ if and only if $F := F_1 - F_2$ is absolutely continuous. Here λ is the Lebesgue measure.* ◆

Proof. (a) First suppose $\mu \ll \lambda$. Fix $\varepsilon > 0$. By Lemma 21.2.1(b) and (e), there exists $\delta > 0$ such that $\lambda(A) < \delta$ implies $|\mu|(A) < \varepsilon$.

Let $(a_i,\ b_i), 1 \le i \le n$ be disjoint intervals of $[a,\ b]$, such that $\sum_{i=1}^{n}(b_i - a_i) < \delta$. Consider the set $A := \cup_{i=1}^{n}(a_i,\ b_i]$. Then $\lambda(A) < \delta$.

Hence

$$\begin{aligned}
\sum_{i=1}^{n} |f(b_i) - f(a_i)| &= \sum_{i=1}^{n} |\mu(a_i,\ b_i]| \quad \text{(by definition of } \mu) \\
&\le \sum_{i=1}^{n} |\mu|(a_i,\ b_i] \quad \text{(since } |\mu(A)| \le |\mu|(A) \text{ for any } A) \\
&= |\mu|(A) \quad \text{(since } \mu\{b_i\} = 0 \text{ for all } i) \\
&\le \varepsilon.
\end{aligned}$$

(b) Now let F be absolutely continuous, and hence continuous. First observe that for any b,

$$\mu\{b\} = \lim_{n\to\infty} \mu(b - 1/n,\ b] = \lim_{n\to\infty} [F(b) - F(b - 1/n)] = 0\,.$$

To show that for any Borel set A, $\lambda(A) = 0$ implies $\mu(A) = 0$, fix $\varepsilon > 0$. Choose $\delta > 0$ as in the definition of absolute continuity of f.

Let γ denote any of the measures μ_1, μ_2 or λ. By Theorem 5.2.4(b),

$$\gamma(A) = \inf\{\gamma(V) : V \supseteq A,\ V \text{ open}\},\ i = 1, 2\,, \tag{22.5}$$

As finite intersection of open sets is open, using (22.5), get a single sequence $\{V_n\}$ open, $V_n \supseteq A$, such that

$$\lambda(V_n) \to \lambda(A) = 0 \text{ and } \mu(V_n) \to \mu(A)\,.$$

Choose a large n, such that for all $k \ge n$, $\lambda(V_k) < \delta$.

Write $V_n = \cup_{i=1}^{\infty}(a_i,\ b_i)$ (disjoint union). Then

$$\begin{aligned}|\mu(V_n)| = |\sum_{i=1}^{\infty}\mu(a_i,\ b_i)| &\le \sum_{i=1}^{\infty}|\mu(a_i,\ b_i)| \\ &= \sum_{i=1}^{\infty}|\mu(a_i,\ b_i]| \quad (\text{since } \mu\{b_i\}=0) \\ &= \sum_{i=1}^{\infty}|f(b_i) - f(a_i)| \le \varepsilon\,.\end{aligned}$$

Since ε was arbitrary, $\lim \mu(V_n) = 0 = \mu(A)$. ∎

Exercise 22.5.2 Absolute continuity of F on any bounded interval is extended in a natural way to $F : \mathbb{R} \to \mathbb{R}$. Suppose F_1 and F_2 are **bounded** distribution functions on $\mathbb{R}$ with Lebesgue-Stieltjes measures μ_1 and μ_2. Let $F := F_1 - F_2$, and $\mu := \mu_1 - \mu_2$. Show that F is absolutely continuous if and only if $\mu \ll \lambda$.

Exercise 22.5.3 Let f be Lebesgue integrable on $\mathbb{R}$. Show that F defined below in Eqn. 22.6, is absolutely continuous. [Theorem 22.5.2 has more information.]

$$F(x) := \int_{(-\infty,\ x]} f d\lambda,\ x \in \mathbb{R}\,. \tag{22.6}$$

Let $F : [a,\ b] \to \mathbb{R}$ be a function. Let $\pi = \{a = x_0 < \cdots < x_n = b\}$ be a *partition* of $[a,\ b]$. Then we define,

$$V_\pi(F) = \sum_{i=1}^{n}|F(x_i) - F(x_{i-1})|\,.$$

Lemma 22.5.1 (Absolutely continuous $\Rightarrow$ bounded variation). Any absolutely continuous $F : [a,\ b] \to \mathbb{R}$ is of bounded variation. ♦

Proof. Let F be absolutely continuous. Fix $\varepsilon > 0$. Let δ be as in Definition 22.5.1. Let π be a partition of $[a,\ b]$. Let τ be a refinement of π, consisting of sub-intervals of π, each with length less than $\delta/2$. Let $a = x_0 < x_1 < \cdots < x_n = b$ be the end-points of the intervals that make up τ. Let $i_0 = 0$, and i_1 be the largest integer such that $x_{i_1} - x_{i_0} < \delta$. Let i_2 be the largest integer such that $x_{i_2} - x_{i_1} < \delta$. Continuing this process, suppose it terminates at $x_{i_r} = x_n$. By construction $x_{i_k} - x_{i_{k-1}} \geq \delta/2$ for all $k = 1, \ldots, r-1$.

Hence,

$$\frac{(r-1)\delta}{2} \leq b-a \quad \text{so} \quad r \leq 1 + \frac{2(b-a)}{\delta} =: M\,.$$

By absolute continuity, $V_\tau(F) \leq M\varepsilon$. But $V_\pi(F) \leq V_\tau(F)$ since τ is a refinement of π. Hence F is of bounded variation on $[a,\ b]$. ■

Lemma 22.5.2. Let $F : [a,\ b] \to \mathbb{R}$ be absolutely continuous. Then $F = F_1 - F_2$ where F_1 and F_2 are non-decreasing on $[a,\ b]$, and are absolutely continuous. ♦

Proof. Define,

$$F_1(x) := V_F[a,\ x],\quad F_2(x) = F(x) - F_1(x),\ a \leq x \leq b.$$

By Exercise 22.4.1, F_1 is non-decreasing.

We now show that F_2 is also non-decreasing. Suppose $x_1 < x_2$. Then

$$\begin{aligned} F_2(x_2) - F_2(x_1) &= V_F(x_2) - V_F(x_1) - (F_2(x_2) - F_2(x_1)) \\ &= V_F[x_1,\ x_2] - (F_2(x_2) - F_2(x_1)) \\ &\geq V_F[x_1,\ x_2] - |F_2(x_2) - F_2(x_1)| \\ &\geq 0 \ \ (\text{by definition of } \ V_F[x_1,\ x_2]). \end{aligned}$$

We now show that F_1 is absolute continuous. Fix $\varepsilon > 0$. Choose $\delta > 0$ as in Definition 22.5.1 Let $\{(a_i,\ b_i)\}_{1\leq i\leq n}$ be disjoint intervals such that $\sum_{i=1}^n (b_i - a_i) < \delta$. Let π_i be a partition of $[a_i,\ b_i]$.

By absolute continuity of F, $\sum_{i=1}^n V_{\pi_i,F}[a_i,\ b_i] \leq \varepsilon$. Taking supremum successively over possible $\pi_1, \ldots, \pi_n$, $\sum_{i=1}^n V_F[a_i,\ b_i] \leq \varepsilon$. In other words, $\sum_{i=1}^n [F_1(b_i) - F_1(a_i)] \leq \varepsilon$. Therefore F_1 is absolutely continuous. Hence F_2 is also so. ■

Theorem 22.5.2 (Fundamental theorem for Lebesgue integrals). *Let $F : [a,\ b] \to \mathbb{R}$ where a, b are finite. The following are equivalent.*

(i) F is absolutely continuous on $[a,\ b]$.

(ii) There is a Lebesgue integrable function $f : [a,\ b] \to \mathbb{R}$ *such that*

$$F(x) - F(a) = \int_{[a,\ x]} f(t)\lambda(dt), \quad a \le x \le b. \tag{22.7}$$

(iii) F *is a.e differentiable,* F' *is Lebesgue integrable and*

$$F(x) - F(a) = \int_{[a,\ x]} F'(t)\lambda(dt), \quad a \le x \le b.$$

◆

Proof. (i) implies (ii): Let F be absolutely continuous. Then by Lemma 22.5.2(b), $F = F_1 - F_2$ where F_1 and F_2 are non-decreasing absolutely continuous (hence continuous) functions. It is enough to prove the result in the case $F_2 = 0$. Let μ be the Lebesgue-Stieltjes measure corresponding to F. Then $\mu \ll \lambda$ by Theorem 22.5.1. By Radon-Nikodym Theorem 21.1.1, there is a λ-integrable function f such that $\mu(A) = \int_A f d\lambda$ for all Borel subsets of $[a,\ b]$. Take $A = [a,\ x]$ to get (22.7).

(ii) implies (i): Suppose (22.7) holds. We can assume $f \ge 0$. Otherwise, we can split $f = f^+ - f^-$. Define μ on the Borel subsets of $[a,\ b]$ by $\mu(A) := \int_A f d\lambda$. Then $\mu \ll \lambda$. Let F be the distribution function of μ. Then by Theorem 22.5.1, F is absolutely continuous.

Trivially (iii) implies (ii).

Remains to show that (i) implies (iii). Without loss of generality, assume that $F(a) = 0$. By Lemma 22.5.2, write $F = F_1 - F_2$ where both F_1 and F_2 are non-decreasing and absolutely continuous. By Theorem 22.5.1, they give rise to measures that are absolutely continuous with respect to λ. Now Exercise 22.3.1(a) can be applied to complete the proof. ■

Chapter 23

Conditional expectation

In elementary probability theory, conditional probabilities are calculated given a finite number of events or random variables. These are based on appropriate conditional probability distributions. More generally, we may consider calculating conditional probabilities and conditional expectations given σ-fields. In this chapter these ideas are made rigorous with the help of the Radon-Nikodym theorem. The theory of Markov processes and of martingales rest on these notions.

23.1 Conditional expectation

In Section 14.3, we defined the conditional probability $P(A|B)$ and showed that $P(\cdot|B)$ is a probability measure on $(\Omega, \mathcal{A})$. We now present two simple examples to motivate the general definition of conditional probability and expectation.

Example 23.1.1. Let $\Omega := \{1,2\} \times \{1,2\}$ and $\mathcal{A} = \mathcal{P}(\Omega)$. Define a probability on this measurable space by, $P(1,1) := 0.5$, $P(1,2) := 0.1$, $P(2,1) := 0.1$ and $P(2,2) := 0.3$. Define the two *coordinate random variables* X and Z as $X(\omega_1, \omega_2) := \omega_1$ and $Z(\omega_1, \omega_2) := \omega_2$. It can be easily checked that

$$P\{X = 1|Z = 1\} = \frac{P(1,1)}{P(1,1) + P(2,1)} \tag{23.1}$$

$$= \frac{0.5}{0.5 + 0.1} = 5/6\,. \tag{23.2}$$

A. Bose, *Elements of Measure and Probability*, Texts and Readings in Mathematics 88,
https://doi.org/10.1007/978-981-95-2758-8_23

Likewise, $P\{X = 2|Z = 1\} = 1/6$. So, $\mathrm{E}(X|Z = 1)$ is,

$$\mathrm{E}(X|Z = 1) = 1 \times 5/6 + 2 \times 1/6 = 7/6\,.$$

Similarly $\mathrm{E}(X|Z = 2) = 7/4$. We combine (23.1) and (23.2) into a rv Y:

$$Y := \mathrm{E}(X|Z) := \frac{7}{6}\mathbf{1}_{\{Z=1\}} + \frac{7}{4}\mathbf{1}_{\{Z=2\}}.$$

Note that Y is a function of Z. It can be checked that $\mathrm{E}(Y) = \mathrm{E}(X)$. ▲

Example 23.1.2. Let $X, Z : (\Omega, \mathcal{A}, P) \to (\mathbb{R}, \mathcal{B}(\mathbb{R}))$, with distinct values $\{x_i\}$ $\{z_j\}$ with positive probabilities. Further $\mathrm{E}(|X|) < \infty$. Then the usual conditional probability and expectation of X given $Z = z_i$ are:

$$P\{X = x_j|Z = z_i\} := \frac{P\{X = x_j, Z = z_i\}}{P\{Z = z_i\}}, \quad i, j \in \{1, 2, \ldots\},$$
$$\mathrm{E}(X|Z = z_i) := \sum_{j=1}^{\infty} x_j P\{X = x_j|Z = z_i\}, \ i \in \{1, 2, \ldots\}.$$

Consider the above relations from the measure theoretic viewpoint. Consider the partition $A_i := Z^{-1}(z_i)$ of Ω. Then define $Y : \Omega \to \mathbb{R}$ as:

$$Y(\omega) := \mathrm{E}(X|Z = z_i), \ \text{ if } \ \omega \in A_i, \ i \geq 1. \tag{23.3}$$

Then $Y \in \sigma(Z)$, and we express (23.3) compactly as $Y = \mathrm{E}(X|Z)$. It has an additional property. Fix any $A \in \sigma(Z)$. Then $A = \cup_{i \in I} A_i$ for some countable set I. Note that

$$\begin{aligned}
\mathrm{E}(Y\mathbf{1}_A) = \sum_{i \in I} \mathrm{E}(Y\mathbf{1}_{A_i}) &= \sum_{i \in I} \mathrm{E}(X|Z = z_i)P(Z = z_i) \\
&= \sum_{i \in I}\sum_{j=1}^{\infty} x_j P(X = x_j|Z = z_i)P(Z = z_i) \\
&= \sum_{j=1}^{\infty} x_j \sum_{i \in I} P(X = x_j, Z = z_i) \\
&= \sum_{j=1}^{\infty} x_j P(\{X = x_j\} \cap A) = \mathrm{E}(X\mathbf{1}_A).
\end{aligned}$$

So we have, $\mathrm{E}(Y\mathbf{1}_A) = \mathrm{E}(X\mathbf{1}_A)$ for any $A \in \sigma(Z)$. ▲

In Example 23.1.2 the conditioning rv assumed only countably many values. One would have to be careful if this is not the case. We now formalise the two properties of Y in Example 23.1.2.

Definition 23.1.1. (**Conditional expectation**) Let X be an integrable rv on $(\Omega, \mathcal{A}, P)$. Let $\mathcal{G}$ be any sub-σ-field of $\mathcal{A}$. An integrable rv Y is called the *conditional expectation of X given $\mathcal{G}$* (and we write $Y = \mathrm{E}(X|\mathcal{G})$) if, it is measurable with respect to $\mathcal{G}$, and

$$\int_A X\, dP = \int_A Y\, dP, \text{ for all } A \in \mathcal{G}. \qquad \diamond$$

It immediately follows that $\mathrm{E}(Y) = \mathrm{E}(X)$. The following result guarantees the existence and uniqueness of conditional expectation.

Theorem 23.1.1 (**Existence and uniqueness**). *Let X be an integrable rv on $(\Omega, \mathcal{A}, P)$ and $\mathcal{G}$ be a sub-σ-field of $\mathcal{A}$. Then there exists a $\mathcal{G}$-measurable and integrable rv Y which satisfies*

$$\int_A X\, dP = \int_A Y\, dP\,, \textit{ for all } A \in \mathcal{G}. \tag{23.4}$$

If Y_1 and Y_2 are choices of Y that satisfy (23.4), then $Y_1 = Y_2$ a.s. ◆

Proof. Let $X^+ := \max\{X, 0\}, X^- := \max\{-X, 0\}$. Then, $X^+, X^- \geq 0$, and $X = X^+ - X^-$. Clearly, $\int_\Omega X^+ dP \leq \int_\Omega |X| dP < \infty$. Define the measure ν^+ on $(\Omega, \mathcal{A})$ by

$$\nu^+(A) := \int_A X^+ dP,\ A \in \mathcal{G}.$$

Then ν^+ is a finite measure and $\nu^+ \ll P$ on $(\Omega, \mathcal{G})$. By Radon-Nikodym Theorem 21.1.1, there is a $\mathcal{G}$-measurable $Y^+ : \Omega \to [0, \infty)$ such that

$$\int_A Y^+ dP = \nu^+(A),\ \ A \in \mathcal{G}.$$

We can define Y^- analogously. Then, $Y := Y^+ - Y^-$ satisfies all the requirements. Proof of uniqueness is left as an easy exercise. ■

Definition 23.1.2. (**Conditional probability**) Suppose $(\Omega, \mathcal{A}, P)$ is a probability space and $\mathcal{G}$ a sub-σ-field of $\mathcal{A}$. The *conditional probability* of $A \in \mathcal{A}$ given $\mathcal{G}$ is same as $\mathrm{E}(\mathbf{1}_A|\mathcal{G})$. ◇

Exercise 23.1.1 Let $(\Omega, \mathcal{A}, P)$ be a probability space and $\mathcal{G}$ be a sub-σ-field.

(a) Let A_1 and A_2 be two sets from $\mathcal{A}$. Show that:

$$P(A_1 \cup A_2|\mathcal{G}) = P(A_1|\mathcal{G}) + P(A_2|\mathcal{G}) - P(A_1 \cap A_2|\mathcal{G}) \text{ almost surely.}$$

(b) Do you foresee any issues in treating the set function $A \to P(A|\mathcal{G})$ as a "probability measure"? First let $\mathcal{A}$ be countably generated.

Exercise 23.1.2 Let $(\Omega, \mathcal{A}, P)$ be a probability space and X be an integrable rv. What are $\mathrm{E}(X|\mathcal{G})$ for $\mathcal{G} = \mathcal{A}$ and $\mathcal{G} = \{\emptyset, \Omega\}$?

Example 23.1.3. Let X be an integrable rv on $(\Omega, \mathcal{A}, P)$. Let $\{A_i\}$ from $\mathcal{A}$ be a measurable partition of Ω. Let $\mathcal{G} := \sigma(A_1, A_2, \ldots)$. Then due to the structure of $\mathcal{G}$, $\mathrm{E}(X|\mathcal{G})$, say Y, should be constant on each A_n. Moreover, it will not matter what value we assign to Y on the sets A_n for those n's for which $P(A_n) = 0$. Thus we can define $\mathrm{E}(X|\mathcal{G})$ as

$$Y := \sum_{\{n: P(A_n) > 0\}} \mathbf{1}_{A_n} \frac{1}{P(A_n)} \int_{A_n} X \, dP .$$

To verify its correctness, by construction, Y is $\mathcal{G}$-measurable. Moreover,

$$\begin{aligned} \int_\Omega |Y| dP &= \sum_{\{n: P(A_n) > 0\}} \left| \int_{A_n} X dP \right| \\ &\leq \sum_{\{n: P(A_n) > 0\}} \int_{A_n} |X| dP = \int_\Omega |X| dP < \infty . \end{aligned}$$

Fix $B \in \mathcal{G}$. Then, for some $I \subset \mathbb{N}$, $B = \cup_{i \in I} A_i$. Hence,

$$\begin{aligned} \int_B Y dP = &\sum_{\{i \in I: P(A_i) > 0\}} \int_{A_i} Y dP \\ = &\sum_{\{i \in I: P(A_i) > 0\}} \int_{A_i} X dP = \int_B X dP . \end{aligned}$$

▲

Example 23.1.4. Let X, Y be rvs on $(\Omega, \mathcal{A}, P)$ where X is integrable, and Y is discrete taking values $\{y_i\}$. Then, using Example 23.1.3,

$$\mathrm{E}(X|Y) := \mathrm{E}(X|\sigma(Y)) = \sum_{\{i:\ P(Y = y_i) > 0\}} \mathbf{1}_{\{Y = y_i\}} \frac{1}{P(Y = y_i)} \int_{\{Y = y_i\}} X dP .$$

▲

Example 23.1.5. Let (X, Y) be a random vector on $(\Omega, \mathcal{A}, P)$, with a a density $f(\cdot,\cdot)$ on $\mathbb{R}^2$ with respect to the Lebesgue measure. Let X be integrable. We wish to identify $\mathrm{E}(X|Y)$. By Fubini's Theorem 10.3.1,

$$f_Y(y) := \int_{\mathbb{R}} f(x,y)\lambda(dx),\ y \in \mathbb{R},$$

is a pdf of Y. Taking a cue from Example 23.1.3, we focus on the set $\{y \in \mathbb{R} : f_Y(y) > 0\}$. As $\mathrm{E}(|X|) = \int_{\mathbb{R}^2} |x| f(x,y)\lambda(dx)\lambda(dy) < \infty$, by Fubini's Theorem 10.3.1, $\int_{-\infty}^{\infty} |x| f(x,y)dx$ is a measurable function of y, and is finite a.s. λ (Lebesgue measure). Define

$$g(y) := \begin{cases} \dfrac{\int_{\mathbb{R}} x f(x,y)\lambda(dx)}{f_Y(y)} & \text{if } \int_{\mathbb{R}} |x| f(x,y)\lambda(dx) < \infty \text{ and } f_Y(y) > 0\,, \\ 0 & \text{otherwise}\,. \end{cases}$$

We claim that $g(Y) = \mathrm{E}(X|Y)$. As $g : \mathbb{R} \to \mathbb{R}$ is a measurable function, $g(Y) : \Omega \to \mathbb{R}$ is also measurable. We check that $\mathrm{E}(|g(Y)|) < \infty$. Since

$$\lambda\{y \in \mathbb{R} : f_Y(y) > 0 \text{ and } \int_{\mathbb{R}} |x| f(x,y)\lambda(dx) = \infty\} = 0\,,$$

$$\begin{aligned} \mathrm{E}(|g(Y)|) &= \int_{\mathbb{R}} |g(y)| f_Y(y)\lambda(dy) \\ &= \int_{\{y: f_Y(y)>0\}} |g(y)| f_Y(y)\lambda(dy) \\ &= \int_{\{y: f_Y(y)>0\}} \left|\frac{1}{f_Y(y)} \int_{\mathbb{R}} x f(x,y)\lambda(dx)\right| f_Y(y)\lambda(dy) \\ &\le \int_{\{y: f_Y(y)>0\}} \int_{\mathbb{R}} |x| f(x,y)\lambda(dx)\lambda(dy) = \mathrm{E}(|X|) < \infty\,. \end{aligned}$$

Now fix $A \in \sigma(Y)$. Then, $A = Y^{-1}(B)$ for some $B \in \mathcal{B}(\mathbb{R})$. Further,

$$\begin{aligned} \mathrm{E}\left(g(Y)\mathbf{1}_A\right) &= \int_{B \cap \{y: f_Y(y)>0\}} g(y) f_Y(y)\lambda(dy) \\ &= \int_{B \cap \{y: f_Y(y)>0\}} \int_{-\infty}^{\infty} x f(x,y)\lambda(dx)\lambda(dy) \\ &= \mathrm{E}\left(X\mathbf{1}_{\{Y \in B\}}\right) = \mathrm{E}(X\mathbf{1}_A)\,. \end{aligned}$$

This confirms the claim that $g(Y) = \mathrm{E}(X|Y)$ almost surely. ▲

Theorem 23.1.2 (Tower property). *Let X be integrable, and $\mathcal{G}_1 \subset \mathcal{G}_2$ be two sub-σ-fields. Then*

$$\mathrm{E}(X|\mathcal{G}_1) = \mathrm{E}(\mathrm{E}(X|\mathcal{G}_2)|\mathcal{G}_1) \quad \textit{almost surely.} \tag{23.5}$$ ◆

Proof. Let $Y := \mathrm{E}(X|\mathcal{G}_2)$ and $Z := \mathrm{E}(X|\mathcal{G}_1)$. Now, for any $A \in \mathcal{G}_1$,

$$\int_A Z dP = \int_A X dP = \int_A Y dP \quad (\text{since } A \in \mathcal{G}_2).$$

That is, $\mathrm{E}(Y|\mathcal{G}_2) = Z$ almost surely. This proves (23.5). ■

The next theorem shows that the conditional expectation satisfies all the properties that the expectation satisfies.

Theorem 23.1.3 (Conditional expectation: basic properties). *Let $X, Y, \{X_n\}$ be integrable rvs on $(\Omega, \mathcal{A}, P)$, and $\mathcal{G} \subset \mathcal{A}$ be a σ-field. Suppose a and b are real numbers. Then the following hold:*

(a) If $X = a$ a.s., then $\mathrm{E}(X|\mathcal{G}) = a$ a.s.

(b) $\mathrm{E}(aX + bY|\mathcal{G}) = a\,\mathrm{E}(X|\mathcal{G}) + b\,\mathrm{E}(Y|\mathcal{G})$ a.s.

(c) If $X \le Y$ a.s., then $\mathrm{E}(X|\mathcal{G}) \le \mathrm{E}(Y|\mathcal{G})$ a.s.

(d) Almost surely, $|\mathrm{E}(X|\mathcal{G})| \le \mathrm{E}(|X|\,|\mathcal{G})$.

(e) If $\lim_n X_n = X$ a.s., $|X_n| \le Y$, and Y is integrable, then

$$\lim_n \mathrm{E}(X_n|\mathcal{G}) = \mathrm{E}(X|\mathcal{G}) \ \textit{a.s.}$$ ◆

Proof. Proofs of Parts (a) and (b) are left as exercises. Part (b) implies Part (c) by considering $\mathrm{E}(Y - X|\mathcal{G})$. For Part (d), using Parts (b) and (c), it follows that a.s.,

$$\begin{aligned} \mathrm{E}(|X|\,|\mathcal{G}) &\ge \mathrm{E}(X|\mathcal{G}), \quad \text{and} \\ \mathrm{E}(|X|\,|\mathcal{G}) &\ge \mathrm{E}(-X|\mathcal{G}) = -\,\mathrm{E}(X|\mathcal{G}). \end{aligned}$$

This completes the proof of Part (d). To prove Part (e), define

$$Z_n := \sup_{k \ge n} |X_k - X|, \ n \ge 1.$$

Clearly $Z_n \downarrow 0$ a.s. Parts (a)–(c) imply that for $n \ge 1$,

$$|\mathrm{E}(X_n|\mathcal{G}) - \mathrm{E}(X|\mathcal{G})| \le \mathrm{E}(Z_n|\mathcal{G}) \ \text{ a.s.}$$

Therefore, it suffices to show that $\mathrm{E}(Z_n|\mathcal{G}) \to 0$ a.s. By Part (c), $\mathrm{E}(Z_n|\mathcal{G})$ is non-increasing, with a limit, say Z. Moreover,

$$\int_\Omega Z dP \le \int_\Omega \mathrm{E}(Z_n|\mathcal{G}) dP = \int_\Omega Z_n dP .$$

Note that $Z_n \le 2Y$ and hence $\mathrm{E}(Z_n|\mathcal{G}) \le 2Y$. Now DCT implies that $\lim_{n\to\infty} \int_\Omega Z_n dP = 0$. This completes the proof. ■

Example 23.1.6. Let X be an integrable rv on $(\Omega, \mathcal{A}, P)$, and let $\mathcal{G}$ be a sub-σ-field of $\mathcal{A}$ which is independent of $\sigma(X)$. Then, $\mathrm{E}(X|\mathcal{G}) = \mathrm{E}(X)$. To see this, fix any $A \in \mathcal{G}$. Then X and $\mathbf{1}_A$ are independent. The rest of the proof is left as an exercise. ▲

Theorem 23.1.4. *Let $X, Y : (\Omega, \mathcal{A}, P) \to (\mathbb{R}, \mathcal{B}(\mathbb{R}))$, so that X, Y, XY are integrable. Let $\mathcal{G} \subset \mathcal{A}$ be a σ-field, and X be $\mathcal{G}$-measurable. Then,*

$$\mathrm{E}\,(XY|\mathcal{G}) = X\,\mathrm{E}(Y|\mathcal{G}) \quad a.s. \qquad \blacklozenge$$

Proof. The equality is trivial if X is a simple function. In general, one can get $\mathcal{G}$-measurable simple functions s_n such that $|s_n| \le |X|$, $s_n \to X$ almost surely. The proof now follows from the observations that $|s_n Y| \le |XY|$, XY is integrable, and Theorem 23.1.3(e). ■

If X is a rv with $\mathrm{E}(X^2) < \infty$, then it is easy to check that $\mathrm{E}(X)$ minimises $\mathrm{E}(X - x)^2$ over all $x \in \mathbb{R}$. The conditional expectation has a similar interpretation.

Theorem 23.1.5 (Conditional expectation as a minimiser). *Let X be square integrable on $(\Omega, \mathcal{A}, P)$. Then for any sub-σ-field $\mathcal{G}$,*

$$\min_{Y \text{ is } \mathcal{G}-\text{measurable}} \mathrm{E}\left((X-Y)^2\right) = \mathrm{E}\left(\left(X - \mathrm{E}\,(X|\mathcal{G})\right)^2\right). \qquad \blacklozenge$$

Proof. Let $Z := \mathrm{E}\,(X|\mathcal{G})$. We first show that $\mathrm{E}(Z^2) < \infty$. Note that

$$0 \;\le\; (X-Z)^2 = X^2 - Z^2 - 2Z(X-Z) .$$

Therefore, for all $n \ge 1$,

$$X^2 \ge X^2 \mathbf{1}_{\{|Z| \le n\}} \ge Z^2 \mathbf{1}_{\{|Z| \le n\}} + 2Z\mathbf{1}_{\{|Z| \le n\}}(X - Z) .$$

As a consequence,

$$
\begin{aligned}
\mathrm{E}(X^2|\mathcal{G}) &\ge \mathrm{E}\left(Z^2 \mathbf{1}_{\{|Z|\le n\}} + 2Z\mathbf{1}_{\{|Z|\le n\}}(X-Z) \big| \mathcal{G}\right) \\
&= Z^2\mathbf{1}_{\{|Z|\le n\}} + 2Z\mathbf{1}_{\{|Z|\le n\}}\,\mathrm{E}\left((X-Z)|\mathcal{G}\right) \text{ (by Theorem 23.1.4)} \\
&= Z^2\mathbf{1}_{\{|Z|\le n\}}\,.
\end{aligned}
$$

Taking expectation on both sides, it follows that

$$\mathrm{E}\left(Z^2\mathbf{1}_{\{|Z|\le n\}}\right) \le \mathrm{E}\left(\mathrm{E}(X^2|\mathcal{G})\right) = \mathrm{E}(X^2)\,.$$

Letting $n \to \infty$, it follows that $\mathrm{E}(Z^2) \le \mathrm{E}(X^2) < \infty$. If $\mathrm{E}(Y^2) = \infty$,

$$\mathrm{E}(X-Y)^2 = \infty > \mathrm{E}(X-Z)^2\,.$$

Therefore, it suffices to show that

$$\mathrm{E}(X-Z)^2 = \inf_{Y \text{ is } \mathcal{G}-\text{measurable: } \mathrm{E}\,Y^2<\infty} \mathrm{E}\left((X-Y)^2\right).$$

For that purpose, fix a $\mathcal{G}$-measurable Y such that $\mathrm{E}(Y^2) < \infty$. By the Cauchy-Schwarz inequality, $(X-Z)(Z-Y)$ is integrable. Thus,

$$
\begin{aligned}
\mathrm{E}(X-Y)^2 &= \mathrm{E}(X-Z)^2 + \mathrm{E}(Z-Y)^2 + 2\,\mathrm{E}\left((X-Z)(Z-Y)\right) \\
&\ge \mathrm{E}(X-Z)^2 + 2\,\mathrm{E}\left((X-Z)(Z-Y)\right) \\
&= \mathrm{E}(X-Z)^2 + 2\,\mathrm{E}\left((Z-Y)\,\mathrm{E}\left(X-Z|\mathcal{G}\right)\right) \text{ (as } Z-Y \in \mathcal{G}) \\
&= \mathrm{E}(X-Z)^2\,.
\end{aligned}
$$

■

23.2 Regular conditional distribution

Let $X : (\Omega, \mathcal{A}, P) \to (\mathbb{R}, \mathcal{B}(\mathbb{R}))$. Let $\mathcal{G}$ be a sub-σ-field of $\mathcal{A}$. Then for every fixed $A \in \mathcal{A}$, we have defined $P(X \in A|\mathcal{G})$ as a $\mathcal{G}$-measurable function. However, this is defined only almost surely for every fixed A, and it is not clear if we can define a single version of $P(X \in A|\mathcal{G})$ that works for all $A \in \mathcal{A}$. This motivates the following.

Definition 23.2.1. Let X be a real-valued rv defined on $(\Omega, \mathcal{A}, P)$, and $\mathcal{G}$ be a sub-σ-field of $\mathcal{A}$. A function $\mu : \mathcal{B}(\mathbb{R}) \times \Omega \to [0,1]$ is called a **regular conditional distribution** (rcd) of X given $\mathcal{G}$ if,

(a) for all fixed $\omega \in \Omega$, $\mu(\cdot, \omega)$ is a probability measure on $(\mathbb{R}, \mathcal{B}(\mathbb{R}))$, and

(b) for all $A \in \mathcal{B}(\mathbb{R})$, $\mu(A, \cdot)$ is a version of $P(X \in A|\mathcal{G})(\cdot)$. ◇

The following theorem guarantees the existence of an rcd.

Theorem 23.2.1 (rcd, existence). *An rcd of X given $\mathcal{G}$ exists.* ◆

The following result will be used in the proof of the theorem.

Exercise 23.2.1 Let $F : \mathbb{Q} \to [0,1]$ be a non-decreasing function such that $\lim_{n\to-\infty} F(n) = 0$ and $\lim_{n\to\infty} F(n) = 1$. Define

$$G(x) := \inf \{F(r) : r > x, r \in \mathbb{Q}\}\,, \quad x \in \mathbb{R}\,.$$

Show that G is a cdf on $\mathbb{R}$. If in addition, $\lim_{n\to\infty} F\left(r + n^{-1}\right) = F(r)\,,$ $r \in \mathbb{Q}$, then G is the unique cdf on $\mathbb{R}$ which agrees with F on $\mathbb{Q}$.

Proof of Theorem 23.2.1. Since $\mathbb{Q}$ is countable, choose $G : \mathbb{Q} \times \Omega \to \mathbb{R}$ such that, for all $r \in \mathbb{Q}$, $G(r, \cdot)$ is a version of $P(X \le r|\mathcal{G})$. Then, we immediately have the following properties of G for all $r, s \in \mathbb{Q}$ a.s.

$$G(r) \le G(s) \text{ when } r \le s\,, \tag{23.6}$$

$$0 \le G(r) \le 1\,, \tag{23.7}$$

$$\lim_{n\to\infty} G(r + n^{-1}) = G(r)\,, \tag{23.8}$$

$$\lim_{n\to-\infty} G(n) = 0\,, \tag{23.9}$$

$$\lim_{n\to\infty} G(n) = 1\,. \tag{23.10}$$

Let $\Omega_0 \subset \Omega$ be the set on which (23.6)–(23.10) hold. Define for all $r \in \mathbb{Q}$ and $\omega \in \Omega$,

$$F(r,\omega) := \begin{cases} G(r,\omega) & \text{if } \omega \in \Omega_0\,, \\ \mathbf{1}_{\{r\ge 0\}} & \text{if } \omega \in \Omega_0^c\,. \end{cases}$$

Thus, (23.6)–(23.10) hold for every $\omega \in \Omega$, with G replaced by F.

Furthermore, $\Omega_0 \in \mathcal{G}$, and hence,

$$F(r,\omega) = P(X \le r|\mathcal{G})(\omega)\,,\ r \in \mathbb{Q},\ \omega \in \Omega\,, \tag{23.11}$$

is a valid version of the conditional expectation. Define

$$F(x,\omega) := \inf\{F(r,\omega) : r > x, r \in \mathbb{Q}\},\ x \in \mathbb{Q}^c, \omega \in \Omega\,.$$

Then Exercise 23.2.1, along with (23.6)–(23.10) shows that for all $\omega \in \Omega$, $F(\cdot,\omega)$ is a valid cdf, and in particular is measurable.

If $x \in \mathbb{Q}^c$, and $r_n \in \mathbb{Q}$ such that $r_n \downarrow x$, then the above shows that,

$$F(r_n, \omega) \to F(x, \omega)\,, \ \omega \in \Omega\,.$$

Since $F(r_n, \cdot)$ is a version of $P(X \le r_n | \mathcal{G})$, it follows that

$$F(x, \omega) = P(X \le x | \mathcal{G})(\omega), \ \omega \in \Omega\,. \tag{23.12}$$

Fix $\omega \in \Omega$. Let $\mu(\cdot, \omega)$ be the probability measure on $(\mathbb{R}, \mathcal{B}(\mathbb{R}))$

$$\mu((a, b], \omega) = F(b, \omega) - F(a, \omega), \ -\infty < a < b < \infty\,.$$

It is immediate from (23.11) and (23.12) that

$$\mu((a, b], \omega) = P(a < X \le b | \mathcal{G})(\omega), \ \omega \in \Omega\,.$$

Now, using standard arguments as seen, for example, in the proof of Theorem 5.2.2, the above relation can be extended to claim that,

$$\mu(B, \omega) = P(X \in B | \mathcal{G})(\omega)\,, \ \text{for all} \ \ B \in \mathcal{B}, \ \text{and all} \ \ \omega \in \Omega\,. \qquad \blacksquare$$

Conditional expectation of $f(X)$ can be calculated by using an rcd.

Theorem 23.2.2. *Let μ be an rcd of X given $\mathcal{G}$. For any measurable function $f : \mathbb{R} \to \mathbb{R}$ such that $f(X)$ is integrable,*

$$\mathrm{E}\left(f(X) | \mathcal{G}\right)(\omega) = \int_{-\infty}^{\infty} f(x) \mu(dx, \omega) \quad a.s. \tag{23.13}$$ ◆

Proof. Eqn. (23.13) holds if $f(X) = \mathbf{1}_{\{X \in B\}}$, $B \in \mathcal{B}(\mathbb{R})$. Then it holds for all simple functions, and then we can pass to limits, to claim it for all integrable functions $f(X)$. ■

23.3 Conditional Jensen's inequality

We now state and prove the conditional version of Jensen's inequality Theorem 8.2.1.

Theorem 23.3.1 (Conditional Jensen's inequality). *Let ϕ be a convex function, and $X, \phi(X)$ be integrable. Then for any sub-σ-field $\mathcal{G}$, $\phi(\mathrm{E}(X | \mathcal{G})) \le \mathrm{E}\left(\phi(X) | \mathcal{G}\right)$ a.s.* ◆

Proof of Theorem 23.3.1. Let $Y := \mathrm{E}(X|\mathcal{G})$. We first assume that $a \leq Y \leq b$ for some finite a, b. Then Lemma 8.2.1(a) implies that

$$\phi(Y) + (X - Y)\phi'_+(Y) \leq \phi(X)\,. \tag{23.14}$$

Since ϕ is convex, it is continuous, and hence $\phi(Y)$ takes values in the compact set $\phi([a,b])$. By Lemma 8.2.1(a), $\phi'_+(a) \leq \phi'_+(Y) \leq \phi'_+(b)$. Thus, the left side of the inequality (23.14) is integrable. Hence,

$$\begin{aligned}\mathrm{E}\left(\phi(X)|\mathcal{G}\right) &\geq \mathrm{E}\left[\phi(Y) + (X-Y)\phi'_+(Y)|\mathcal{G}\right] \text{ a.s.}\\ &= \phi(Y) + \phi'_+(Y)\,\mathrm{E}(X - Y|\mathcal{G}) \text{ a.s. (as } Y \text{ is } \mathcal{G}\text{-measurable)}\\ &= \phi(Y) \text{ a.s. (since } Y = \mathrm{E}(X|\mathcal{G}))\,,\end{aligned}$$

which is the desired inequality (for bounded Y). To prove the result in general, note that for $n \geq 1$, using (23.14),

$$\phi(Y)\mathbf{1}_{\{|Y|\leq n\}} + (X - Y)\phi'_+(Y)\mathbf{1}_{\{|Y|\leq n\}} \leq \phi(X)\mathbf{1}_{\{|Y|\leq n\}} \text{ a.s}\,.$$

Taking conditional expectation of both sides with respect to $\mathcal{G}$, and going through the same steps as earlier, it follows that,

$$\phi(Y)\mathbf{1}_{\{|Y|\leq n\}} \leq \mathbf{1}_{\{|Y|\leq n\}}\,\mathrm{E}\left(\phi(X)|\mathcal{G}\right)\,.$$

Letting $n \to \infty$, the proof follows. ■

We invite the reader to construct proofs for the following inequalities.

Theorem 23.3.2. (a) (**Conditional Cauchy-Schwarz inequality**) Let $X, Y \in L^2$. Then,

$$[\mathrm{E}(XY|\mathcal{G})]^2 \leq \mathrm{E}(X^2|\mathcal{G})\,\mathrm{E}(Y^2|\mathcal{G}) \text{ a.s.}$$

(b) (**Conditional Hölder's inequality**) Let $X \in L^p$, $Y \in L^q$, where $p, q \in (1, \infty)$, $p^{-1} + q^{-1} = 1$. Then $XY \in L^1$ and

$$|\mathrm{E}(XY|\mathcal{G})| \leq [\mathrm{E}(|X|^p|\mathcal{G})]^{1/p}\,[\mathrm{E}(|Y|^q|\mathcal{G})]^{1/q} \text{ almost surely.}$$ ◆

The next result on uniform integrability of conditional expectations is a crucial tool in the study of martingales.

Theorem 23.3.3. *Let X be integrable and let $\{\mathcal{G}_\alpha\}$ be a family of sub-σ-fields. Then the collection $\{X_\alpha := \mathrm{E}(X|\mathcal{G}_\alpha)\}$ is ui.* ◆

Proof. For $\alpha \in I$, $T > 0$, $\{X_\alpha > T\}$ and $\{X_\alpha < -T\} \in \mathcal{G}_\alpha$. Hence,

$$\begin{aligned} \int_{\{|X_\alpha|>T\}} |X_\alpha|\, dP &= \int_{\{X_\alpha>T\}} X_\alpha\, dP - \int_{\{X_\alpha<-T\}} X_\alpha\, dP \\ &= \int_{\{X_\alpha>T\}} X\, dP + \int_{\{X_\alpha<-T\}} (-X)\, dP \\ &= \int_{\{|X_\alpha|>T\}} |X|\, dP\,. \end{aligned}$$

Fix $\varepsilon > 0$. By integrability of X, get $\delta > 0$ such that

$$\int_A |X|\, dP \le \varepsilon\,, \quad \text{whenever } P(A) \le \delta\,.$$

Let $T := \frac{\mathrm{E}|X|}{\delta}$. By Markov's inequality Lemma 8.1.1, for any $\alpha \in I$,

$$P\{|X_\alpha| > T\} \le T^{-1}\,\mathrm{E}\,|X_\alpha| \le T^{-1}\,\mathrm{E}\,|X| = \delta\,.$$

Thus,

$$\int_{\{|X_\alpha|>T\}} |X_\alpha|\, dP \le \int_{\{|X_\alpha|>T\}} |X|\, dP \le \varepsilon\,.$$ ■

23.4 Conditional independence given a σ-field

Recall the concept of conditional independence given an event, developed in Section 14.4 of Chapter 14. We can now extend this to conditional independence given a σ-field in the natural way.

Definition 23.4.1. Let $(\Omega, \mathcal{A}, P)$ be a probability space and $\mathcal{G}$ be a sub-σ-field. We say that

(i) events A and B are conditionally independent given $\mathcal{G}$ if,

$$P(A \cap B|\mathcal{G}) = P(A|\mathcal{G})\ P(B|\mathcal{G})\ a.s.$$

(ii) sub-σ-fields $\mathcal{G}_1$ and $\mathcal{G}_2$ are conditionally independent given $\mathcal{G}$ if all $A \in \mathcal{G}_1$ and $B \in \mathcal{G}_2$ are independent given $\mathcal{G}$.

(iii) random variables or vectors X and Y are conditionally independent given $\mathcal{G}$ if, $\sigma(X)$ and $\sigma(Y)$ are conditionally independent given $\mathcal{G}$.

Finally, we define conditional independence given a random variable or vector X by using $\mathcal{G} := \sigma(X)$. ◇

23.5 Exercises

Exercise 23.5.1 Suppose $\mathcal{G} = \sigma(A_1, A_2, \ldots, A_n)$, where $\{A_i\} \subset \mathcal{A}$ but not necessarily disjoint and let X be an integrable rv. Describe $\mathrm{E}(X|\mathcal{G})$.

Exercise 23.5.2 Let $X : (\Omega, \mathcal{A}, P) \to (\bar{\mathbb{R}}, \mathcal{B}(\bar{\mathbb{R}}))$, and $\int X\,dP = \infty$. Let $\mathcal{G}$ be a sub-σ-field of $\mathcal{A}$. Prove or give counter-example to the following:
(a) There exists a $\mathcal{G}$-measurable function $Y : \Omega \to [0, \infty)$ such that

$$\int_A X\,dP = \int_A Y\,dP\,, \quad \text{for all} \quad A \in \mathcal{G}\,.$$

(b) There exists a $\mathcal{G}$-measurable function $Y : \Omega \to [0, \infty]$ such that

$$\int_A X\,dP = \int_A Y\,dP\,, \quad \text{for all } \mathcal{A} \in \mathcal{G}\,.$$

Exercise 23.5.3 Let $X \in L^2(\Omega, \mathcal{A}, P)$ and $\mathcal{H} \subset \mathcal{G} \subset \mathcal{A}$ be σ-fields. Show that the covariance between $\mathrm{E}(X|\mathcal{H})$ and $\mathrm{E}(X|\mathcal{G}) - \mathrm{E}(X|\mathcal{H})$ is 0.

Exercise 23.5.4 Suppose $X \in L^2(\Omega, \mathcal{A}, P)$ and $\mathcal{G} \subset \mathcal{A}$ is a σ-field.
(a) Show that $\mathrm{Cov}(X, \mathrm{E}(X|\mathcal{G})) \geq 0$.
(b) Show that equality holds above if and only if $\mathrm{E}(X|\mathcal{G}) = \mathrm{E}(X)$ a.s.

Exercise 23.5.5 Suppose $\mathcal{G}_1 \subset \mathcal{G}_2$ are sub-σ-fields and $\mathrm{E}(X^2) < \infty$. Then show that

$$\mathrm{E}\left[(X - \mathrm{E}(X|\mathcal{G}_2))^2\right] \leq \mathrm{E}\left[(X - \mathrm{E}(X|\mathcal{G}_1))^2\right].$$

Exercise 23.5.6 Let $\{X_n\}, X$ be integrable rvs, $X_n \downarrow X$ almost surely. Show that on $\cup_{n=1}^{\infty}\{\mathrm{E}(X_n|\mathcal{G}) < \infty\}$, $\mathrm{E}(X_n|\mathcal{G}) \downarrow \mathrm{E}(X|\mathcal{G})$ almost surely.

Exercise 23.5.7 Suppose $\mathrm{E}(X^2 + Y^2) < \infty$, and $\mathrm{E}(X|Y) = f(Y)$ for a decreasing function f. Show that $\mathrm{Cov}(X, Y) \leq 0$.

Exercise 23.5.8 (Conditional Markov's inequality) Show that if $a > 0$ then

$$P(|X| \geq a|\mathcal{G}) \leq a^{-2}\,\mathrm{E}(X^2|\mathcal{G}) \text{ a.s.}$$

Exercise 23.5.9 Let $X \in L^2(\Omega, \mathcal{A}, P)$ and $\mathcal{G}$ a sub-σ-field. Show that
(a) $\mathrm{Var}(X) = \mathrm{Var}\left(\mathrm{E}(X|\mathcal{G})\right) + \mathrm{E}\left((X - \mathrm{E}(X|\mathcal{G}))^2\right)$. Hence

$$\mathrm{Var}\left(\mathrm{E}(X|\mathcal{G})\right) \leq \mathrm{Var}\left(X\right)\,. \tag{23.15}$$

(b) Using (23.15), deduce the Rao-Blackwell theorem: *Let $\{P_\theta\}$ be probability distributions such that $P_\theta \ll \mu$ for all θ, for some fixed measure μ and for which T is a sufficient statistic. Let X be any unbiased estimator of $f(\theta)$, so that $\mathrm{E}_\theta(X) = f(\theta)$, with finite variance. Then,*

$$\mathrm{E}\left[\,(\mathrm{E}(X|\sigma(T)) - f(\theta))^2\,\right] \le \mathrm{E}\left[(X - f(\theta)^2\right].$$

Exercise 23.5.10 Let $X \in L^2(\Omega, \mathcal{A}, P)$ and $\mathcal{G} \subset \mathcal{A}$ be a σ-field. Let $Z := \mathrm{E}(X|\mathcal{G})$. Suppose $Y \in L^2(\Omega, \mathcal{G}, P)$ is such that

$$\mathrm{E}\left[(X - Y)^2\right] = \mathrm{E}\left[(X - Z)^2\right].$$

Show that $Y = Z$ almost surely.

Exercise 23.5.11 Let $X \sim U(0,\ 1)$. Fix a positive integer k. Compute $\mathrm{E}(X|Y)$ where $Y = kX - [kX]$ and $[x]$ denotes the integer part of x.

Exercise 23.5.12 Let $(X, Y)'$ be a Gaussian vector X, Y have 0 means, and $\mathrm{E}(Y^2) > 0$. Show that

$$\mathrm{E}(X|Y) = \rho Y,$$

where $\rho = \mathrm{E}(XY)/\,\mathrm{E}(Y^2)$.

Exercise 23.5.13 Let X and Y be independent rvs, and $f : \mathbb{R}^2 \to \mathbb{R}$ be bounded Borel measurable. Define $g : \mathbb{R} \to \mathbb{R}$ by

$$g(y) := \int_{-\infty}^{\infty} f(x, y)\, P_X(dx)\,,\ y \in \mathbb{R}\,.$$

Show that

$$\mathrm{E}\left(f(X, Y)|Y\right) = g(Y) \text{ a.s.}$$

Exercise 23.5.14 Let $X : (\Omega, \mathcal{A}, P) \to (\mathbb{R}, \mathcal{B}(\mathbb{R}))$ be integrable. Let $\{\mathcal{G}_n\} \subset \mathcal{A}$ be a non-decreasing sequence of σ-fields. Let Y be such that $\mathrm{E}(X|\mathcal{G}_n) = Y$ a.s., for all $n \ge 1$. Show that

$$\mathrm{E}\left(X|\sigma(\cup_{n=1}^{\infty}\mathcal{G}_n)\right) = Y \text{ a.s.}$$

Exercise 23.5.15 Prove or give counter-examples to the following.

(a) If X is integrable and independent of the σ-field $\mathcal{G}$, then for any σ-field $\mathcal{H}$, $\mathrm{E}\left(X|\sigma(\mathcal{G} \cup \mathcal{H})\right) = \mathrm{E}\left(X|\mathcal{H}\right)$ a.s.

(b) Let X be an integrable rv and $\mathcal{G}, \mathcal{H}$ be σ-fields such that $\sigma(\mathcal{H} \cup \sigma(X))$ is independent of $\mathcal{G}$. Then $\mathrm{E}(X|\sigma(\mathcal{G} \cup \mathcal{H})) = \mathrm{E}(X|\mathcal{H})$ a.s.

Exercise 23.5.16 Let X, Y be integrable rvs on $(\Omega, \mathcal{A}, P)$ such that,

$$\mathrm{E}(X|\sigma(Y)) = Y \quad \text{and} \quad \mathrm{E}(Y|\sigma(X)) = X \quad \text{a.s.}$$

Show that $X = Y$ almost surely.
Hint: First show that $\int_{[Y \le c]} (X - Y) dP = 0$, for all $c \in \mathbb{R}$. Then show that $\int_{[X \le c, Y \le c]} (X - Y) dP \le 0$.

Exercise 23.5.17 Let X be an integrable rv. Show that whenever $\mathcal{G}$ is a trivial σ-field, $\mathrm{E}(X|\mathcal{G}) = \mathrm{E}(X)$ a.s.

Exercise 23.5.18 Let $X, Y : (\Omega, \mathcal{A}, P) \to (\mathbb{R}, \mathcal{B}(\mathbb{R}))$ be integrable. Let $\mathcal{G} \subset \mathcal{A}$ be a σ-field. Suppose $Y \in \sigma(\mathcal{G})$, $X - Y$ is independent of $\mathcal{G}$, and $\mathrm{E}(X) = \mathrm{E}(Y)$. Then show that $\mathrm{E}(X|\mathcal{G}) = Y$ a.s.

Exercise 23.5.19 Let $X : (\Omega, \mathcal{A}, P) \to (\mathbb{R}, \mathcal{B}(\mathbb{R}))$, $\mathcal{G} \subset \mathcal{A}$ be a σ-field. Let $Y \in \sigma(\mathcal{G})$ and $\phi : (\mathbb{R}^2, \mathcal{B}(\mathbb{R}^2)) \to (\mathbb{R}, \mathcal{B}(\mathbb{R}))$ be bounded. If μ is the rcd of X given $\mathcal{G}$, then show that

$$(\mathrm{E}(\phi(X,Y)|\mathcal{G}))(\omega) = \int_{\mathbb{R}} \phi(x, Y(\omega)) \mu(dx, \omega) \quad \text{for almost all} \quad \omega.$$

Exercise 23.5.20 Consider the coin tossing space.

(a) Let X_1 be the number of tosses required to obtain the first head.
(i) Find $\mathrm{E}(X)$ using conditioning on $A := \{\text{First toss is head}\}$.
(ii) Calculate the probability distribution of X.
[This is known as the *geometric distribution* with parameter p, and is written as $X_1 \sim Geo(p)$.]

(b) Let X_k be the number of tosses required to get exactly k heads. Show that $X_k \overset{D}{=} Y_1 + \cdots + Y_k$ where $\{Y_i\}$ are iid $Geo(p)$ rvs.
[This distribution is known as the *negative binomial distribution* with parameters k and p and is written as $X_k \sim NegBin(k, p)$.]

Definition 23.5.1. (**Kernel**) Let $h : (\mathbb{R}^m, \mathcal{B}(\mathbb{R}^m)) \to (\mathbb{R}, \mathcal{B}(\mathbb{R}))$, be is symmetric in its arguments. That is, $h(x_1, \ldots, x_m) = h(x_{\pi(1)}, \ldots, x_{\pi(m)})$ for all permutations $(\pi(1), \ldots, \pi(m))$ of $(1, \ldots, m)$. Then it is called a *kernel of order m*. ◇

Exercise 23.5.21 (*U*-**statistics CLT**) Let $\{X_i\}$ be iid rvs. The U-statistics with kernel h is defined as:

$$U_n := \binom{n}{m}^{-1} \sum_{1 \le i_1 < i_2 < \cdots < i_m \le n} h(X_{i_1}, \ldots, X_{i_m}).$$

(a) Check that if $m = 1$, $h(x) = x$, then U_n is the *sample mean,*

$$\bar{X}_n = (X_1 + \cdots + X_n)/n.$$

(b) Check that if $m = 2$, $h(x_1, x_2) = (x_1 - x_2)^2/2$, then U_n is the *sample variance*

$$s_n^2 = (n-1)^{-1} \sum_{i=1}^{n} (X_i - \bar{X}_n)^2.$$

(c) If $\theta := \mathrm{E}[h(X_1, \ldots, X_m)]$ is finite, show that $\mathrm{E}(U_n) = \theta$ for every n.

(d) Suppose $\mathrm{Var}[h(X_1, \ldots, X_m)] < \infty$. Prove the following U-statistics CLT:

$$n^{1/2}(U_n - \theta) \Rightarrow N(0,\ \sigma^2) \text{ as } n \to \infty,$$

where

$$\sigma^2 = m^2\, \mathrm{Var}[\mathrm{E}(h(X_1, \ldots, X_m)|X_1)].$$

Hint: Use Lemma 17.10.1(b) and Theorem 17.8.1; first assume that h is a bounded function and use the method of cumulants. Then remove this restriction by truncation and Mallows metric.

(e) Suppose U_{n1} and U_{n2} are U-statistics with kernels h_1 and h_2, of orders m_1 and m_2 respectively. Formulate and prove a CLT for the vector $n^{1/2}(U_{n1} - \theta_1,\ U_{n2} - \theta_2)'$, where $\theta_i = \mathrm{E}(U_{ni})$, $i = 1, 2$.

(f) If $\{X_i\}$ are iid with $P\{X_i = \pm 1\} = 1/2$ then show that for the U-statistics in (b), $\sigma^2 = 0$.

(g) Suppose $\{X_i\}$ are iid with mean 0 and variance 1 and finite fourth moment. Show that $n^{1/2}(\bar{X}_n,\ s_n^2 - 1)'$ converges to a bivariate Gaussian distribution. Identify the parameters of the limit distribution.

Bibliography

W. J. Adams. *The Life and Times of the Central Limit Theorem*, volume 35 of *History of Mathematics.* American Mathematical Society, 2nd edition, 2009.

J. Alexopoulos. de La Vallée Poussin's theorem and weakly compact sets in Orlicz spaces. *Quaest. Math.*, 17:231–248, 1994.

N. Alon and J. H. Spencer. *The Probabilistic Method.* John Wiley & Sons, 2008.

R. Ash and C. A. Doléans-Dade. *Probablility & Measure Theory.* Elsevier Science, 2000.

K. B. Athreya and S. N. Lahiri. *Measure Theory and Probability Theory.* Springer, New York, 2006.

S. Athreya and V. Sunder. *Measure & Probability.* CRC Press, 2019.

Z. D. Bai and J. W. Silverstein. *Spectral Analysis of Large Dimensional Random Matrices.* Springer, 2nd edition, 2010.

B. Bercu, B. Delyon, and E. Rio. *Concentration Inequalities for Sums and Martingales.* Springer International Publishing, 2015.

S. Bernstein. On a modification of Chebyshev's inequality and of the error formula of Laplace. *Ann. Sci. Inst. Sav. Ukraine, Sect. Math*, 1924.

S. Bernstein. *Theory of Probability* (In Russian). Moscow, 1927.

S. Bernstein. On certain modifications of Chebyshev's inequality. *Doklady Akademii Nauk SSSR*, 17(6):275–277, 1937.

A. Bose, *Elements of Measure and Probability*, Texts and Readings in Mathematics 88,
https://doi.org/10.1007/978-981-95-2758-8

K. Bhaskara Rao and M. Bhaskara Rao. *Theory of Charges: A Study of Finitely Additive Measures.* Elsevier, 1983.

P. Billingsley. *Probablility & Measure.* John Wiley, 3rd edition, 1995.

V. Bogachev. *Measure Theory, Vol. I and II.* Springer, 2007.

A. Bose. *Random Matrices and Non-commutative Probability.* CRC Press, Boca Raton, 2021.

A. Bose, A. Chakrabarty, and R. S. Hazra. *A Little Book of Martingales.* Hindustan Book Agency, 2024. International edition by Springer.

L. Breiman. *Probability.* Society for Industrial and Applied Mathematics. Orginally published in 1968 by Addison-Wesley, 1992.

T. Carleman. *Les Fonctions Quasi Analytiques.* Gauthier-Villars, Paris, 1926.

C. Castaing and R. de Fitte. $\mathfrak{G}$-uniform scalar integrability and strong law of large numbers for Pettis integrable functions with values in a separable locally convex space. *J. Theoret. Probab.*, 13:93–134, 2000.

R. Cerf and P. Petit. A short proof of Cramér's theorem in $\mathbb{R}$. *The American Mathematical Monthly*, 118(10):925–931, 2011.

D. Chafai. de La Vallée Poussin On uniform integrability. `https://djalil.chafai.net/blog/2014/03/09/de-la-vallee-poussin-on-uniform-integrability/`, 2014. Accessed 16-March-2025.

Y. S. Chow and H. Teicher. *Probability theory: independence, interchangeability, martingales.* Springer Science & Business Media, 2003.

K. L. Chung. *A Course in Probability Theory.* Academic Press, Inc., 3rd edition, 2001.

D. L. Cohn. *Measure Theory.* Birkhäuser, 2013.

A. Dembo and O. Zeitouni. *Large Deviation Techniques and Applications*, volume 38. Springer, 2nd edition, 1998. Applications of Mathematics.

F. den Hollander. *Large Deviations*, volume 14. American Mathematical Society, 2000. Fields Institute Monographs.

J.-D. Deuschel and D. W. Stroock. *Large Deviations.* AMS Chelsea Publishing, 1989.

R. M. Dudley. *Real Analysis and Probability.* Cambridge University Press, 2002.

R. Durrett. *Probability: Theory and Examples*, volume 49. Cambridge University Press, 2019.

M. Elekes and T. Keleti. Is Lebesgue measure the only sigma-finite invariant Borel measure? *Math. Anal. Appl.*, 321:445–451, 2006.

R. S. Ellis. *Entropy, Large Deviations, and Statistical Mechanics.* Springer, 2006.

N. Etemadi. An elementary proof of the strong law of large numbers. *Z. Wahr. verw Gebiete*, 55:119–122, 1981.

W. Feller. *An Introduction to Probability Theory and Its Applications, Vol. I.* John Wiley & Sons, 3rd edition, 1968.

T. Ferguson. *Mathematical Statistics.* Academic Press, New York, 1967.

H. Fischer. *A History of the Central Limit Theorem.* Sources and Studies in the History of Mathematics and Physical Sciences. Springer, 2010.

G. B. Folland. *Real Analysis: Modern Techniques and their Applications.* John Wiley & Sons, 1999.

G. B. Folland. *Fourier Analysis and its Applications.* American Mathematical Society, 2009.

A. Goswami and B. V. Rao. *Measure Theory: for analysis and probability.* Springer-Verlag, 2025.

G. R. Grimmett and D. R. Stirzaker. *Probability and Random Processes.* Cambridge University Press, 2002.

A. Gut. *Probability: A Graduate Course.* Springer, 2013.

P. R. Halmos. *Measure Theory.* Springer, 1950.

T.-C. Hu and A. Rosalsky. A note on the de La Vallée Poussin criterion for uniform integrability. *Stat. Probab. Letts.*, 81:169–174, 2011.

L. Isserlis. On a formula for the product-moment coefficient of any order of a normal frequency distribution in any number of variables. *Biometrika*, 12(1-2):134–139, 1918.

D. Khoshnevisan. *Probability.* American Mathematical Society, 2007.

R. Lang. A note on the measurability of convex sets. *Arch. Math.*, 47: 90–92, 1986.

M. Ledoux. *The Concentration of Measure Phenomenon.* Mathematical Surveys and Monographs. American Mathematical Society, 2001.

M. Loéve. *Elementary Probability Theory.* Springer, 1977.

E. Lukacs. *Characteristic Functions.* Charles Griffin & Co. Ltd., 1970.

E. Lukacs and R. G. Laha. *Applications of Characteristic Functions.* Charles Griffin & Co. Ltd., 1964.

J. Marcinkiewicz. Sur une propertie de la loi deGauss. *Math. Zeitschr.*, 44:612–618, 1938. Reprinted in *Collected Papers*, Pantswowe wydawnictwo, Naukowe, Warsawa, 1964.

J. Martin and M. Milman. Modes of convergence: interpolation methods. I. *J. Approx. Theory*, 111:91–127, 2001.

C. McDiarmid. On the method of bounded differences. *Surveys in Combinatorics*, 141(1):148–188, 1989.

P. Meyer. *Probability and Potentials.* Blaisdell Publishing Co. Ginn and Co. Waltham Massachusetts, 1966.

J. Neveu. *Mathematical Foundations of the Calculus of Probability.* Holden-Day Series in Probability and Statistics, 1965.

K. R. Parthasarathy. *Introduction to Probability and Measure.* Hindustan Book Agency, 2005.

B. Ramachandran. *Advanced Theory of Characteristic Functions.* Statistical Publication Society, Calcutta, 1967.

S. Resnick. *A Probability Path.* Springer, New York, 2005.

M. Riesz. Surle probleme des moments. Troisieme note (in French). *Ark.for matematik. Astronomi och Fysik*, 16:1–52, 1923.

M. Rivera. The de La Vallée Poussin theorem for vector valued measures. *Rocky Mountain J. Math.*, 30:715–722, 2000.

J. S. Rosenthal. *A First Look at Rigorous Probability Theory.* World Scientific, 2nd edition, 2006.

G. G. Roussas. *An Introduction to Measure-Theoretic Probability.* Elsevier, 2nd edition, 2014.

H. L. Royden and P. M. Fitzpatrick. *Real Analysis.* Pearson, 4th edition, 2015.

G. R. Shorack. *Probability for Statisticians.* Springer, 2nd edition, 2017.

M. Steele. *Probability Theory and Combinatorial Optimization.* CBMS-NSF Regional Conference Series in Applied Mathematics. Society for Industrial and Applied Mathematics, 1997.

E. M. Stein and R. Shakarchi. *Real Analysis: Measure Theory, Integration and Hilbert Spaces.* Princeton Lectures in Analysis. Princeton University Press, 2005.

T. Tao. The central limit theorem. `https://terrytao.wordpress.com/2010/01/05/254a-notes-2-the-central-limit-theorem/`, 2010. Accessed 16-March-2025.

T. Tao. *An Introduction to Measure Theory.* American Mathematical Society, 2011.

L. Tonelli. Sull'integrazione per parti. *Atti della Accademia Nazionale dei Lincei*, 5(18):246–253, 1909.

S. R. S. Varadhan. *Large Deviations*, volume 27. Courant Institute of Mathematical Sciences, New York, 2016. Courant Lecture Notes.

G. C. Wick. The evaluation of the collision matrix. *Physical Review*, 80 (2):268–272, 1950.

D. Williams. *Probability with Martingales.* Cambridge Mathematical Textbooks. Cambridge University Press, 1991.

Author Index

A. Bose, *Elements of Measure and Probability*, Texts and Readings in Mathematics 88,
https://doi.org/10.1007/978-981-95-2758-8

Subject Index

A. Bose, *Elements of Measure and Probability*, Texts and Readings in Mathematics 88,
https://doi.org/10.1007/978-981-95-2758-8